5 STEPS TO A 5™

500 AP Biology Questions to Know by Test Day

Third Edition

Mina Lebitz MS

New York Chicago San Francisco Athens London Madrid Mexico City
Milan New Delhi Singapore Sydney Toronto

1 2 3 4 5 6 7 8 9 QFR 24 23 22 21 20 19

ISBN 978-1-260-44203-8
MHID 1-260-44203-9

e-ISBN 978-1-260-45218-1
e-MHID 1-260-45218-2

CONTENTS

ABOUT THE AUTHOR

Mina Lebitz worked in biochemistry research before transitioning to teaching and tutoring at both the high school and college level. As a teacher, she designed curricula for several science courses, mentored students with research projects, and was awarded the *New York Times* Teachers Who Make a Difference Award. She spent several years as the senior science tutor at one of the most prestigious private tutoring and test prep agencies in the United States. Mina continues to develop science curricula, working with students and their parents to help them achieve their academic goals, while continuing to learn everything she can about what she loves most, science.

INTRODUCTION

Congratulations! You've taken a big step toward AP success by purchasing *5 Steps to a 5: 500 AP Biology Questions to Know by Test Day*. We are here to help you take the next step and earn a high score on your AP Exam so you can earn college credits and get into the college or university of your choice.

This book gives you 500 AP-style multiple-choice and free-response questions that cover all the most essential course material. Each question has a detailed answer explanation. These questions will give you valuable independent practice to supplement both your regular textbook and the groundwork you are already covering in your AP classroom. This and the other books in this series were written by expert AP teachers who know your exam inside and out and can identify crucial exam information and questions that are most likely to appear on the test.

You might be the kind of student who takes several AP courses and needs to study extra questions a few weeks before the exam for a final review. Or you might be the kind of student who puts off preparing until the last weeks before the exam. No matter what your preparation style is, you will surely benefit from reviewing these 500 questions that closely parallel the content, format, and degree of difficulty of the questions on the actual AP exam. These questions and their answer explanations are the ideal last-minute study tool for those final few weeks before the test.

Remember the old saying "Practice makes perfect." If you practice with all the questions and answers in this book, we are certain you will build the skills and confidence needed to do great on the exam. Good luck!

—Editors of McGraw-Hill Education

NOTE FROM THE AUTHOR

Dear Student,

Thank you for choosing this book to help prepare you for your AP Biology exam. Knowledge is perishable, so consistent practice is crucial to your success on this exam. Practice problems and questions are the most efficient and effective way to test and reinforce your knowledge, comprehension, and application of biology for the AP Biology exam. Your mistakes will identify the areas that need improvement.

This book is meant as a **comprehensive review** and is most helpful *after* you've learned at least half of the AP curriculum. These questions here are intended to develop and refine your reasoning skills as well as your knowledge of biology. Although the questions range from fairly simple and straightforward to complex and sophisticated, I chose to write most of the questions on the more challenging side so *many of these questions will take more time to answer than you'll have on the AP Biology exam.* It's impractical to spend 7 minutes on a single multiple-choice question during the actual AP exam, but it's time well spent on many of the questions in this book. Like the AP Biology exam, the questions in this book will contain data and experiments that you've likely never seen or heard of. That's the point: *These types of questions are designed to test and sharpen your reasoning skills, not your specific knowledge of biology facts.* If you put the work into understanding each of the questions in this book, you will be very well-prepared for your exam.

The College Board has made a small sample of practice questions available to students. I strongly recommend that you practice with all of the questions they have made available. They will help you get a feeling for *exactly* what the questions will be like on test day.

In order to make the book a more useful learning tool, I had to make some formatting compromises. All of the questions in this book are "AP Biology–like," but there are some notable differences. *Some answer explanations will give away the answers to later questions, and some questions will require answers from previous questions.* The actual AP Biology exam will not contain questions like this. To avoid inadvertently getting answers to later questions, *answer all the grouped questions before looking at the answers.* If a question requires an answer from a previous question, wait until you get to that question and then check *the answer* but *not the explanation. Unlike the AP Biology exam, I have interspersed free-response questions, grid-in questions, and multiple-choice questions.*

Here are some other tips for using this book in the most effective way possible and generally preparing for the AP Biology exam:

✓ **If you guess on a question,** you should mark that question for later review. If you guessed correctly, you may not remember to review the question and answer to clarify and reinforce the information. Remember, when the topic comes up again, you may not guess as well!

✓ In the answer section of this book, *italicized notes refer to specific information regarding the AP Biology curriculum.* **You should familiarize yourself with the information in these notes.** They are there to streamline what you need to know for the exam.

✓ Be aware that, at the time of publication, **the College Board has released a full-length practice test** that is *available only to teachers.* Make sure you ask your teacher to administer it to your class. Check with your teacher and the College Board website to make sure you have access to all released exams and questions from actual AP Biology exams.

✓ **The first administration of the new AP Biology exam was in May of 2013**, so in addition to the free-response questions in the course syllabus and the released practice exam, you can find all the free-response questions from 2013 and later on the website with the scoring rubrics.

✓ **Familiarize yourself with as many scoring rubrics as you can,** even if they are from exams administered prior to 2013. They are all free to download on the College Board website. Even the free-response rubrics from older exams will help you get acquainted with the nuances of earning points on the free-response questions.

✓ Make sure you **review your labs**! Biology is both a laboratory and a field science, so expect to see a significant number of questions related to your labs on the AP exam.

✓ **Familiarize yourself with the formula sheet** you'll be provided with. It contains many equations, formulas, and a chi-square table. This will help limit your memorization load and the time you spend looking for the relevant information on the sheet.

✓ **Be familiar with the format of the exam *before* you sit for it**. On test day, let your brain focus solely on the biology.

SECTION I:

⇒ 63 multiple choice
⇒ 6 grid-in questions
⇒ 90 minutes
⇒ No calculator

SECTION II:

⇒ 2 long free-response questions
⇒ 6 short free-response questions
⇒ 10-minute reading period
⇒ 80 minute answer period
⇒ Any type of calculator

Remember, I want to know what you think! Your comments will help improve later editions for future students.

THANK YOU and best of luck in your AP Biology class and on the exam!

☺ Mina

P.S. Here are some helpful websites:

- Full course description with practice questions https://secure-media.collegeboard.org/digitalServices/pdf/ap/ap-biology-course-and-exam-description-effective-fall-2015.pdf
- A quick overview of the AP Biology course and exam http://media.collegeboard.com/digitalServices/pdf/ap/13b-7589-AP-Biology-ADA-v0.1.pdf
- The official College Board AP Biology Lab manual http://apcentral.collegeboard.com/apc/members/courses/teachers_corner/218954.html

CHAPTER 1

Evolution, Diversity and Unity

Big Idea 1: The process of evolution drives the diversity and unity of life.

MULTIPLE-CHOICE QUESTIONS

1. Which of the following true statements best supports the claim that organisms are linked by lines of descent from a common ancestor?

(A) All organisms have the capacity to grow and reproduce.

(B) All organisms use oxygen for cellular respiration in mitochondria, an organelle that was once a free-living prokaryote.

(C) Different species have specific traits and adaptations that allowed them to succeed in diverse environments.

(D) All organisms share a genetic code that allows one organism to express a gene from another organism.

2. Which explanation is *least* consistent with evolutionary theory regarding why humans lack a tail?

(A) Humans didn't need a tail.

(B) Primate ancestors without tails were more fit than those with tails.

(C) In the population of the common ancestor of apes and humans, those that had shorter or no tails left behind more offspring than those with full-sized tails.

(D) Tails were a disadvantage to our ancestors.

3. Which of the following phylogenies correctly illustrates the data in the following table?

	Tuna	Lamprey	Turtle	Dog	Frog
Hair				X	
Hinged jaws	X		X	X	X
Vertebral column	X	X	X	X	X
Four walking legs			X	X	X
Amniotic egg			X	X	

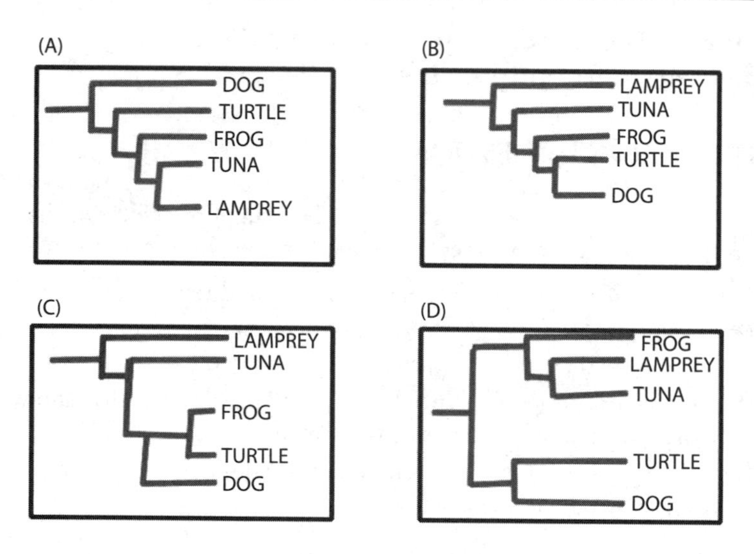

Question 4 refers to the data below showing the evolution of drug resistance in *Moraxella catarrhalis*, a bacteria that causes middle ear infections in children. (Data adapted from Levin and Anderson, 1999.)

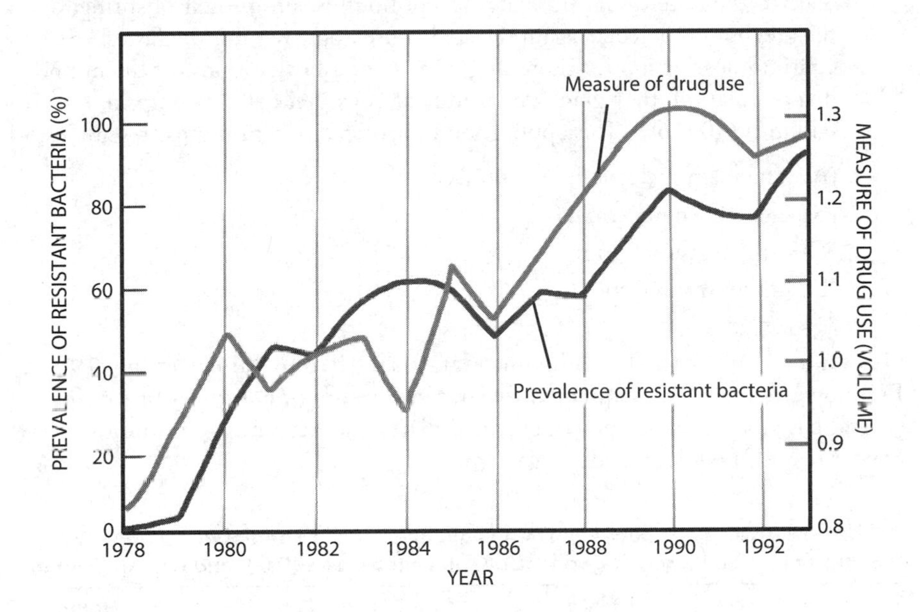

4. Which of the following statements is an accurate conclusion based on the data and modern evolutionary theory?

   (A) As the number of resistant strains increases, a greater volume of antibiotics is required to kill them.

   (B) An increase in the prevalence of resistant bacteria directly caused an increase in the use of antibiotics.

   (C) Lowering the volume of antibiotic use would result in a decreased prevalence of resistant strains.

   (D) It is highly probable that increased use of antibiotics directly caused an increase in the prevalence of resistant bacterial strains.

5. The greyhound dog was originally used to hunt the fastest of game, fox and deer. The breed dates to ancient Egypt. Early breeders of greyhound dogs were interested in dogs with the greatest speed. Breeders carefully selected the fastest dogs from a group of hounds. From their offspring, the greyhound breeders again selected those dogs that ran the fastest. By continuing selection for those dogs that ran faster than most of the hound dog population, they gradually produced dogs that could run up to 64 km/h (40 mph). This application of artificial selection most resembles:

(A) stabilizing selection.
(B) genetic engineering.
(C) directional selection.
(D) disruptive selection.

Question 6 refers to the following data table (from Bailey, et al., 1991). Pseudogenes are DNA sequences that resemble functional genes but differ in several base pairs and are not transcribed. They can accumulate mutations that appear to have no effect on the organism.

Percent divergence between DNA sequences of the ψη-globin pseudogene among orangutan (*Pongo*), gorilla (*Gorilla*), chimpanzee (*Pan*), and human (*Homo*)

	Gorilla	**Pan**	**Homo**
Pongo	3.39	3.42	3.30
Gorilla		1.82	1.69
Pan			1.56
Homo			0.38*

* Percent divergence between two human sequences.
**Divergence between *Homo* and other species calculated using the average of these two sequences.

6. Which of the following statements is the *least* accurate interpretation of the data above?

(A) *Pongo* and *Pan* may have diverged the longest time ago.
(B) *Pan* and *Pongo* most likely diverged further back in time than *Pongo* and *Gorilla.*
(C) *Pan* may be the oldest of the four genera represented.
(D) *Pan* and *Homo* likely share the most recent common ancestor.

7. Which of the following true statements is the best single piece of evidence that nonhuman apes such as chimpanzees, bonobos, and gorillas are more closely related to humans than any other mammal?
   (A) Apes and humans have similar anatomy. They both have nails instead of claws, lack a tail and full facial hair, and have forward-facing eyes.
   (B) Apes and humans have very similar behaviors and social structures, and they both have the ability to use tools.
   (C) Apes have 24 pairs of chromosomes and humans have 23.
   (D) Humans have a greater percentage of DNA sequences in common with apes than with any other group of organisms.

Question 8 refers to the data below showing the relationship between two genotypes (lines I and II) and their respective phenotype for a single gene with two alleles. Organisms with the same genotypes for a particular trait do not always show the same phenotype due to factors in the environment or the expression of other genes. The *y*-axis shows the phenotype for the three genotypes. Notice that A_1A_1 and A_2A_2 phenotypes are the same, but the phenotypes for the heterozygotes are different.

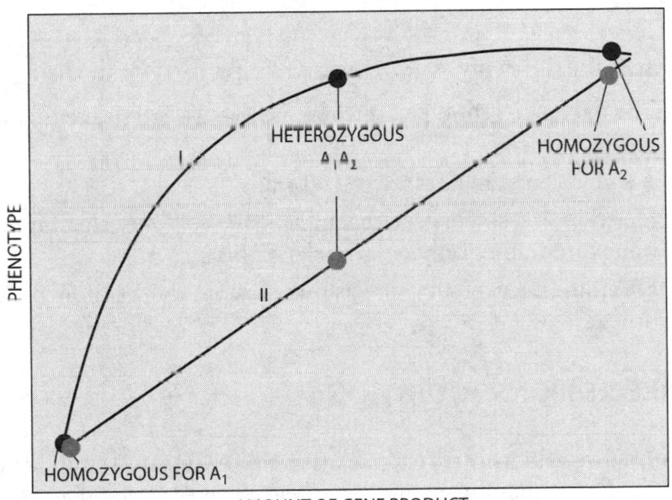

8. Which of the following is true concerning the relationship between the A_1 and A_2 alleles?
   (A) A_1 is dominant to A_2 in both cases.
   (B) A_1 is dominant to A_2 in curve I.
   (C) A_2 is dominant to A_1 in both cases.
   (D) A_2 is dominant to A_1 in curve II.

Question 9 refers to the following data showing the relative fitness in *Saccharomyces cerevisiae*, a strain of yeast. Relative fitness is the fitness of a genotype relative to a reference genotype, which is assigned the value of 1. Each point represents a strain in which a *single gene* has been deleted. (Data taken from Cooper, et al., 2007.)

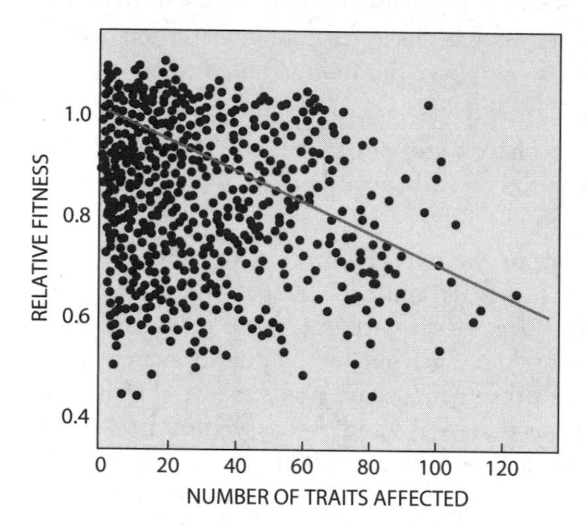

9. Which of the following statements is *not* supported by the data?

(A) The reference genotype, in which no genes were deleted, has the greatest fitness.

(B) A single gene can affect over 100 traits.

(C) Deletion of a single gene that affects one or a very small number of traits can dramatically reduce relative fitness.

(D) Deletion of a gene that affects one or more traits can increase relative fitness.

LONG FREE-RESPONSE QUESTION

10. Terrestrial vertebrates are tetrapods (have four limbs). Tetrapods include amphibians, reptiles, birds, and mammals. Snakes are reptiles, but they lack limbs.

- **Consider** the evolutionary origins of the tetrapods and **explain** why all terrestrial vertebrates are tetrapods.
- **Explain** how snakes fit into this picture.
- **Describe** a piece of evidence that would support the purported relationship between snakes and the rest of the terrestrial vertebrates.

Question 11 refers to the method of replica plating, as shown below, an experimental technique pioneered by Joshua and Esther Lederberg in 1952. The *Escherichia coli* culture used in the experiment was derived from a single cell. Cells from the culture are then spread onto an agar plate. Each colony (there were many more than shown below) is the result of the divisions of a single bacterial cell.

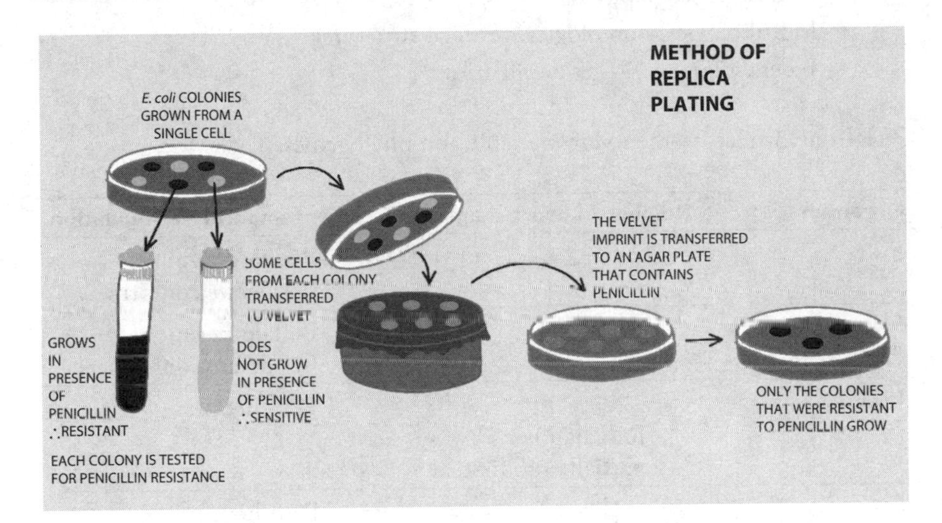

MULTIPLE-CHOICE QUESTIONS

11. The results of this experiment demonstrate that:

(A) the mutation for penicillin resistance was induced by a previous exposure to penicillin.

(B) the mutation for penicillin resistance arose spontaneously, before exposure to penicillin.

(C) the mutations in *E. coli* that are caused by exposure to penicillin do not promote resistance to penicillin.

(D) mutations in the penicillin-resistant colonies caused a sensitivity to penicillin.

LONG FREE-RESPONSE QUESTION

12. Many lines of evidence support biological evolution. **State** one example of supporting evidence from each of the following categories and **explain** how it supports the theory of evolution.

- Fossils/biogeography
- Morphological homologies, vestigial structures
- Biochemical and genetic similarities

Question 13 refers to the following table and phylogenies.

Organism	Number of heart chambers	Body temperature regulation
Fish	2	Poikilothermic (moderate conformer)
Frog	3	Poikilothermic (moderate conformer)
Lizard	3 3rd chamber partially subdivided	Ectothermic
Crocodile	4	Ectothermic
Bird	4	Endothermic
Mammal	4	Endothermic

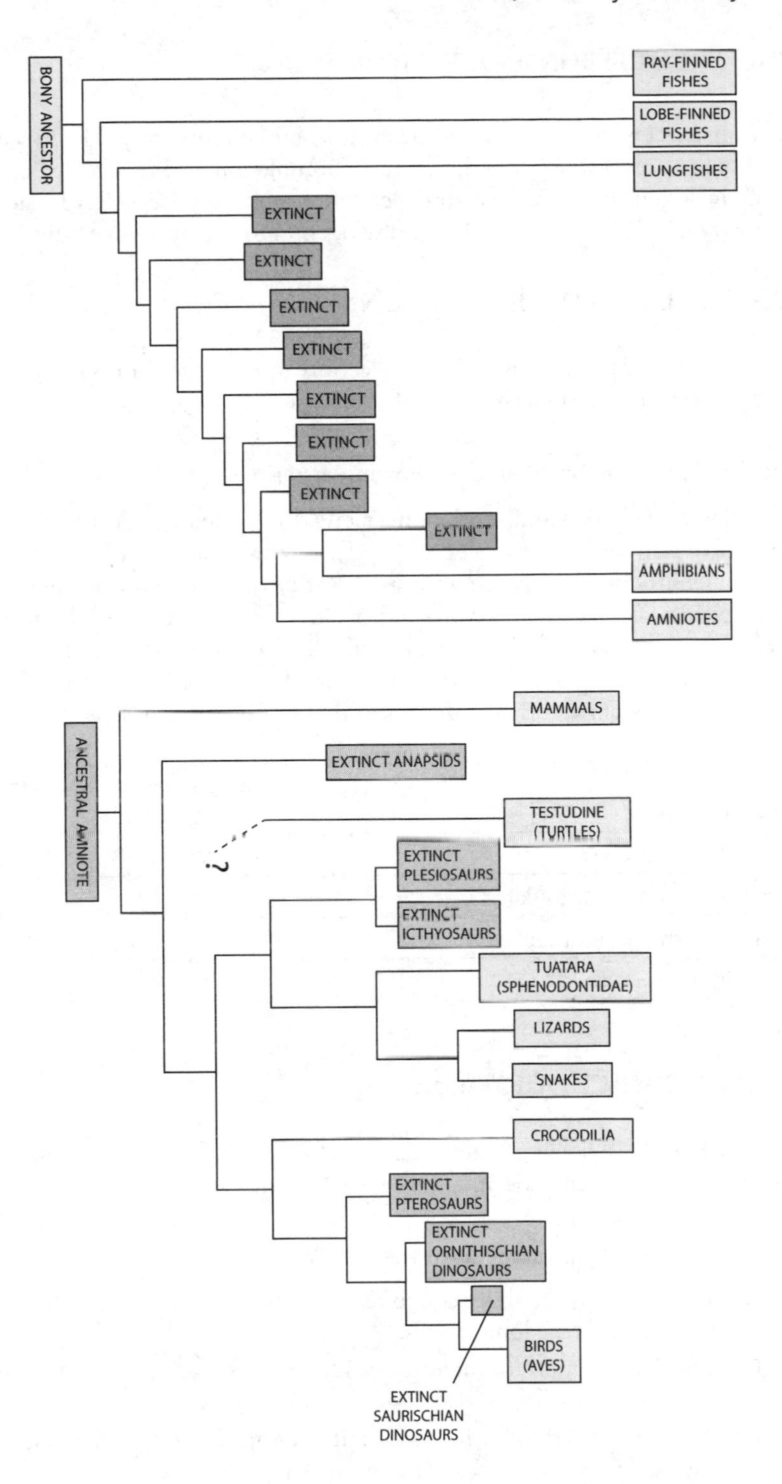
BONY ANCESTOR
RAY-FINNED FISHES
LOBE-FINNED FISHES
LUNGFISHES
EXTINCT
EXTINCT
EXTINCT
EXTINCT
EXTINCT
EXTINCT
EXTINCT
EXTINCT
AMPHIBIANS
AMNIOTES
ANCESTRAL AMNIOTE
MAMMALS
EXTINCT ANAPSIDS
TESTUDINE (TURTLES)
?
EXTINCT PLESIOSAURS
EXTINCT ICTHYOSAURS
TUATARA (SPHENODONTIDAE)
LIZARDS
SNAKES
CROCODILIA
EXTINCT PTEROSAURS
EXTINCT ORNITHISCHIAN DINOSAURS
BIRDS (AVES)
EXTINCT SAURISCHIAN DINOSAURS

SHORT FREE-RESPONSE QUESTION

13. **Construct** a hypothesis about the evolution of endothermy. The hypothesis should address whether endothermy is homologous (i.e., it evolved once) or analogous (i.e., it evolved independently at least twice) in birds and mammals. Refer to the table and phylogenies to **support** your claim.

SHORT FREE-RESPONSE QUESTION

14. Evolution occurs by both natural selection and genetic drift. **Compare** and **contrast** these two mechanisms of evolution.

Questions 15–17 are based on the following experiment.

Male widowbirds have significantly longer tails than female widowbirds. In this experiment, captive male widowbirds had their tails shortened (by cutting), artificially lengthened (by gluing on longer feathers), or left with normal length tails (by either keeping their natural tail or cutting and replacing the tail with feathers the same length as the normal tail). The birds were allowed to establish and defend nesting sites and the average number of nests is shown in the following table. The number of nests is an indicator of the number of mates per male.

Group of males	Average number of nests per male
Artificially lengthened	2.0
Normal length, unmanipulated	1.0
Normal length, cut and replaced	1.0
Artificially shortened	0.5

MULTIPLE-CHOICE QUESTIONS

15. Which of the following statements most likely explains the difference in tail length between male and female widowbirds?

(A) Males with the longest tails leave behind more offspring than females with the longest tails.

(B) Females with the shortest tails leave behind more offspring than females with the longest tails.

(C) Positive selection for increased tail length only acts on the males in the population.

(D) Negative selection for increased tail length only acts on the females in the population.

16. Which of the following hypotheses is best supported by the data above?

(A) Increased tail length improves the reproductive success of male widowbirds.

(B) There is no difference in the reproductive success between normal length tails that are naturally occurring and those that are cut and replaced.

(C) Males with longer tails mate with a greater number of females.

(D) Males with shortened tails leave behind less offspring than males with longer tails.

17. Which of the following statements most accurately explains the purpose of having two groups with normal tail length?

(A) The normal length, cut and replaced group is a control for the artificially shortened group.

(B) The normal tail length is the most common in the wild, so two groups are needed to represent them in the experiment.

(C) The purpose is to test the effect of artificial tails on female choice.

(D) The purpose is to control for the effect artificial tails have on bird flight.

Questions 18 and 19 refer to the following table showing the characteristics of the four major plant groups.

Plant group	Gametophyte or sporophyte dominant	Vascular tissue?	Requires water for reproduction?	Pollen and seeds?	Fruit and flowers?
Mosses	Gametophyte	No	Yes	No	No
Ferns	Sporophyte	Yes	Yes	No	No
Gymnosperms	Sporophyte	Yes	No	Yes	No
Angiosperms	Sporophyte	Yes	No	Yes	Yes

18. Which of the following statements most accurately describes the trend in plant evolution indicated in the table above?

(A) Plants evolved mechanisms for inhabiting many terrestrial habitats by evolving mechanisms of water transport and reproduction that do not require water.

(B) Plants were only able to survive on land once they evolved vascular tissue.

(C) Pollen and seeds evolved so that plants could live on land without drying out.

(D) Flowers and fruits evolved so plants could store energy on land without having to rely on seeds.

19. One aspect of the lifestyle of plants gives rise to a whole suite of related challenges. Which aspect of plant life is the root cause of the other challenges listed here?

(A) avoiding predation (herbivory)
(B) choice of location
(C) mate choice
(D) sessile existence

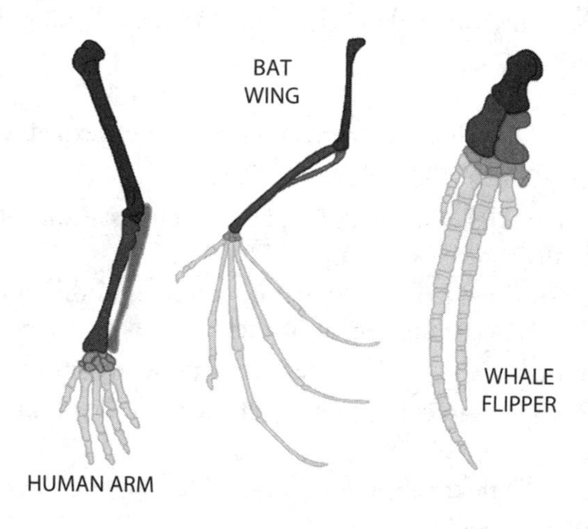

20. The structural similarities between the forelimbs shown in the figure above are evidence that:

(A) divergent evolution of a common ancestor occurred.
(B) humans are more closely related to bats than to whales.
(C) convergent evolution produces homologous structures.
(D) vertebrates show a limited number of body plans.

21. Insects first evolved approximately 479 million years ago. The first flying insects evolved around 406 million years ago. Birds evolved from therapod dinosaurs approximately 160 million years ago. The functional similarity between the wings of a bird and the wings of an insect is:

(A) the result of divergent evolution.
(B) an example of analogous structures.
(C) the result of the convergent evolution of two common ancestors.
(D) the direct result of adaptive radiation.

22. The frequency of a particular trait in a wild population increased over three generations. Which of the following most accurately explains the allele frequency change?

(A) The trait produced by the allele was adaptive.
(B) Evolution by natural selection occurred.
(C) The allele was created by a mutation that was selected for over the two generations.
(D) Unless there is evidence that natural selection occurred, the null hypothesis, that genetic drift is the cause of the change, should be assumed.

Question 23 refers to the following graph showing the distribution of allele frequencies between several populations of size *N*. Each curve represents the distribution of allele frequencies among different populations as genetic drift proceeds over time. Each population began with the same initial allele frequency. The *y*-axis shows the changes in the probability that the allele will have a particular frequency (allele frequencies are represented on the *x*-axis).

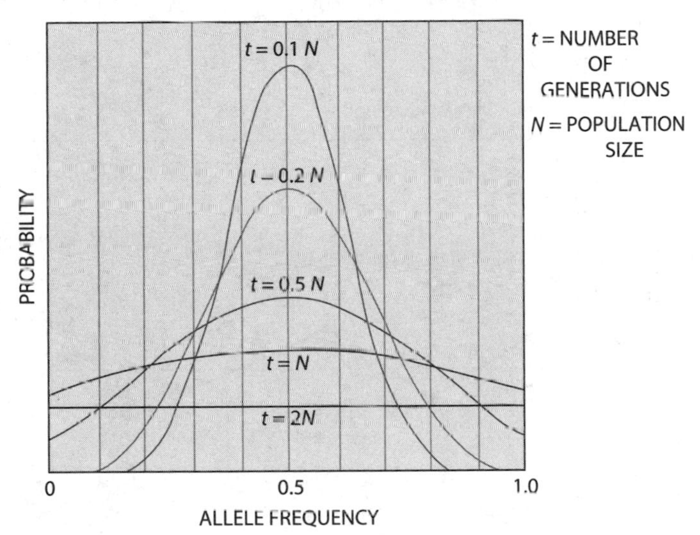

23. Which of the following statements accurately summarizes the graph? Assume genetic drift is the only force acting to change allele frequencies.

(A) Smaller populations are most likely to maintain allele frequencies of 0.5 as a result of genetic drift.
(B) Over the course of many generations, the allele frequencies drift toward 0 (the allele is lost) or 1 (the allele is fixed) and at $t = 2N$, all allele frequencies between 0 and 1 are equally likely.
(C) If there is no selective pressure on an allele, its frequency is dependent only on population size.
(D) An allele is more likely to become fixed in a large population.

Question 24 refers to the following figure and data. *Perissodus microlepis* is a scale-eating fish whose population in under frequency-dependent selection. The frequency of right- and left-mouthed individuals oscillates over time but balancing selection (due to frequency dependence) keeps the frequency of each phenotype to approximately 50%. The trait is determined by two alleles by simple Mendelian inheritance. Open circles indicate the frequency of left-mouthed individuals among adults that reproduced in the 3 years that data was collected.

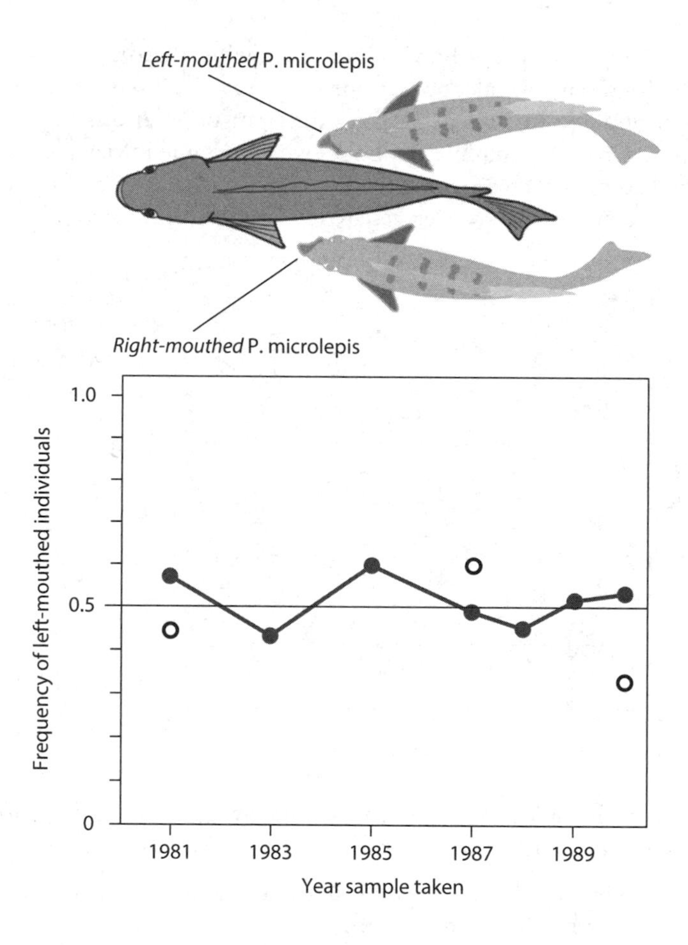

GRID IN

24. Calculate the change in frequency of left-mouthed individuals between 1985 and 1988. Report your answer as a whole number percent. Include a negative sign, if appropriate.

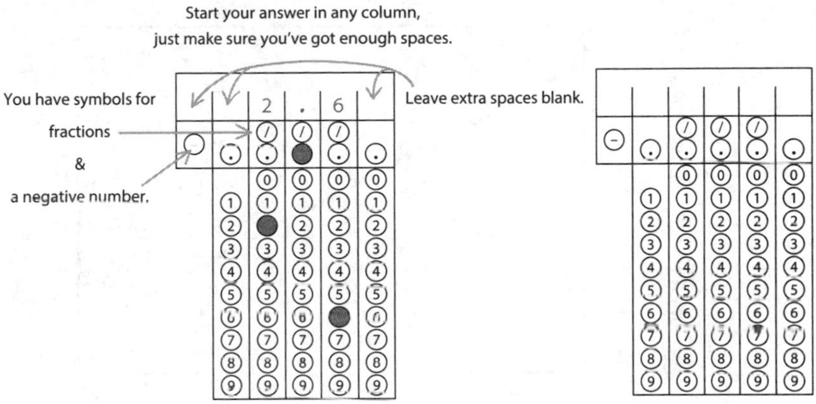

Start your answer in any column, just make sure you've got enough spaces.

You have symbols for fractions & a negative number.

Leave extra spaces blank.

MULTIPLE-CHOICE QUESTIONS

25. Which of the following best explains the role of variation in natural selection?

(A) Variation is not necessary for evolution, but increases the speed at which evolution occurs.

(B) Variation impedes natural selection but is an inevitable consequence of sexual reproduction.

(C) Environmental pressures applied to a population of genetically different individuals may result in divergent evolution.

(D) Variations result in new species, regardless of the selective pressures present.

Questions 26–28 refer to the following graph, which shows the relationship between first mean flowering dates and spring temperatures for six species. Data was gathered over several years. Each dot represents the first flowering date of a particular species for a given year. (Data adapted from Ellwood and colleagues, 2013.)

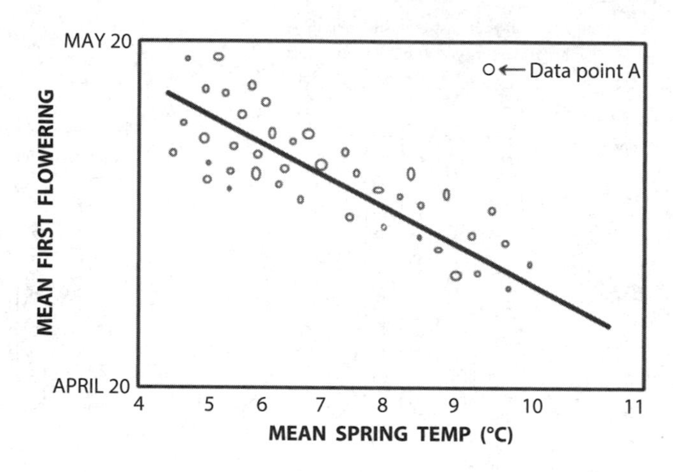

26. Which of the following statements most accurately describes the data above?

(A) As mean spring temperature increases, the time required for flowering decreases.

(B) Plants flower earlier when spring temperatures are higher.

(C) Shorter winters tend to promote early flowering.

(D) Species that flower early in the spring prefer higher temperatures.

27. What is the relevance of data point A?

(A) Point A is a clear outlier and is most likely due to an error in data collection.

(B) Point A represents a species that does not adhere to the trend expressed in the graph and may be interesting to study for this reason.

(C) Point A represents a species that does not adhere to the trend expressed in the graph and therefore should not be included in the study.

(D) Point A represents a species that does not adhere to the trend expressed in the graph, so the trend may not be biologically relevant.

28. Which of the following can be reasonably inferred from the data above?

(A) Increased temperature negatively affects plant flowering.
(B) Insects are more active when the temperature is higher.
(C) Plant reproduction is responsive to environmental cues.
(D) The timing of plant flowering is genetically controlled.

SHORT FREE-RESPONSE QUESTION

29. Flowering is an energetically expensive process for a plant. For many species of angiosperm, in order to reproduce sexually, flowering must occur within a particular window of time. **Describe** how the environment applies selective pressure to this trait (the timing of flowering). Consider both biotic and abiotic factors. Include a **description** of the selective pressures this may put on another organism that interacts with the plant.

Questions 30 and 31 refer to the following graphs, each of which shows the genotype frequencies of the gametes of the F_1 generation that resulted from the random mating between the two parent populations, one with the genotype $A_1A_1B_1B_1$ (only) and the other with genotype $A_2A_2B_2B_2$ (only). The two loci, A and B, each have two alleles, 1 and 2.

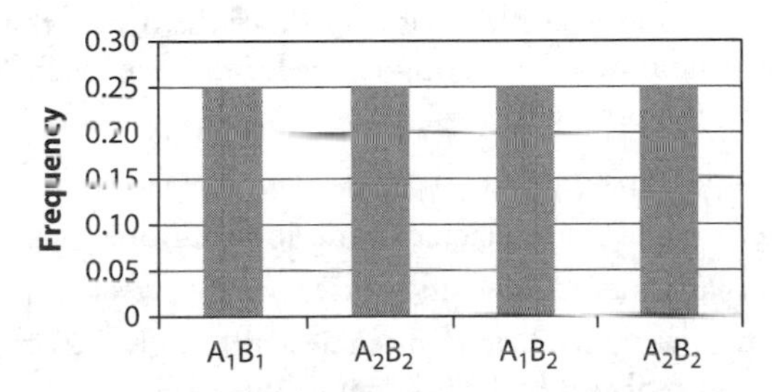

MULTIPLE-CHOICE QUESTIONS

30. Which of the following statements is true if the mating between the two populations produced offspring with the gametic frequencies shown above?

(A) Genes A and B must be on different chromosomes.
(B) Independent assortment is not acting during this time.
(C) The alleles for genes A and B segregate randomly during meiosis.
(D) The traits for genes A and B were not acted upon by natural selection in the production of this generation.

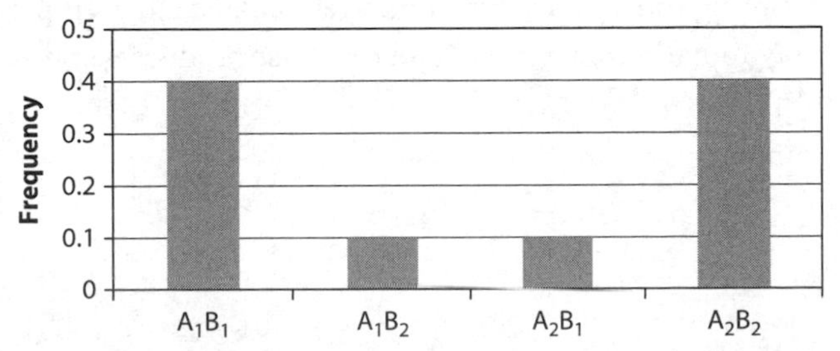

31. Which of the following statements is true if the mating between the two populations produced offspring with the gametic frequencies shown above?

(A) Allele A_1 is dominant to A_2 and allele B_1 is dominant to B_2.
(B) Genes A and B assort independently.
(C) The traits for genes A and B were acted upon by natural selection in the production of this generation.
(D) Genes A and B are located relatively close together on the same chromosome.

LONG FREE-RESPONSE QUESTION

32. Can natural selection perfect a species? **Explain. Support** your choice with examples from nature.

MULTIPLE-CHOICE QUESTIONS

33. A biologist created a computer program to simulate a large population of insects. The program specifies that no mutations occur, there is no natural selection or gene flow, and mating is random. After several generations, the population in the program:

(A) evolved, but may not yet have speciated.
(B) will get smaller due to lack of genetic variation.
(C) will get larger and larger since there are no selective pressures.
(D) will have the same allele frequencies, although the original members have been replaced by later generations.

34. In an experiment, lab mice have their tails cut off soon after birth for 18 generations, but all the offspring of the 19th generation have tails of the same average length as the first generation. Which of the following correctly explains this observation?

(A) This observation is explained by natural selection.
(B) Offspring can inherit characteristics acquired by the parent in their lifetime but only if there is a selective pressure to maintain the trait.
(C) Offspring usually have little or no resemblance to their parents.
(D) Only mutations in gametes and cells that produce gametes affect offspring.

35. Which of the following organisms would be considered the most fit?

(A) a young female bear with five baby cubs and three males competing to mate with her
(B) an old, male bear whose two offspring each have two offspring
(C) a young male bear with three offspring who has just killed an unrelated bear before he was able to reproduce
(D) an adolescent female bear who has three mature males (each with two offspring) competing to mate with her

36. In northern elephant seals (*Mirounga angustirostris*), only a few large males compete for the much smaller females. Each of these males will father many offspring by many females. Most of the males in the population do not get the opportunity to breed. From a genetic point of view:

(A) the rate of genetic drift would be higher than expected for a population of equal size.
(B) the frequency of heterozygotes in the population is higher than expected for a population of equal size.
(C) males and females will eventually become reproductively isolated due to the dramatic difference in body size.
(D) small males will devise a new reproductive strategy.

37. Which of the following is *not* a mechanism by which new genes are generated in organisms?

(A) Existing genes are modified by mutations in its DNA sequence.
(B) Two or more segments of a gene are recombined.
(C) Gene duplication and mutation create new alleles.
(D) DNA nucleotide chains are randomly assembled by DNA polymerase without a template strand.

Questions 38–42 refer to the following graph of allele frequencies in a hypothetical population of tropical insects. Their color is determined by a single gene with two alleles. M (red) is dominant to m (black). Allele frequencies over 35 years are summarized as follows.

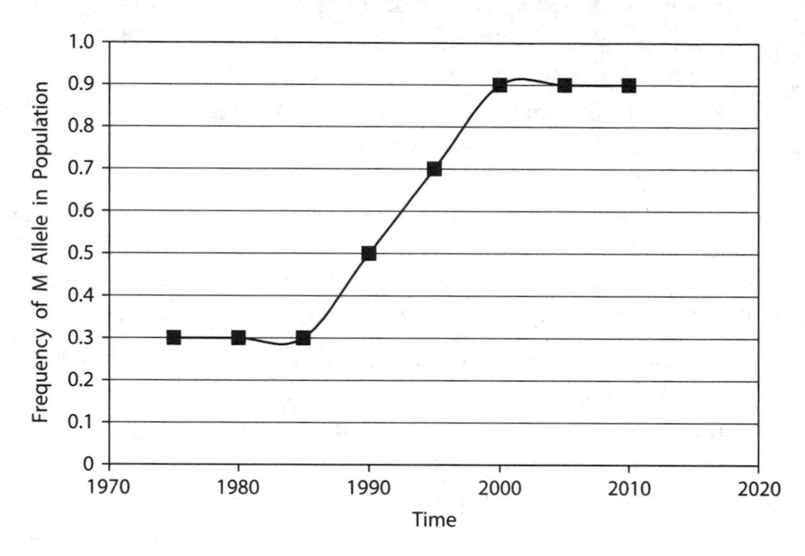

38. During which period of time(s) was the insect population in the Hardy–Weinberg equilibrium for M allele?

(A) 1975–2010
(B) ~1985–2000
(C) 2000–2010 only
(D) 1975–1985 and 2000–2010

GRID INS

39. In 1980, what percentage of **red** insects are heterozygous? Report your answer as a whole number.

40. In 2005, what percentage of the population is expected to be homozygous? Report your answer as a whole number.

41. In 2005, what percentage of the population is expected to be red? Report your answer as a whole number.

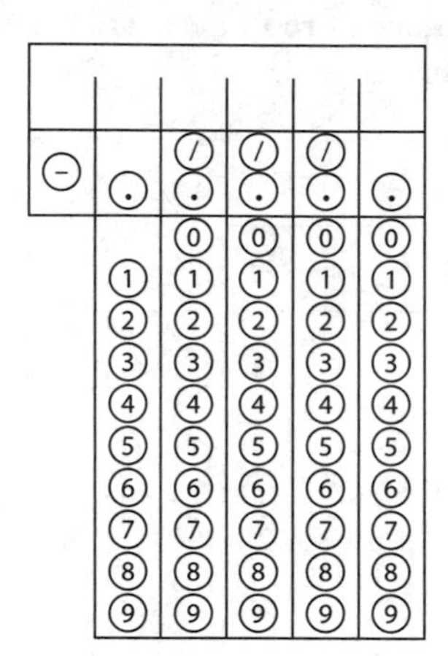

Questions 42 and 43 refer to experiments performed by Lenski and colleagues (2000) who studied the role of new, beneficial mutations in bacteria for 20,000 generations. Each population was initiated with a single individual and was therefore genetically uniform at the start. Samples taken at different times from an evolving population were stored and their fitness directly compared later upon revival. (Bacteria can be frozen, during which time they remain alive but do not undergo reproduction and therefore no genetic change occurs in the population. Later, they can be revived.) The results are shown as follows.

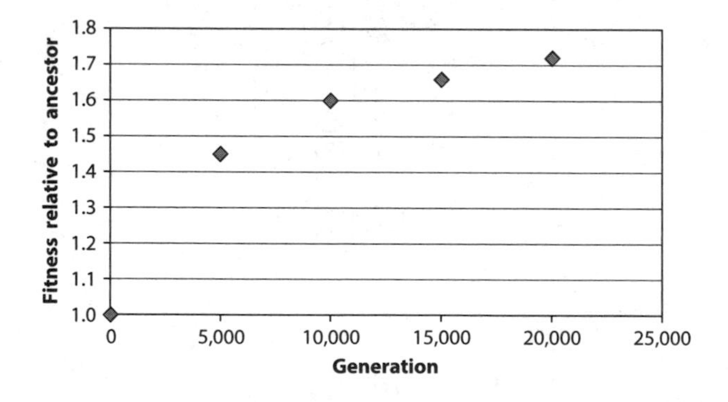

MULTIPLE-CHOICE QUESTIONS

42. Which of the following statements most accurately summarizes the results?

(A) The increase in adaptation was due to natural selection acting on new, advantageous alleles created by mutations.
(B) Most advantageous mutations occurred after approximately 10,000 generations.
(C) Most mutations are harmful, so natural selection most likely acted on pre-existing variation.
(D) After 20,000 generations, the bacteria are no longer able to reproduce with the bacteria from the initial population.

Question 43 also refers to the following information.

In a similar experiment, the medium in which the bacteria were grown contained both glucose and citrate. Normally, bacteria cannot use citrate for energy. After 33,000 generations, one population increased dramatically in density because a mutation enabled it to use citrate. Billions of mutations had occurred in the previous populations.

43. Which of the following statements is a logical hypothesis based on these observations?

(A) Mutations are always beneficial for bacterial populations.
(B) Several mutations are necessary to create a new, useful allele.
(C) Natural selection limits the rate of the mutational process.
(D) The evolution of new characteristics may rely on rare combinations or sequences of mutational events.

LONG FREE-RESPONSE QUESTION

44. Explain the mechanism(s) by which new alleles are created.

Questions 45 and 46 refer to the following graph.

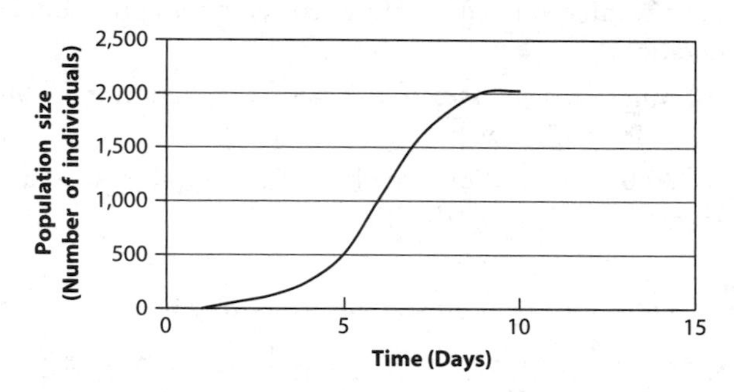

GRID IN

45. Use the graph to calculate the mean rate of population growth per day between days 4 and 8. Give your answer to the nearest tens place.

−	.	/ .	/ .	/ .	.
		0	0	0	0
	1	1	1	1	1
	2	2	2	2	2
	3	3	3	3	3
	4	4	4	4	4
	5	5	5	5	5
	6	6	6	6	6
	7	7	7	7	7
	8	8	8	8	8
	9	9	9	9	9

SHORT FREE-RESPONSE QUESTION

46. The previous graph shows the typical growth pattern when an initially small population (at Time 0) is introduced into a new area. **Explain** why the population grows rapidly and why that rapid growth eventually slows.

Questions 47–49 refer to the data below. The alleles show complete dominance (the heterozygote has the dominant phenotype). The population from which the data were taken has random mating. Assume all individuals are diploid.

Generation	Number of individuals with the *dominance* phenotype	Number of individuals with the *recessive* phenotype
1	880	120
2	800	200
3	740	260
4	710	290
5	660	340
6	650	350
7	655	345

GRID INS

47. Estimate the number of recessive alleles in generation 1 of the population. Report your answer as a whole number.

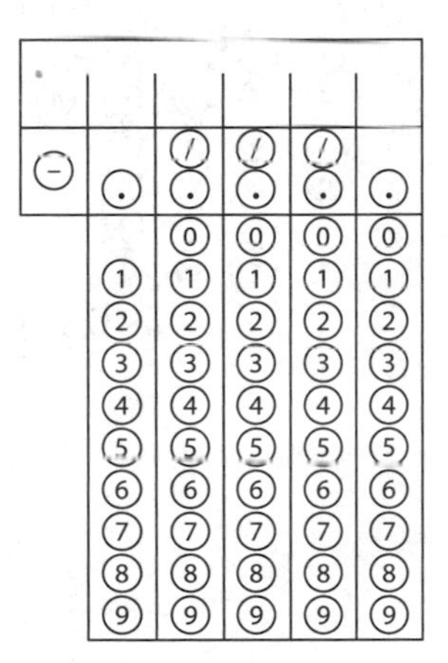

48. What is the change in the recessive allele frequency between generations 1 and 7? Report your answer as a whole number.

49. What percentage of the population in generation 5 is heterozygous? Report your answer as a whole number.

Question 50 refers to the following sequence data. (**Important note**: It is unlikely the AP Biology exam would include a question with as much sequence data as in this question. It would take too long to answer. However, one long question instead of a several short, easy questions makes the analysis more useful, challenging, and illuminating. **Be patient, it's a tough problem.** You may want to make copies of the sequence data and even cut out the individual sequences so you can line them up and sort them out spatially.)

Species	Sequence	Age of oldest fossil (mya = millions of years ago)
A	A A C G C T T A A G	75 mya
B	C T T A C T T C C G	
C	G T T A C T T C C G	150 mya
D	G T T A C C T C C G	
E	A A C G A T T A A T	
F	A A C G T T T A A T	
G	A A T G C T T C A G	100 mya, extinct
H	A T T G C T A C A G	400 mya, extinct
I	A T T A C T T C A G	
J	A T T G C T T C A G	200 mya, extinct
K	A A T G C T T A A G	150 mya, extinct
L	A T T A C T T C C G	120 mya, extinct
M	A A C G C T T A A T	

LONG FREE-RESPONSE QUESTION

50. Construct a phylogeny using the preceding sequence data. **State** three assumptions you used in constructing the phylogeny.

Question 51 refers to an experiment done with four identical populations of 1,000 houseflies (*Musca domestica*) taken from a natural population that was established in a laboratory. Each population was subject to a random bottleneck of either 1, 4, or 16 mating pairs of flies. When the population reached approximately 1,000 flies, each population was subject to the same initial bottleneck (1, 4, or 16 mating pairs) as the previous population. Allele frequencies at four loci were estimated for each population after each recovery (population = 1,000 individuals) from the bottleneck. A diagram of the experimental procedure follows. (Adapted from McCommas and Bryant, 1990.)

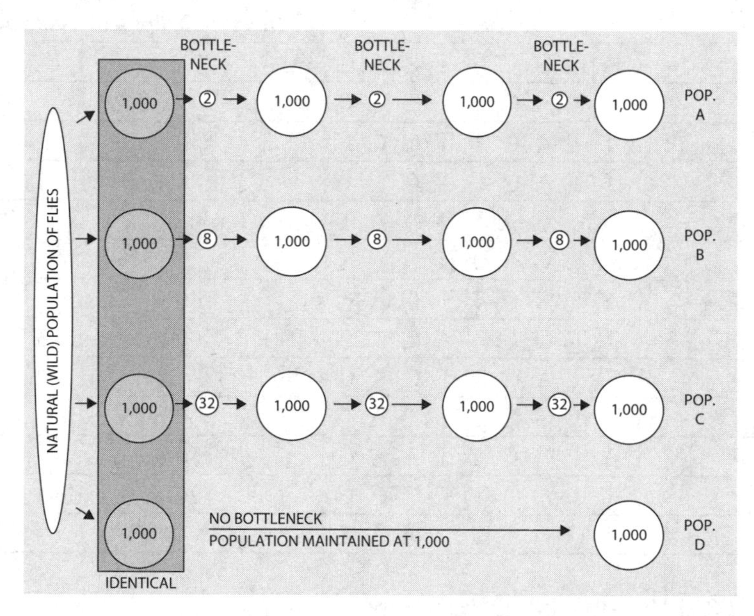

MULTIPLE-CHOICE QUESTIONS

51. Which of the following states the expected results based on the current model of genetic drift?

(A) Heterozygosity decreased after each bottleneck. Population A experienced the greatest rate in reduction of heterozygotes.

(B) Heterozygosity increased after each bottleneck. Population A had the greatest heterozygosity.

(C) In all populations except D, the allele frequencies for at least one of the four loci became fixed or lost.

(D) Population D was the least genetically similar to the wild population the flies were obtained from.

Questions 52 and 53 refer to the following data comparing the rate of recessive, harmful allele removal versus the frequency of the allele for 4 alleles under different strengths of selection. The *x*-axis shows the frequency with which the deleterious allele occurs in the population. The *y*-axis shows the rate of removal of the allele from the population as a function of its frequency.

s = strength of selection
0 = neutral, 1 = lethal

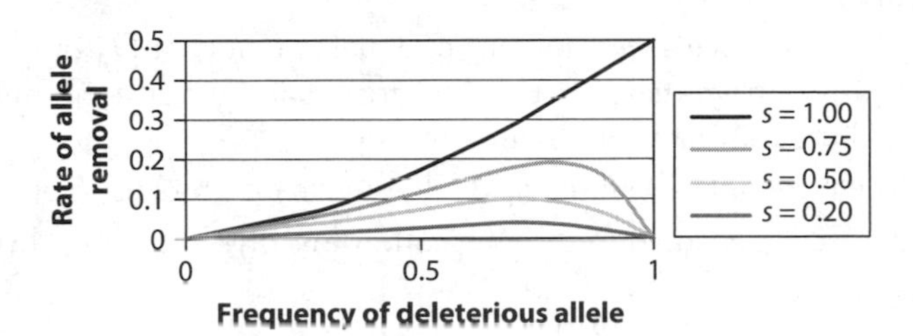

52. The rate of evolution is greatest in which of the following situations?

(A) $s = 1$
(B) $s = 2$
(C) $s = 3$
(D) $s = 4$

SHORT FREE-RESPONSE QUESTION

53. State two factors that affect the time required for a dominant allele to replace a recessive allele in a population. **Explain** how these factors affect the time it takes for one allele to replace the other and **state** if (and how) it would be different for a recessive allele to replace a dominant allele. Assume *no gene flow.*

MULTIPLE-CHOICE QUESTIONS

54. Which of the following most accurately states the effect of selection on the frequency of dominant versus recessive alleles?

(A) Dominant alleles are subject to selection even at low frequencies.

(A) Lethal, recessive alleles' frequencies reach zero if selected against for a long enough time.

(B) Dominant alleles are more affected by negative than positive selection.

(C) Recessive alleles are more subject to selection than dominant alleles because there is only one genotype for the recessive phenotype.

55. Fragments from a small meteorite that fell to Earth approximately one hundred million years after Earth formed were found to contain more than 80 amino acids. Which of the following true statements is the best evidence that the amino acids were *not* contaminants from organisms from Earth?

(A) The proportions of amino acids found in the meteorite were similar to those produced in the Miller–Urey experiment (showing that organic molecules can be synthesized from inorganic precursors).
(B) The amino acids were present in equal amounts of D- and L-isomers (mirror image enantiomers), but organisms on Earth can only make and use L-isomers (with rare exceptions).
(C) There are only 20 amino acids that make up proteins in organisms.
(D) Molecules exposed to extreme conditions often undergo chemical changes.

56. Which of the following true statements does *not* support the idea that RNA was likely the first genetic material?

(A) Cells typically contain about 8 times more RNA than DNA.
(B) RNA plays a central, informational role in protein synthesis.
(C) RNA can replicate by Watson–Crick base pairing rules.
(D) Single-stranded RNA molecules can take on a variety of three-dimensional shapes and can carry out several catalytic functions.

Question 57 refers to the following diagram.

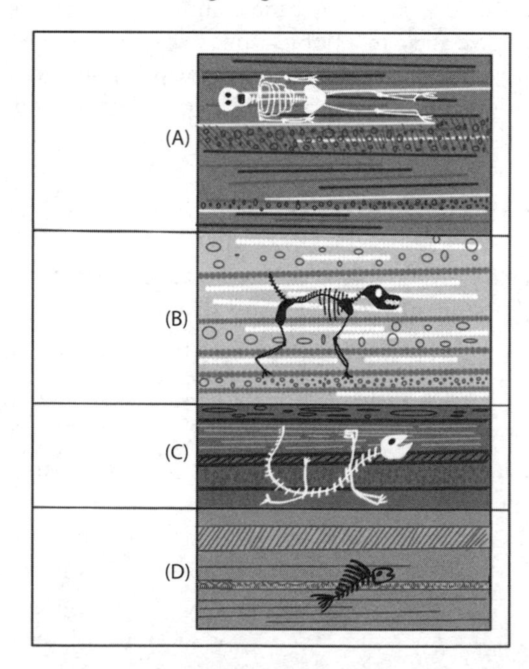

57. In which layer would scientists be most likely to find the fossils of vertebrates making the transition from sea to land?

(A) layer D
(B) between layers D and C
(C) layer C
(D) between layers C and B

58. When planning a fossil hunt for marine fossils of the Devonian period (354 million years ago), which of the following is the most crucial aspect of planning?

(A) Make sure you pack the equipment used for radioactive dating.
(B) Locate sedimentary layers of 354 million years ago in coastal regions.
(C) Find sites where sedimentary rocks of approximately 354 million years ago are exposed and accessible.
(D) Collect fossils of various ages and place them in order of relative age.

Question 59 is based on the graph below. Iridium is one of the least abundant elements in the Earth's crust but is often found at levels of 4–5 parts per million (notice graph is parts per billion) in meteorites that have hit the planet.

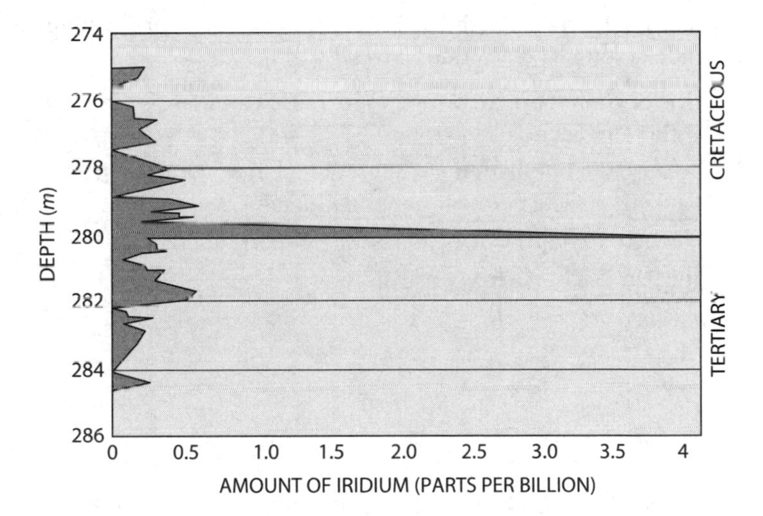

59. Which of the following hypotheses is *not* supported by the evidence that follows (hypothesis: evidence)?

(A) *Rapid evolution/punctuated equilibrium:* different compositions of taxa are represented above and below the iridium rich layer.
(B) *Mass extinction:* many more fossils are present directly below compared to directly above the iridium rich layer.
(C) *Iridium layer poisoned some plant life:* the composition of plant biomass in the taxa above the iridium rich layer is different from the composition of plant biomass below the iridium-rich layer.
(D) *Iridium-rich cloud blocked out the sun:* greater biomass of plant life directly above the iridium layer.

Question 60 refers to the following information.

Ashkenazi Jewish populations in Eastern Europe have a relatively high frequency of the allele that causes Tay-Sachs disease in which homozygotes for the trait die at an early age. The frequency of the allele is approximately 0.03 in the Ashkenazi Jewish population but only about 0.002 in the general population. Grandparents of Tay-Sachs carriers die from proportionally the same causes as grandparents of noncarriers.

60. Which of the following is the most likely explanation for the higher allele frequency despite its deleterious effects?

(A) One or more bottlenecking events caused an increase in the frequency of the Tay-Sachs allele.
(B) Because the population is a subset of a larger population, the Tay-Sachs allele frequency appears higher.
(C) The heterozygous case confers resistance to tuberculosis.
(D) The Tay-Sachs allele is a transposon.

Question 61 refers to the following data. Allozymes are variant forms of an enzyme coded for by different alleles at the same locus. Allozyme heterozygotes have two different alleles for a particular enzyme. Allozyme heterozygosity may or may not affect the phenotype. Genotypes that are heterozygous at several or many loci often appear to be fitter than more homozygous genotypes. (Data adapted from Soulé, 1976.)

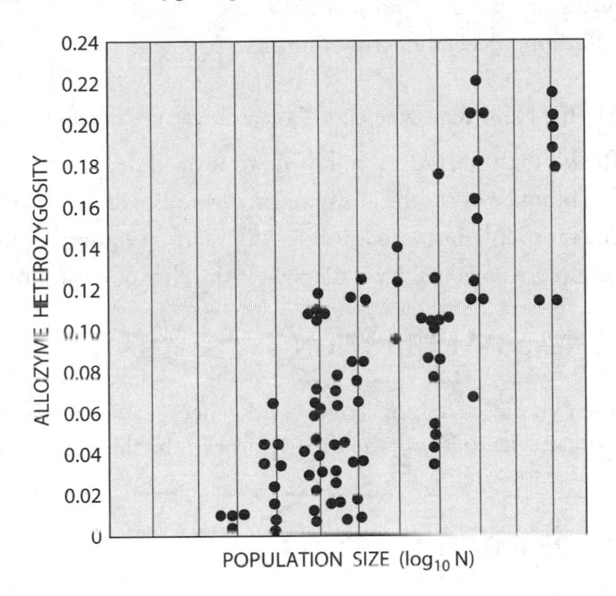

61. Which of the following true statements is *least* illustrative of the data above?

(A) European populations of the ruddy duck are thought to stem from 7 individuals that escaped captivity. The North American species has 23 haplotypes compared with 1 haplotype found in European ducks.

(B) In locations where *Plasmodium*, the protist that causes malaria, is a significant threat. Heterozygotes for the sickle-cell trait have an adaptive advantage: they are resistant to malaria.

(C) Frequency-dependent selection can confer greater fitness to a rare phenotype in some cases and a more common phenotype in other cases.

(D) Larger populations tend to exhibit greater genetic diversity than small populations.

Questions 62 and 63 refer to an experiment in which researchers collected data on the number and colors of moths eaten by birds, in two different forests. The following data were obtained.

	Moths eaten by birds	
	Light	**Dark**
Unpolluted woods	26	164
Polluted woods	43	15

62. Which of the following statements is the most reasonable inference of the data in the table?

(A) Dark moths are the preferred food source of birds.
(B) Dark moths are rare in polluted woods.
(C) Birds use vision to find food.
(D) Pollution affects the taste of moths.

63. Which of the following statements is a reasonable prediction based on the data?

(A) The biomass of birds in polluted woods will decrease.
(B) The biomass of moths in unpolluted woods will decrease.
(C) The ratio of light to dark moths will increase in unpolluted woods.
(D) Polluted woods will have the lowest number of dark moths.

LONG FREE-RESPONSE QUESTION

64. Describe two pieces of biochemical and genetic evidence from extant and extinct organisms that support the hypothesis that all organisms on Earth share a common origin of life.

SHORT FREE-RESPONSE QUESTION

65. Use the data below to **construct** a cladogram. **State** two hypotheses of the cladogram.

Plant	Produces seeds	Vascular tissue	Pollen	Flowers	Chlorophyll a
Fern		X			X
Moss					X
Conifer	X	X	X		X
Monocot	X	X	X	X	X
Dicot	X	X	X	X	X

Question 66 refers to the following graph. The effects of 665 newly arisen single mutations in laboratory cultures of *Pseudomonas fluorescens* were studied. Of the 665 mutations, 28 were shown to have a beneficial effect, measured by an increased growth relative to the ancestral genotype. The fitness effect of the 28 beneficial mutations measured as their *increase* in growth rate relative to the ancestral genotype is shown. If a particular genotype has a fitness effect of 0.07, it means the population with that genotype grows 7% faster than populations of the ancestral genotype. (Adapted from Kassen and Bataillon, 2006.)

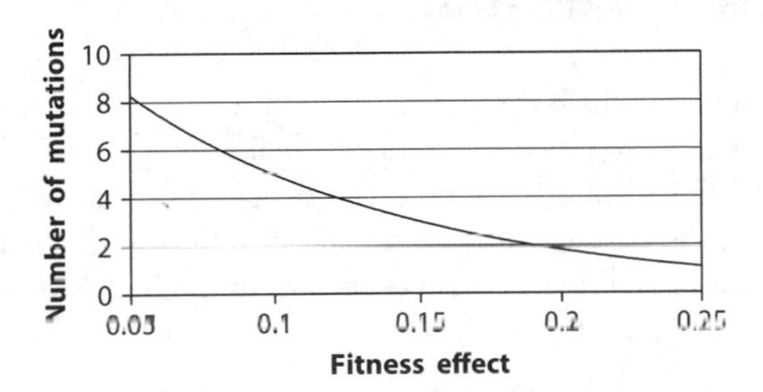

MULTIPLE-CHOICE QUESTIONS

66. Which of the following most accurately summarizes the results of the experiment?

(A) *Pseudomonas* with three mutations grow 15% faster than *Pseudomonas* with the ancestral genotype.

(B) Most of the mutations conferred a small fitness advantage and a small number conferred a relatively large fitness advantage.

(C) The lower the number of mutations in *Pseudomonas*, the greater the fitness effect.

(D) The majority of the mutations in *Pseudomonas* are deleterious, a small number conferred a slight advantage, and very few conferred a significant advantage.

Questions 67 refers to the following information.

The presumed ancestor of all placental mammals is thought to be a furry tailed, tree-climber and insect eater that weighed between 6 and 245 grams. It had a highly folded brain and three pairs of molars on each jaw. The placental mammals are believed to have diversified after the K/T extinction. No one has ever found fossils of placental mammals from before the K/T extinction event 65 million years ago. Genetic studies, however, place the origin of the placental mammals at 100 million years ago.

LONG FREE-RESPONSE QUESTION

67. Did placental mammals first appear before or after the K/T extinction? **State** the type of data that would convincingly support each of the two alternate hypotheses. Include a **description** of what would need to be measured or observed and the expected difference between pre-extinction and post-extinction diversification.

MULTIPLE-CHOICE QUESTIONS

68. Many bacteria obtain resistance to an antibiotic (such as ampicillin) by carrying a plasmid that contains a gene that confers resistance to the antibiotic. The plasmid is passed on to other bacteria through conjugation and can be passed onto offspring during binary fission. If bacteria already resistant to ampicillin were not exposed to ampicillin, which of the following would most likely be the effect over several generations?

(A) The gene for the ampicillin would eventually mutate to confer resistance to the new antibiotic.
(B) The plasmid containing the gene for ampicillin resistance would disappear from the population.
(C) Transmission of the plasmid containing the ampicillin resistance would increase as those bacteria without the plasmid would be killed.
(D) The transmission of the plasmid conferring ampicillin resistance would decrease but would not disappear from the population.

Question 69 refers to the following data. The figure shows the effects of single mutations on the fitness of experimental cultures of bacteria. The effects of 665 newly arisen single mutations in laboratory cultures of *Pseudomonas fluorescens* were studied. The fitness of ancestral genotype = 0.95. (Adapted from Kassen and Bataillon, 2006.)

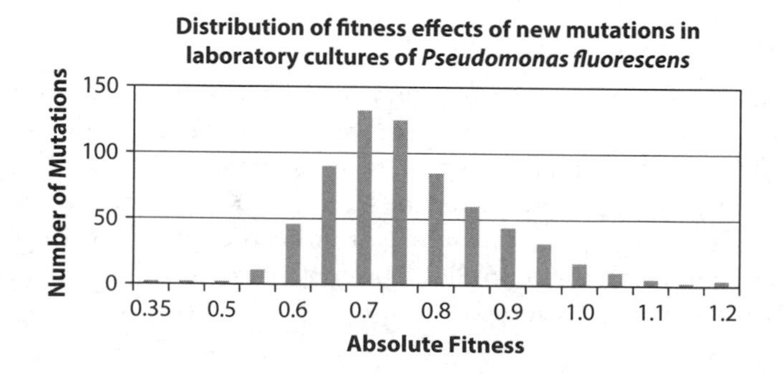

69. Which of the following statements most accurately summarizes the effect of mutations on fitness in *Pseudomonas fluorescens?*

(A) The majority of mutations have a small, deleterious effect on fitness.
(B) The majority of mutations have a large, deleterious effect on fitness.
(C) *Pseudomonas* individuals with about 25 mutations have the same absolute fitness as the ancestral genotype.
(D) *Pseudomonas* with approximately 125 mutations have the greatest absolute fitness.

SHORT FREE-RESPONSE QUESTION

70. Viruses are not considered alive, yet they can evolve. **Explain** how a nonliving entity can evolve by natural selection. **State** the minimum the requirements for natural selection to occur in a biological entity.

Questions 71 and 72 refer to the following chi-square (χ^2) table.

	Degrees of freedom							
P	**1**	**2**	**3**	**4**	**5**	**6**	**7**	**8**
0.05	3.84	5.99	7.82	9.49	11.07	12.59	14.07	15.51
0.01	6.64	9.32	11.34	13.28	15.09	16.81	18.48	20.09

MULTIPLE-CHOICE QUESTIONS

71. Two groups of genetically identical plants were grown in two different environments. At 12 weeks, the average height of one group was larger than the other. Which of the following statements most accurately describes the statistically significant difference in plant height between the two groups?

(A) A σ^2 value of 3.84 means the null hypothesis is correct and the apparent difference in height between the two groups of plants was a result of chance.
(B) A σ^2 value of 3.84 means that there is a 95% probability that the difference in height observed between the two groups is due to their different environments.
(C) A σ^2 value of 5.99 means that there is a 5% probability that the difference in height observed between the two groups is due to chance.
(D) A σ^2 value of 6.64 means that there is a 99% chance that the null hypothesis is incorrect.

72. A cross between two plants of genotypes AaBb × AaBB produces seemingly unusual results. A student performs a chi-square analysis to determine if the results are significant. Which of the following is the *lowest* χ^2 value that would indicate significance? Assume simple dominance inheritance pattern.

(A) 3.84
(B) 5.80
(C) 6.20
(D) 7.82

Questions 73 and 74 refer to the following diagram of the greenish warbler, *Phylloscopus trochiloides.* Each circle represents a species. A is believed to be the ancestral species, with radiation moving out (B → D → F → G and C → E → H).

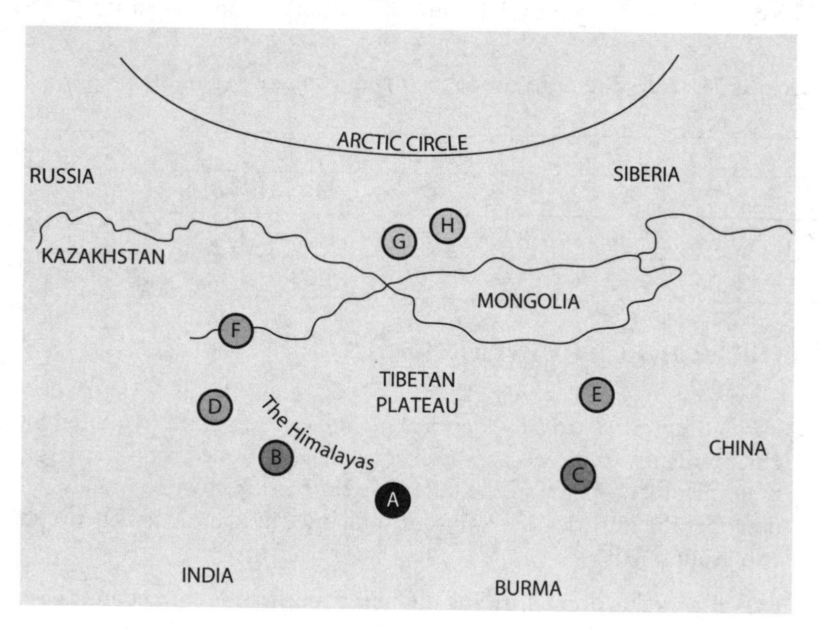

73. Which pairs of species are most likely to be *not* able to breed?

(A) A and H
(B) B and C
(C) D and E
(D) G and H

74. Of the following true statements, which is the *weakest* piece of evidence to support the proposed radiation?

(A) Plumage patterns and wing bar patterns change gradually from species A → G and A → H. Groups G and H differ most in their feather appearance.

(B) Species E and D have the same number of DNA sequence differences (compared to species A) in a specific, homologous region of DNA.

(C) Species A has the simplest song. Songs of the bird groups on each side of species A increase in complexity northward. Groups G and H have the most complex songs.

(D) Groups F and G can interbreed, groups C and E can interbreed, but groups G and H cannot interbreed.

Question 75 refers to the following information and diagram below.

Cabbage, brussels sprouts, cauliflower, kale, broccoli, and kohlrabi are all descended from a common ancestor, *Brassica oleracea*, the wild mustard plant. Breeders produced these varieties by selecting variations in different parts of the plant.

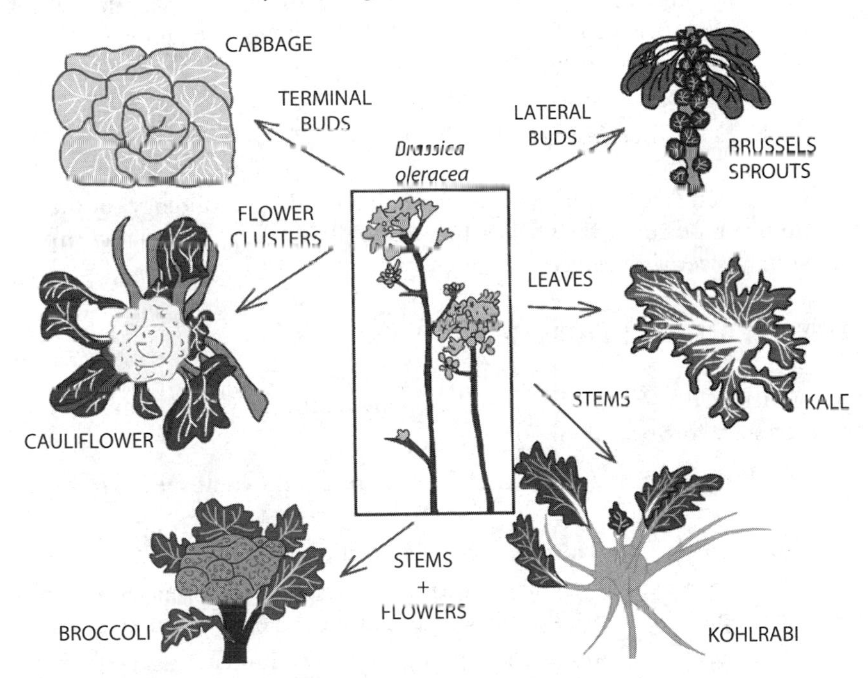

LONG FREE-RESPONSE QUESTION

75. **Select** one of the traits and describe how artificial selection for this trait could produce the new plant. **Compare** and **contrast** the mechanisms and consequences of artificial and natural selection.

MULTIPLE-CHOICE QUESTIONS

76. Mules are the offspring of male donkey ($2n = 62$) and a female horse ($2n = 64$). Mules have 63 chromosomes and are typically sterile, although rarely they may produce offspring with a donkey or horse. Which of the following statements most accurately explains this observation?

(A) Organisms with an odd number of chromosomes are sterile.
(B) Hybrid sterility occurs only when the nonhybrid populations or species have a different number of chromosomes.
(C) The genes for hybrid sterility are not expressed in nonhybrid individuals.
(D) The genes that contribute to hybrid sterility do not affect the fertility of the parent species; the effect likely stems from the interaction in hybrid individuals.

LONG FREE-RESPONSE QUESTION

77. Can genetic engineering (using recombinant DNA technology) of the human food supply be safely achieved? **Explain** your position and **support** your answer with evidence.

MULTIPLE-CHOICE QUESTIONS

78. Which of the following best explains why reproductive isolation is necessary for speciation?

(A) Pre- and post-zygotic mechanisms prevent different species from interbreeding.
(B) Gene flow is needed to prevent speciation.
(C) Isolation introduces different selective pressures, which differentially affect disparate gene pools or allow genetic drift to fix or eliminate certain alleles.
(D) Adaptive radiation isolates populations by eliminating gene flow.

Question 79 refers to the following data, showing the strength of prezygotic reproductive isolation among allopatric and sympatric pairs of *Drosophila* populations plotted against genetic distance.

Allopatric refers to a population or species that occupies a particular geographic region that is separate and different from that of another population or species.

Sympatric refers to two or more species or populations that occupy the same geographic location. Genetic distance is measured by the degree of difference in allozyme (one of several forms of an enzyme coded for by different alleles at a particular locus) or allele frequencies or in DNA sequences that are used as a molecular clock to estimate the relative time of divergence between pairs of populations. (Data adapted from Coyne and Orr, 1997.)

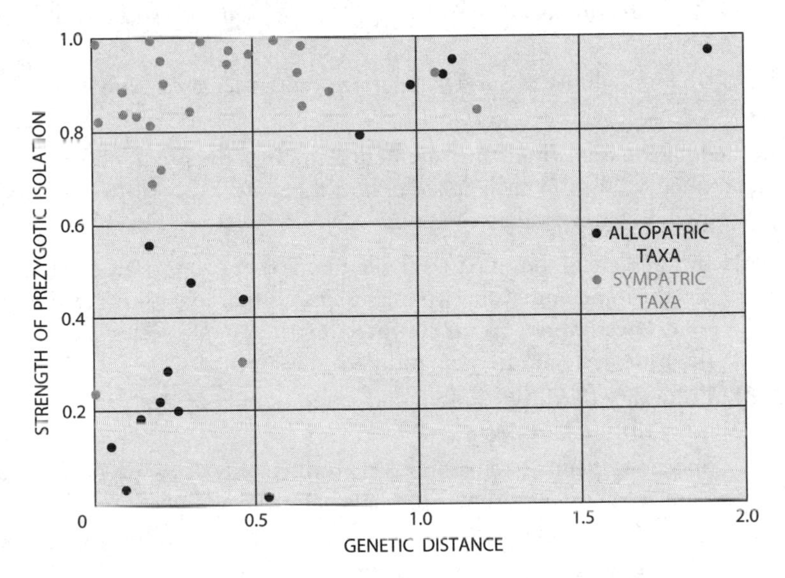

79. The data indicate that:

(A) prezygotic isolation is a stronger barrier to gene exchange than postzygotic isolation.

(B) prezygotic isolation is stronger among allopatric pairs compared with sympatric pairs of taxa.

(C) reproductive isolation evolved to prevent hybridization in allopatric taxa.

(D) prezygotic isolation mechanisms are stronger in sympatric taxa, whereas postzygotic isolation mechanisms are stronger in allopatric taxa.

Questions 80 and 81 refer to the age schedules of reproduction. **Semelparous** species have a life history in which individuals, especially females, reproduce only once. **Iteroparous** species have a life history in which individuals reproduce more than once.

80. In which of the following situations would semelparous reproduction be suboptimal?

(A) The rate of reproductive output increases exponentially with body mass.
(B) The rate of body mass growth decreases as the individuals increase in size.
(C) Reproduction is likely to fail in some years and be successful in others.
(D) The probability of survival decreases with increased body size.

81. Which of the following true statements supports the prediction that reproductive effort in each episode should be lower in iteroparous species as compared with semelparous species?

(A) Inflorescences (floral parts) make up a lower proportion of plant weight in perennial species (plant species that live for more than two that flower every year) as compared with annual species (complete their life cycle in one year and then die) of plants.
(B) Fecundity (fertility) is often correlated with body mass in species that grow throughout life.
(C) Species of plants like bamboos reproduce only after several years but produce massive numbers of seeds before they die.
(D) Reproducing at an early age may increase the risk of death, decrease growth, or decrease subsequent fecundity.

SHORT FREE-RESPONSE QUESTION

82. Inflorescence (clusters of flowers) make up a lower proportion of plant biomass in perennial species (reproduce every year) as compared with annual species (reproduce only once). **Explain** this observation with reasoning that is consistent with evolutionary theory.

Questions 83–85 refer to an experiment performed by Resnick and colleagues in Trinidad (1990–2002). Species of cichlid, *Crenicichla*, prey on large, sexually mature guppies in some streams. In others and above waterfalls, there are no *Crenicichla* and predation on the guppies is much lower or absent.

MULTIPLE-CHOICE QUESTIONS

83. Which of the following is a reasonable prediction?

(A) Predation by *Crenicichla* should select for an overall increase in body mass of the guppies.
(B) Predation by *Crenicichla* should favor the evolution of early sexual maturation and reproduction.
(C) Lack of predation by *Crenicichla* should favor the production of many, smaller offspring earlier in life.
(D) Lack of predation by *Crenicichla* should select for guppies that can choose when to reproduce based on environmental conditions.

84. Which of the following is an acceptable method to estimate reproductive effort (reproductive investment)?

(A) weight of embryos relative to the weight of the mother
(B) weight of embryos relative to the weight of the father
(C) the average weight of an individual embryo relative to the weight of the mother
(D) the total number of embryos that live to sexual maturity

Question 85 refers to an experiment in which guppies were relocated from their home stream to a new, nearby stream. Their home stream contained a predator, *Crenicichla*, which feeds on large, sexually mature guppies. The new stream did not contain *Crenicichla*. After several generations, researchers took guppies from both the home stream and the new stream and reared their offspring in a laboratory.

85. Which of the following would be expected from the populations relieved of predation on large adults?

(A) They acclimated to the laboratory environment by delaying maturation but tended to have a greater number of smaller offspring.
(B) They acclimated to the laboratory environment by reproducing earlier and at a smaller body size, and their reproduction became semelparous.
(C) They evolved delayed maturation at larger adult sizes and tended to have fewer, larger offspring and lower reproductive effort.
(D) They evolved earlier maturation at smaller sizes and tended to have a larger number of smaller offspring with a greater reproductive effort.

86. In mammals, *Drosophila*, and many other animals, a greater number of new mutations enter the population through sperm as compared to the number that enter the population through eggs. Which of the following most *directly* and *accurately* accounts for this observation?

(A) For each fertilization event, there are millions of sperm for each egg.
(B) More cell divisions occurred in the germ line before spermatogenesis as compared with oogenesis in individuals of the same age.
(C) Eggs are larger and less subject to mutation.
(D) Sperm are produced at a faster rate than eggs, increasing the probability that DNA polymerase will make a mistake that is uncorrected by proofreading.

SHORT FREE-RESPONSE QUESTION

Question 87 refers to the following data.

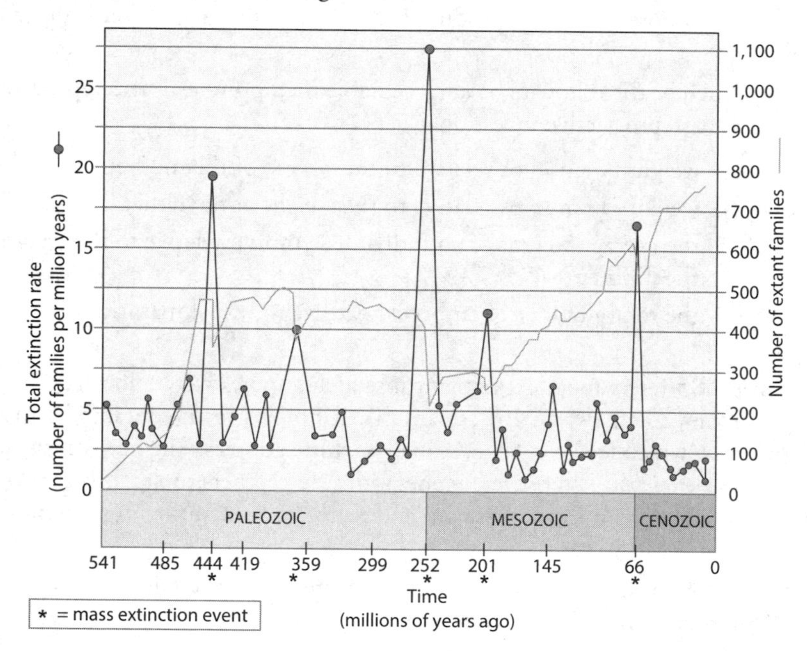

87. The end-Permian mass extinction (252 million years ago) resulted in the extinction of approximately 96% of marine species. **Explain** how this can be true even though the light-grey line on the graph shows only a 50% drop at that time.

Question 88 refers to **vicariance**, a process by which the geographical range of an individual taxon, or a whole biota, is split into discontinuous parts by the formation of a physical or biotic barrier to gene flow or dispersal.

Gondwana is one of two supercontinents that formed from the breakup of Pangea. Gondwana included Antarctica, South America, Africa, Madagascar, and Australia, the Arabian Peninsula, and the Indian subcontinent. The vicariance hypothesis proposes that the breakup of Gondwana isolated the descendants of a common ancestor of the *Cichlidae*, a large family of freshwater fishes.

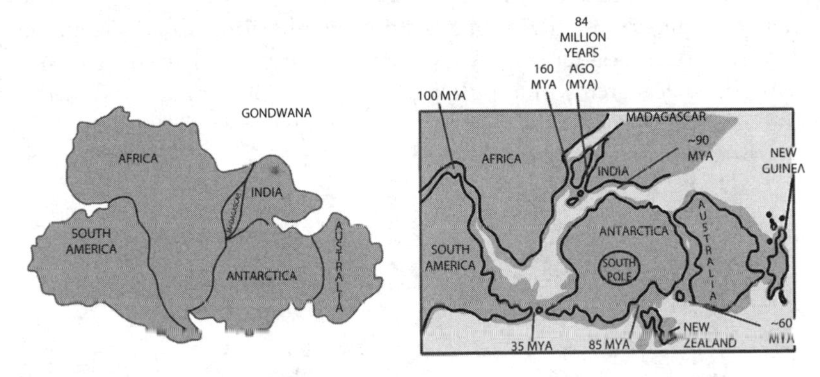

MULTIPLE-CHOICE QUESTIONS

88. Which of the following observations, if true, supports the vicariance hypothesis for *Cichlidae* diversification?

(A) The time of origin for each clade (using DNA sequence data) matches the times of separation of these land masses.

(B) The greatest DNA sequence similarities among the fish closest to the oldest breakup site (160 million years ago) and the greatest number of descendant species at the site of the newest breakup (35 million years ago).

(C) The migration trail matches the path of the rifting plates.

(D) The greatest number of species at the newest breakup site.

Question 89 refers to the following information.

The Isthmus of Panama, the narrow strip of land connecting North and South America, formed approximately 3 million years ago. Armadillo fossils from the Tertiary period (66–2.6 million years ago) are found only in South America. They begin to appear in North America during the Pliocene (5.22–2.6 million years ago).

89. Which of the following statements is the most accurate conclusion of this observation?

(A) It provides definitive proof that armadillos evolved in South America and later dispersed into North America

(B) It supports the hypothesis that the separate land masses caused a diversification process leading to speciation and migration.

(C) It suggests that armadillos dispersed into North America from South America.

(D) It proves that a taxon proliferates in one area before appearing in another.

Questions 90–94 refer to the following information regarding jaw evolution in vertebrates.

Agnathans are jawless fish (like the lampreys). Some agnathans have teeth that developed from the first pharyngeal arch. Development of teeth from this structure requires expression of Hox genes.

Gnathostomes are jawed vertebrates. Some gnathostome fish lineages, like the cichlids, have both oral and pharyngeal jaws. Pharyngeal jaws develop from a different pharyngeal arch without Hox gene input (expression). In fish with both oral and pharyngeal jaws, teeth may be present on either or both. Note that jaws and teeth are two separate traits.

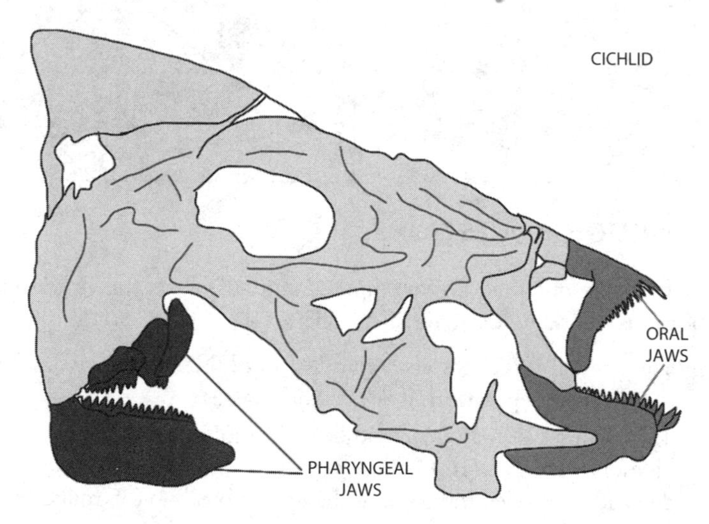

90. Which of the following hypotheses is a reasonable explanation for the jaw and tooth development in gnathostomes?

(A) Loss of Hox gene expression is one of the genetic factors involved in the appearance of the oral jaw.

(B) The loss of the pharyngeal jaw is due to loss of Hox gene expression.

(C) Development of teeth requires Hox gene input regardless of the presence or absence of a jaw.

(D) Jaws could not evolve until teeth evolved.

91. A 2009 study (by Fraser) found that tooth number and position were highly correlated in the oral and pharyngeal jaws of the cichlids in Lake Malawi. This observation taken in context with the previous information suggests that:

(A) the two sets of teeth share the same regulatory network.
(B) development of the first teeth (in agnathans) was under Hox control, but the network came under different control later in the evolution of gnathostomes.
(C) the development of the two sets of toothed jaws require the same gene expression but in different areas.
(D) Hox gene expression indirectly regulates the development of the jaw through the development of teeth and associated structures.

Questions 92–94 refer to the many diverse species of cichlids endemic (unique) to the Great African Lakes, Victoria (~26,830 square miles), Tanganyika (~12,700 square miles), and Malawi (~8,683 square miles). These lakes are located on the Great Rift Valley that formed approximately 25 million years ago as the African and Eurasian continental plates collided. The following diagram shows the lakes.

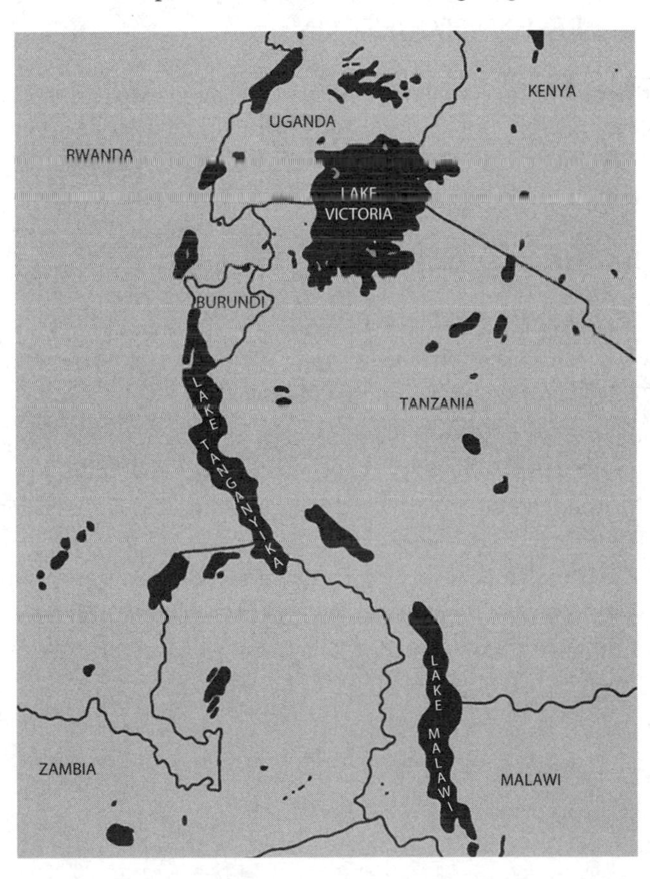

Lake Tanganyika is the oldest lake, approximately 12–14 million years old. Lake Malawi is approximately 1–2 million years old. Lake Victoria has dried up at least three times since it formed about 400,000 years ago. The last dry-out was 17,300 years ago.

Each of the three lakes has hundreds of cichlid species and much of the diversification of the cichlids involves diet. Each lake contains cichlid species that are specialists in plankton feeding, grazing algae from rock; predators of other fish; scavengers; food robbers; and fish-egg eaters (and more!). The cichlids vary greatly in coloration, body form, teeth, and jaws that appear to correspond with their feeding habits. The double-jaw innovation may have primed them for dietary versatility.

LONG FREE-RESPONSE QUESTION

92. Each of the lakes is home to hundreds of different species of cichlids. **Explain** how fish living in the same lake can speciate. Include in your discussion a **hypothesis** of the mechanism of reproductive isolation.

SHORT FREE-RESPONSE QUESTION

93. The three lakes are completely separate, yet many of the species from one lake resemble species from a different lake. **Explain** how convergent evolution and adaptive radiation could produce the large numbers of species with similarities to species that diverged (first).

MULTIPLE-CHOICE QUESTIONS

94. Several phylogenies have been proposed for the cichlids. Which of the following methods of phylogeny construction would be the *least* useful in the construction or evaluation of a phylogeny?

(A) *Maximum parsimony:* the phylogeny that requires the minimum number of evolutionary changes assumed for all traits is preferred.

(B) Morphological, molecular, developmental, and if applicable, behavioral data are considered.

(C) The complete sequence of the genomes of all the species involved are compared.

(D) Mathematical models are applied to the data to identify the tree that most likely produced the observed data.

95. The amount of genetic variation within populations of hunter-gatherers in southern Africa is much greater than in non-African populations. The genetic divergence among these African populations as compared to non-African populations suggests that:

(A) the environment in Africa has not changed much in the past 200,000 years.
(B) populations in southern Africa have been subjected to more inbreeding.
(C) the members of the population that initially left Africa were all related by a common ancestor not more that 3 or 4 generations prior to their departure.
(D) the populations that colonized the rest of Earth carried the genetic information from their populations of origin almost intact and few genetic differences have occurred since then.

GRID IN

Question 96 refers to the following data.

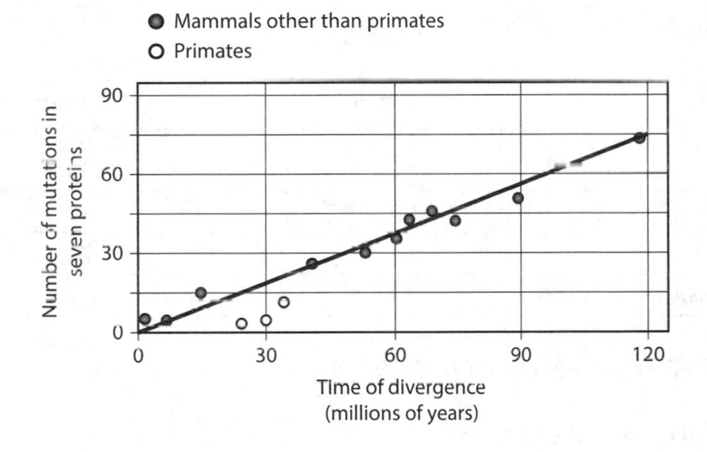

96. **Estimate** the difference in the number of mutations (in seven proteins) accumulated between non-primate mammals and primates in a 10-million-year period. Report your answer with a whole number.

Question 97 refers to the following data.

Species	GAPDH gene % similarity to humans	GAPDH protein % similarity to humans
Chimpanzee	99.6	100.0
Dog	91.3	95.2
Fruit fly	72.3	76.7
Roundworm	68.2	74.3

SHORT FREE-RESPONSE QUESTION

97. **Explain** how the percent difference in the nucleotide sequence of the gene could be different from the percent difference in amino acid sequence of the protein. **State** whether the protein sequence could have a greater percent difference than the gene sequence. **Justify** your response.

SHORT FREE-RESPONSE QUESTION

98. **Explain** why reproductive isolation must be maintained for one species to become and remain distinct from another.

LONG FREE-RESPONSE QUESTION

99. **Coevolution** (reciprocal adaptation) is an evolutionary process in which an adaptation in one species leads to an adaptation in another species with which it interacts. Use an **example** of your choice to illustrate coevolution. **Identify** the trait or traits that coevolved in each organism and **describe** the selective pressure that promoted it.

SHORT FREE-RESPONSE QUESTION

100. The ability to learn is adaptive for animals, particularly those that can move. Give an **example** of animal learning that is adaptive and **explain** how natural selection could act on that behavior.

Free Energy and Homeostasis

Big Idea 2: Biological systems utilize free energy and molecular building blocks to grow, to reproduce, and to maintain dynamic homeostasis.

MULTIPLE-CHOICE QUESTIONS

101. Which of the following best describes the use of free energy by organisms?

(A) Exergonic processes are always exothermic.
(B) Processes that release heat and increase the disorder of a system are able to perform biological work.
(C) Only exergonic processes can be performed by organisms.
(D) Cells use catalysts to convert endergonic processes into spontaneous (exergonic) processes.

Questions 102–104 refer to the following graphs.

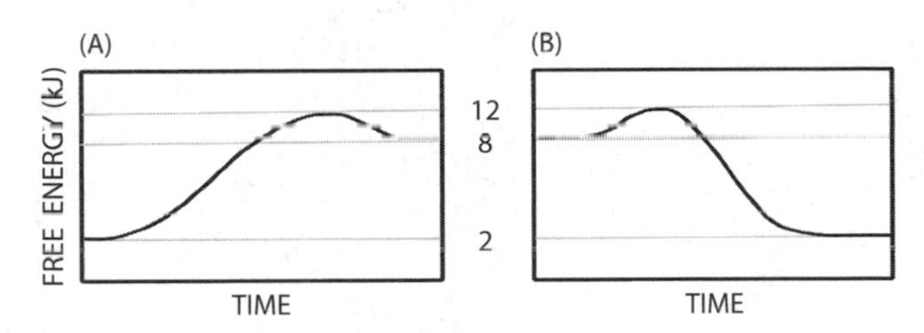

102. Which of the following is *not* true regarding the processes represented above?

(A) Reaction A can do biological work.
(B) The reaction profile of ATP synthesis is more similar to reaction A than B.
(C) Reactions A and B have the same free energy change, just in opposite directions.
(D) The activation energy of reaction A is greater than the activation energy of reaction B.

GRID INS

103. Calculate the free energy change (ΔG) of the process represented in graph B. Indicate a negative number if applicable.

Start your answer in any column, just make sure you've got enough spaces.

You have symbols for fractions & a negative number.

2 . 6

Leave extra spaces blank.

104. Calculate the activation energy (E_A) of the process represented in graph A. Indicate a negative number, if applicable.

MULTIPLE-CHOICE QUESTIONS

105. The contraction of skeletal muscle occurs by the action of the proteins actin and myosin. In an experiment done in the 1940s by Szent-Györgyi, a solution of actin was mixed with a solution of myosin. The viscosity (thickness, texture, resistance to flow) of the resulting solution was dramatically increased, indicating the formation of an acto-myosin complex. The viscosity of the solution was lowered by the addition of ATP (adenosine triphosphate). Which of the following statements is an accurate conclusion based on the results of this experiment?

(A) ATP increases the affinity of myosin for actin.
(B) The release of actin from myosin is coupled to ATP hydrolysis.
(C) Actin has ATP-ase activity.
(D) Conformational changes drive the hydrolysis of ATP during muscle contraction.

Question 106 refers to the chemical equation for photosynthesis and the following table.

$$6\ CO_2 + 12\ H_2O + \text{light energy} \rightarrow C_6H_{12}O_6 + 6\ O_2 + 6\ H_2O$$

Several experiments were performed with radiolabeled atoms to trace (follow) specific atoms through the photosynthetic process. The results are summarized in the following table.

Radioactively labeled atom	Molecule in which the radioactive atom was provided to plant	Molecule in which the radioactive atom was recovered
C	CO_2	$C_6H_{12}O_6$
O	$H_2\mathbf{O}$	$\mathbf{O}_2$
O	$C\mathbf{O}_2$	$C_6H_{12}\mathbf{O}_6$

106. Which of the following descriptions of the process of photosynthesis most accurately accounts for the data in the table?

(A) Two carbon dioxide molecules are required for each water molecule to synthesize glucose.
(B) Carbon and oxygen atoms from carbon dioxide are incorporated into organic molecules.
(C) Hydrogen from water is combined with carbon dioxide to form organic molecules.
(D) The oxygen atoms from carbon dioxide are released as oxygen gas.

Question 107 refers to the following chemical equation for cellular respiration and table.

$$C_6H_{12}O_6 + 6\ O_2 \rightarrow 6\ CO_2 + 6\ H_2O + \text{energy}$$

Several experiments were performed with radiolabeled atoms to trace (follow) specific atoms through the process of cellular respiration. The results are summarized in the following table.

Radioactively labeled atom	Molecule in which the radioactive atom was provided to animal	Molecule in which the radioactive atom was recovered
C	$\mathbf{C}_6H_{12}O_6$	$\mathbf{C}O_2$
O	$C_6H_{12}\mathbf{O}_6$	$C\mathbf{O}_2$
O	$\mathbf{O}_2$	$H_2\mathbf{O}$

107. Which of the following statements most accurately accounts for the data?

(A) Glucose is the source of atoms in expired carbon dioxide.
(B) ATP is synthesized from the atoms in glucose and oxygen.
(C) The oxygen atoms in glucose are reduced to form water.
(D) The hydrogen atoms in water come from oxygen gas.

108. A student placed a plant in a sealed container and measured the amount of carbon dioxide and oxygen present. For the first several days, she kept the plant in the light and found that the oxygen concentration increased. For the next several days, she kept the plant in the dark and observed that the oxygen concentration decreased. Which of the following most accurately explains why the oxygen concentration decreased in the dark?

(A) The plant used the oxygen for photosynthesis.
(B) The plant used the oxygen for cellular respiration.
(C) The plant produced more carbon dioxide, displacing the oxygen.
(D) The plant consumed more carbon dioxide, resulting in more oxygen use.

109. Suppose a student attempts to grow a plant in a sealed container under a green light. If the container, initially containing air from the atmosphere, is completely sealed, which of the following is most likely to occur over a 24-hour period?

(A) The oxygen concentration will increase.
(B) The carbon dioxide concentration will increase.
(C) The carbon dioxide concentration will decrease.
(D) No change is expected because plants do not use green light.

110. Which of the following is the most likely reason expired milk typically tastes sour and develops curds (becomes clumpy)?

(A) Yeast produces ethanol and CO_2, which causes the lactose sugars to polymerize.
(B) The proteins and sugars in the milk are being converted into fats (lipids).
(C) Bacteria produce lactic acid, which lowers the pH and denatures the proteins.
(D) Bacteria produce ethanol, which denatures the sugars and proteins.

111. ATP serves as the cellular energy source for all organisms because:

(A) it is a negatively charged molecule.
(B) its synthesis is exergonic (spontaneous).
(C) it is stable enough to store for long periods of time.
(D) its hydrolysis provides energy that can do cellular work.

Questions 112 and 113 refer to the following answer choices.

(A)

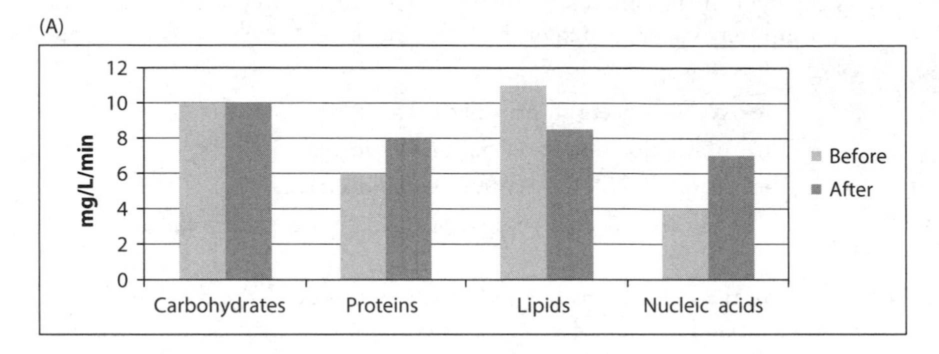

(B)

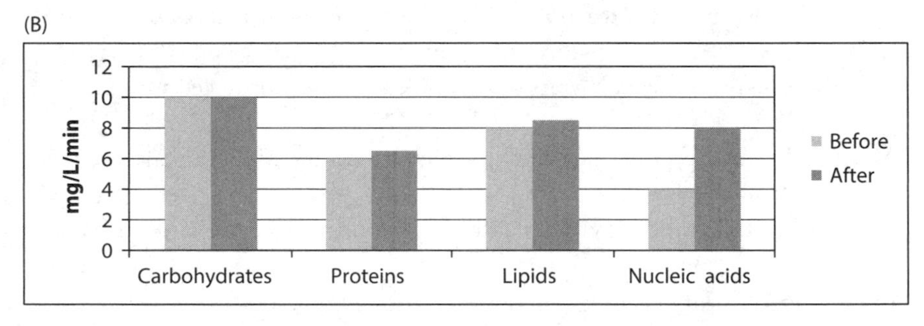

(C)

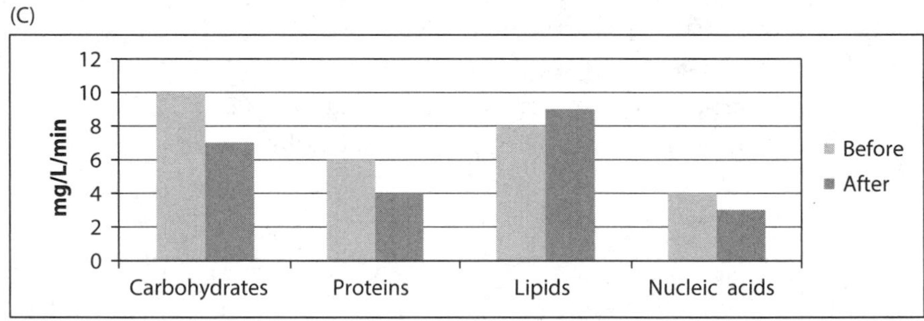

(D)

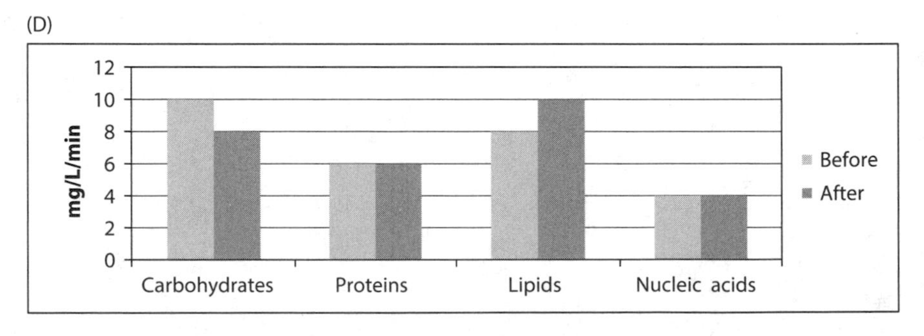

112. A culture of autotrophic algae was supplemented with additional phosphates. The rate of synthesis of various organic compounds was measured before and after the addition of phosphates to the culture. Which graph illustrates the most likely result of the experiment?

113. A culture of autotrophic algae was supplemented with additional nitrogen. The rate of synthesis of various organic compounds was measured before and after the addition of nitrates to the culture. Which graph illustrates the most likely result of the experiment?

SHORT FREE-RESPONSE QUESTION

114. **Biological oxygen demand (BOD)** is the amount of dissolved oxygen needed by the organisms in a particular body of water to decompose the organic material present in a specified amount of time. **Explain** how increasing the concentration of phosphorus in a lake creates a biological oxygen demand (BOD). **State** one human activity that is likely to result in increased organic material in a body of water.

LONG FREE-RESPONSE QUESTION

115. **Design an experiment** to test whether nitrogen or phosphorus is the limiting factor in a freshwater community of algae and small fish in a laboratory aquarium.

- **State** the conditions of the experiment, the variable or variables that will be measured, and how it (they) will be measured.
- **Sketch** a graph of the expected results and **briefly explain** how you will make the final determination.
- **State** one source of potential error in measurement or data collection.

Questions 116–118 refer to the following diagram of the breakdown of the carbon chain of glucose during cellular respiration.

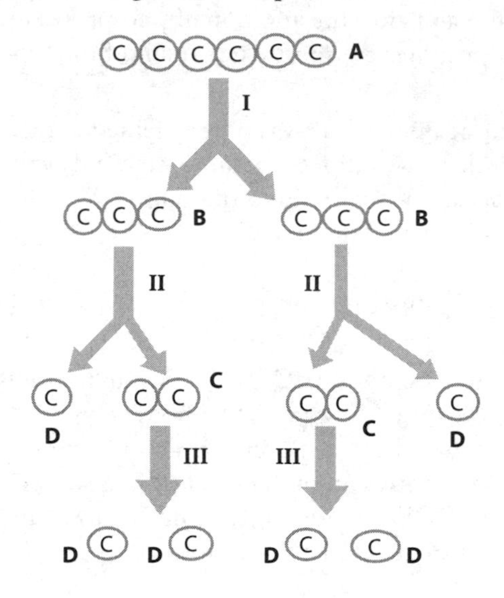

MULTIPLE-CHOICE QUESTIONS

116. Which of the following is represented by the process labeled I?

(A) aerobic respiration in the mitochondria
(B) the splitting of glucose into two ATP and two pyruvate molecules
(C) the conversion of glucose into two lactate molecules by fermentation
(D) a series of enzyme-mediated reactions that occur in the cytosol

117. Which of the following is *not* required by process II?

(A) transport of a metabolic intermediate into the mitochondria
(B) the direct participation of oxygen
(C) the transfer of hydrogen atoms from one of the carbons to a coenzyme
(D) enzymes

118. Which of the following is the identity of molecule D?

(A) carbon dioxide
(B) glucose
(C) pyruvic acid
(D) ATP

Questions 119–122 refer to the following diagram of the chemiosmotic production of ATP in mitochondria and chloroplast.

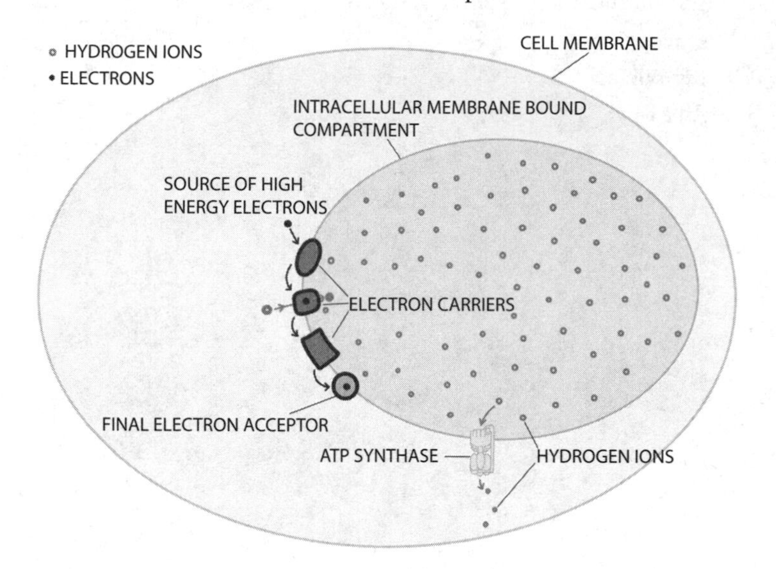

119. Which of the following is the direct source of high-energy electrons in plants?

(A) the sun
(B) glucose
(C) NADPH
(D) chlorophyll

120. In which of the following would the high-energy electrons generated in the light reactions of photosynthesis be found at the end of the photosynthesis?

(A) glucose
(B) oxygen
(C) water
(D) the sun

121. Which of the following is the source of high-energy electrons in animals?

(A) glucose
(B) oxygen
(C) water
(D) the sun

122. In which of the following molecules would electrons be found after they have been released from the mitochondrial electron transport chain?

(A) glucose
(B) oxygen
(C) water
(D) ATP synthase

LONG FREE-RESPONSE QUESTION

123. The following diagrams represent the cytochrome complexes of the mitochondrial electron transport chain and the photosystems of thylakoids.

- **Compare** and **contrast** the free energy changes in electrons during electron transport in cellular respiration and photosynthesis.
- **State** the reason the final electron acceptors differ.

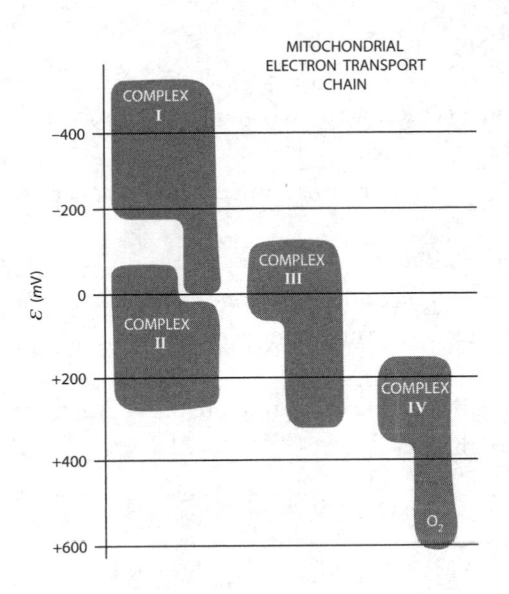

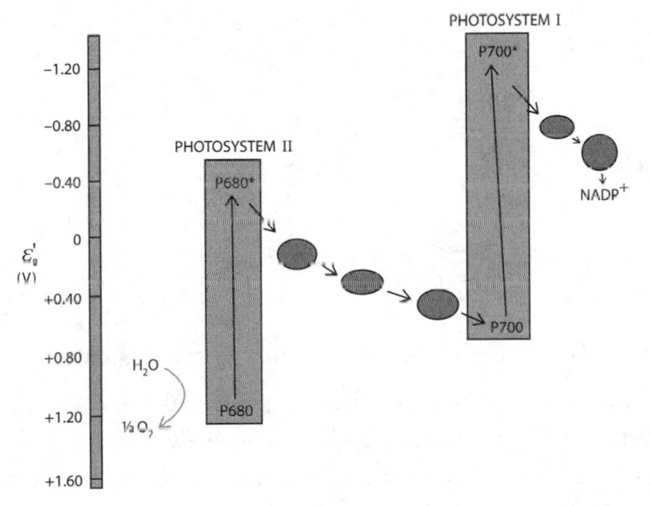

ALCOHOL

ALDEHYDE

CARBOXYLIC ACID

CARBON DIOXIDE (CO_2)

MOST REDUCED
LEAST OXIDIZED

LEAST REDUCED
MOST OXIDIZED

MULTIPLE-CHOICE QUESTIONS

124. Which of the following statements most accurately models the process of chemiosmosis?

(A) When water flows through a dam, it turns a water wheel that can do work.
(B) Electron movement in a wire causes a lightbulb to glow.
(C) A pulley changes the direction of work being done.
(D) Combustion of gasoline causes pressure changes in an engine.

Question 125 refers to the following diagrams of the structure of ATP (left) and the energy profile of its hydrolysis (right).

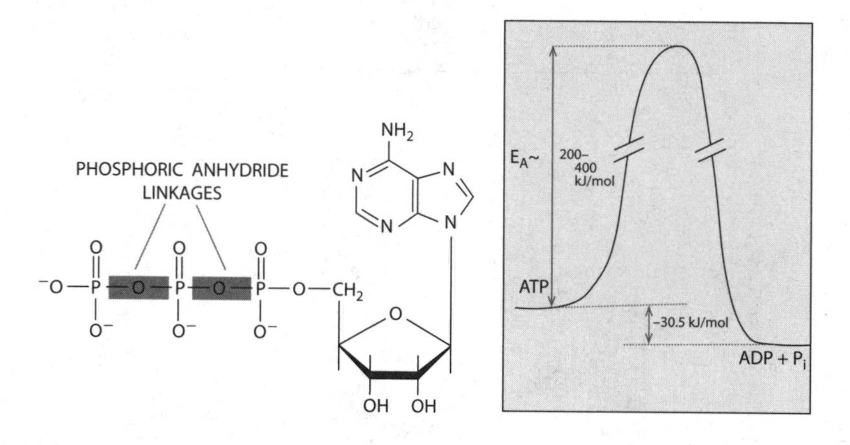

125. Which of the following statements is *not* supported by the diagrams?

(A) ATP is a stable molecule because it has a large activation energy (E_A).
(B) The products of ATP (ADP and P_i) hydrolysis are more stable than ATP.
(C) The products of ATP hydrolysis are more entropic than ATP (due to increased number of particle types in solution).
(D) The energy of the transition state during the transition from ATP $\rightarrow$ ADP + P_i is high, which makes the standard free energy change of the reaction high.

126. Organisms exhibit great diversity in the way they obtain or generate nutrients. Consider an organism that uses high-energy electrons from H_2S and CO_2 from its aquatic habitat to synthesize organic molecules in the absence of light. To which group does this organism belong?

(A) a bacteria
(B) an animal
(C) a plant
(D) a fungus

LONG FREE-RESPONSE QUESTION

127. **State** and **describe** the energy transformations involved in the chemiosmotic production of ATP.

Process	Free-energy change
Electrons are passed along an electron transport chain.	Exergonic
Protons are pumped into a membrane-bound compartment.	Endergonic
Protons are transported down their concentration gradient.	Exergonic
ADP + P → ATP	Endergonic

LONG FREE-RESPONSE QUESTION

128. The **endosymbiotic theory** (also known as symbiogenesis) is an evolutionary theory that explains the origin of eukaryotic cells from prokaryotic cells. According to this theory, the mitochondria and plastids (including the chloroplast) were once free-living bacteria that were engulfed by a primitive eukaryote about 1.5 billion years ago (see the following diagram).

State four pieces of evidence that support the endosymbiotic theory of the mitochondria (or chloroplast). For each piece of evidence, **provide a brief evaluation** (less than one sentence) of the strength of the evidence.

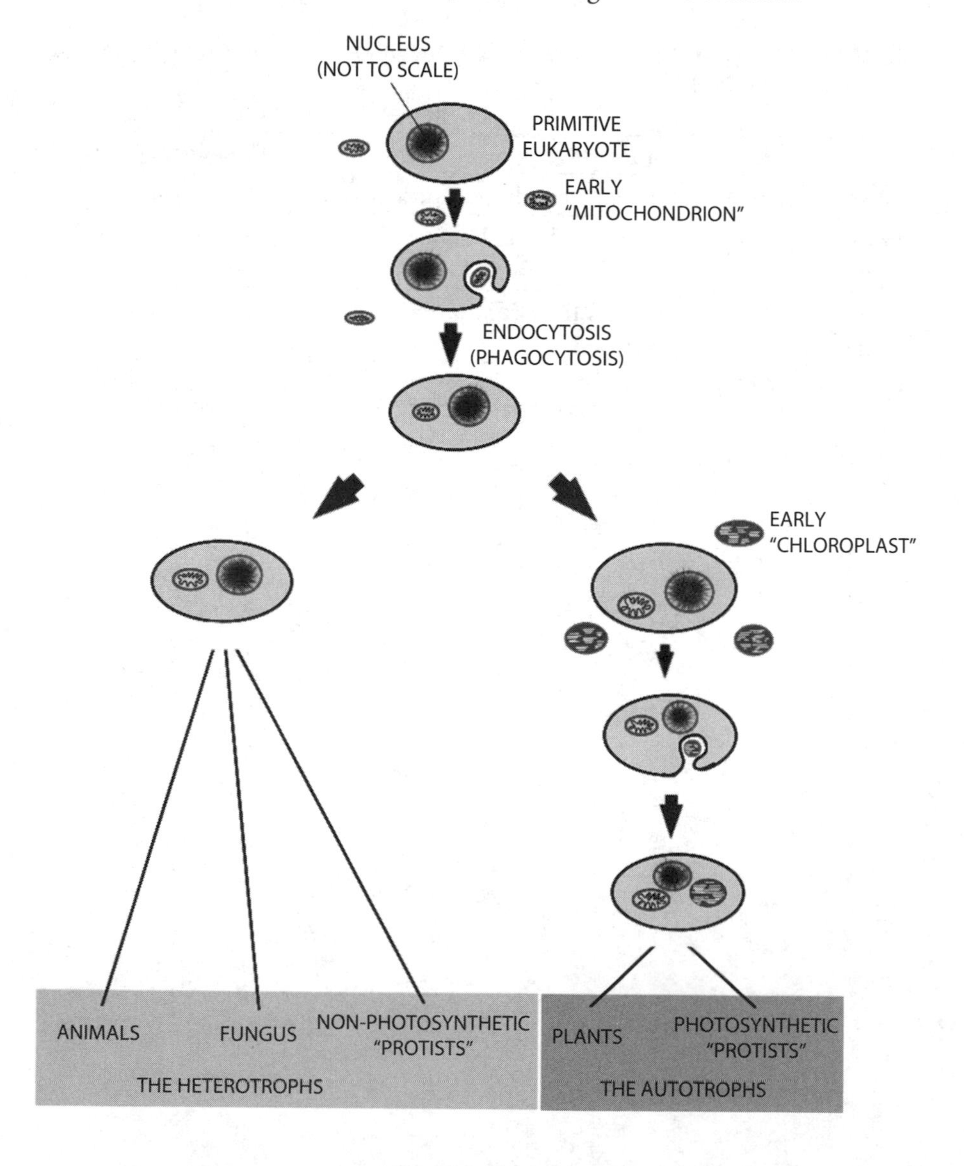

Questions 129–131 refer to the following table.

Location	Substrate	Product	NADH	$FADH_2$	ATP
Cytosol	Glucose	2 pyruvate	2	0	2
Mitochondria	Pyruvate	Acetyl-CoA + CO_2	1	0	0
Mitochondria	Acetyl-CoA	2 CO_2	3	1	1
Mitochondria	Palmitoyl-CoA	8 acetyl-CoA	7	7	0

GRID INS

129. According to the table, how many NADH are produced from the complete oxidation of 1 molecule of palmitoyl-CoA to 16 carbon dioxide molecules?

−					
		/	/	/	
	.	.	.	.	.
		0	0	0	0
	1	1	1	1	1
	2	2	2	2	2
	3	3	3	3	3
	4	4	4	4	4
	5	5	5	5	5
	6	6	6	6	6
	7	7	7	7	7
	8	8	8	8	8
	9	9	9	9	9

130. How many acetyl-CoA molecules can be produced from 1 glucose molecule?

131. How many pyruvate molecules need to be completely oxidized to carbon dioxide to produce 24 NADH molecules?

Questions 132 and 133 refer to the following diagram of the Calvin cycle.

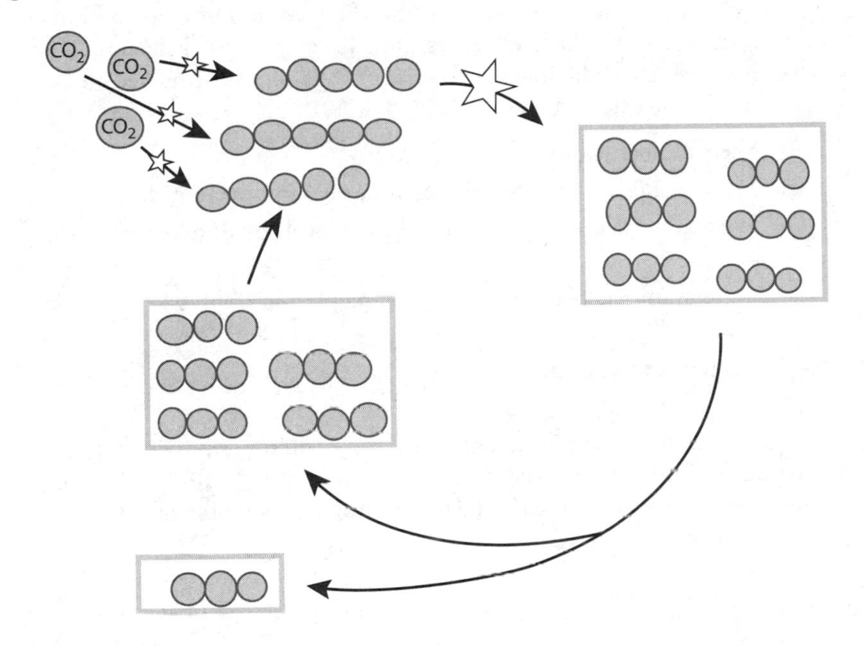

MULTIPLE-CHOICE QUESTIONS

132. Which of the following statements most accurately describes the cycle depicted in the diagram?

(A) Mitochondria oxidize organic compounds into carbon dioxide and water.

(B) Inorganic carbon is absorbed from the environment by autotrophs that used it to build carbohydrates, proteins, lipids, and nucleic acids.

(C) Autotrophs use light to excite photons for the synthesis of ATP.

(D) Organic carbon is stored in molecules that can be used to produce CO_2 by heterotrophs.

133. Which of the following is true regarding the process?

(A) It requires energy and a source of hydrogen atoms.

(B) It is spontaneous and does not require enzymes.

(C) It is performed by heterotrophs.

(D) The energy requirements are met by the oxidation of glucose.

134. Glucose-6-phosphatase is an enzyme that removes the phosphate from glucose-6-phosphate and is necessary for the secretion of glucose from cells that store glycogen. The liver expresses this enzyme, but skeletal muscle does not. Which of the following is the most likely consequence of this pattern of gene expression?

(A) Skeletal muscle cannot store glycogen.
(B) Skeletal muscle can store glycogen but cannot break it down.
(C) Skeletal muscle can only change blood levels by uptake from the blood.
(D) Skeletal muscle plays no role in the regulation of blood glucose levels.

LONG FREE-RESPONSE QUESTION

135. The body of an average 70-kilogram person contains approximately 50 grams of ATP. One mole (6.02×10^{23} molecules) of ATP provides ~50 kJ of free energy under cellular conditions. One mole of ATP has a mass of 0.551 kg (551 g).

Calculate the number of times each ATP is cycled (one cycle = synthesis and hydrolysis) per day for an average 70 kg person consuming a 2,800 kcal diet (11,700 kJ) whose metabolic efficiency is 40%. Report your answer to the nearest hundred.

MULTIPLE-CHOICE QUESTIONS

136. Which of the following statements correctly describes how cells transport water into and out of cells?

(A) Aquaporins are active transporters that allow water to be moved across a membrane regardless of the concentration gradient.
(B) Water moves passively by osmosis from an area of higher solute concentration to an area of lower solute concentration.
(C) Most water crosses the membrane by facilitated diffusion moving from areas of lower solute concentration to areas of higher solute concentration.
(D) Water crosses the cell membrane mostly by simple diffusion based on the needs of the cell.

Questions 137 and 138 refer to the following diagram.

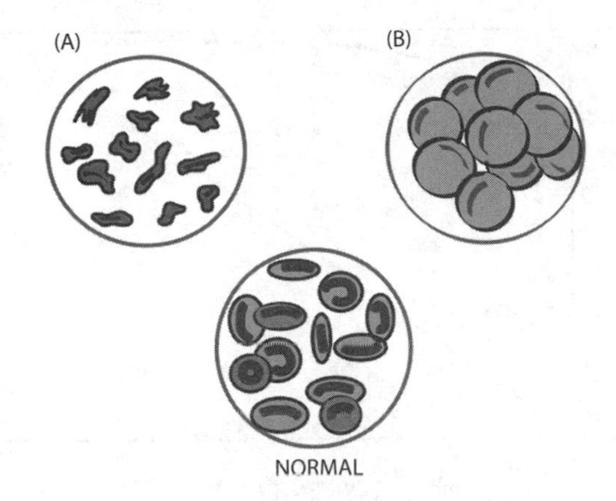

137. A sample of blood was viewed under a microscope with a few drops of physiological saline (0.9% sodium chloride). Which of the following predicts the view under the microscope with the correct reasoning?

(A) A, because the RBCs are permeable to water.
(B) A, because the RBCs are permeable to salt.
(C) B, because the RBCs are permeable to water.
(D) Normal, because there is no net movement of water in or out of the cell.

138. Which of the following conditions is most likely to produce view A with the correct reason?

(A) Pure water, because it will cause the water inside the cell to diffuse out.
(B) Pure water, because it will cause the solutes inside the cell to diffuse out.
(C) 10% sodium chloride, because it will cause the water inside the cell to diffuse out.
(D) 10% sodium chloride, because the salts will diffuse inside the cell.

139. Which of the following best explains why animal cells do *not* have cell walls?

(A) Animals tightly regulate the composition of their extracellular fluids.
(B) Animals have bones, muscles, and connective tissue for support.
(C) Animal cells have complex cytoskeletons for shape, support, and intracellular trafficking.
(D) Animal cells exchange large quantities of nutrients and wastes across their surfaces.

Question 140 refers to the following graph.

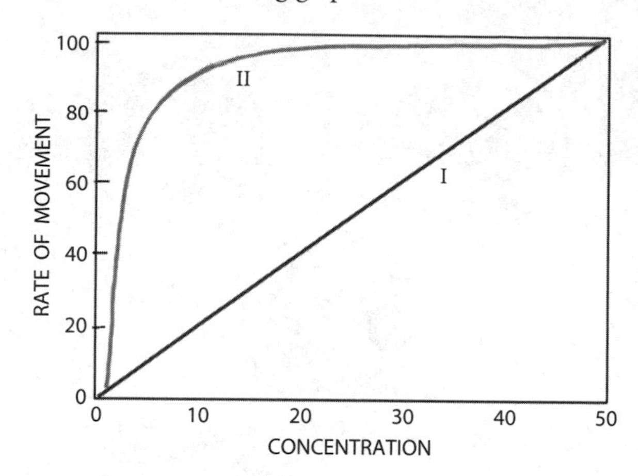

140. Which of the following correctly identifies the two curves in the graph?

(A) Graph I represents simple diffusion; graph II represents facilitated diffusion.

(B) Graph I represents facilitated diffusion; graph II represents simple diffusion.

(C) Graph I represents simple diffusion; graph II represents active transport.

(D) Graph I represents facilitated diffusion; graph II represents active transport.

Questions 141 and 142 are based on the graph of glucose transport rate versus concentration for three different glucose transporters, each with a different K_M. The K_M of a transporter is the concentration at which ½ the maximum rate of transport is achieved.

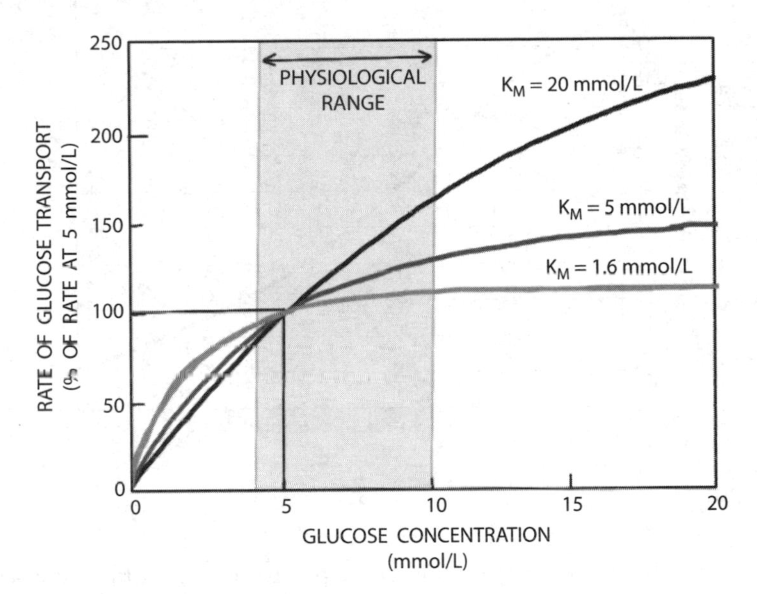

141. The neurons of the brain use glucose at a relatively constant rate. To achieve this, the rate of uptake must be independent of the plasma glucose concentration (within physiological range). Which of the following K_M values would be expected for the glucose transporters on neurons?

(A) 1.6 mmol/L
(B) 5 mmol/L
(C) 20 mmol/L
(D) >20 mmol/L

142. The functioning of pancreatic β-cells and liver cells requires that glucose concentrations equilibrate across the membrane (the intracellular glucose concentration is equal to the plasma glucose concentration). Which of the following K_M values would be expected for the glucose transporters on pancreatic β and liver cells?

(A) <1 mmol/L
(B) 1–2 mmol/L
(C) 2–5 mmol/L
(D) >5 mmol/L

Questions 143 and 144 refer to the following data taken from osmosis experiments on potato cores.

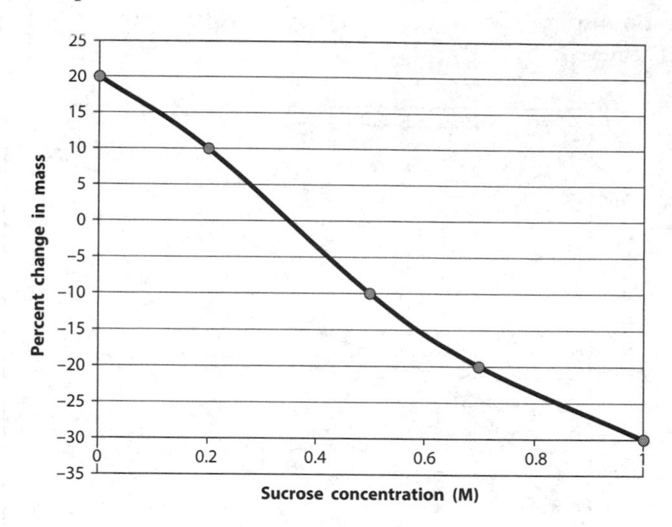

GRID IN

143. What is the solute concentration of the potato cells? Record your answer to two decimal places (hundredths place).

LONG FREE-RESPONSE QUESTION

144. Design an experiment to determine which part of a plant—the stem, the leaf, or the root—contains the greatest percentage of water by mass.

- **Describe** the general procedure.
- **State what** will be measured.
- **State how** it will be measured.
- **State the results** that would support your hypothesis.
- **Explain how** the data support it.

Questions 145–147 refer to the process of sodium and potassium transport. See the following equation.

$$\begin{matrix} ATP & & ADP \\ + H_2O & \rightarrow & + H_2PO_4^- \\ + 3\ Na^+_{\text{INSIDE CELL}} & & + 3\ Na^+_{\text{OUTSIDE CELL}} \\ + 2\ K^+_{\text{OUTSIDE CELL}} & & + 2\ K^+_{\text{INSIDE CELL}} \end{matrix}$$

GRID INS

145. How many molecules of ATP are required for the transport of 600 sodium ions out of the cell?

		/	/	/	
−	.	.	.	.	.
		0	0	0	0
	1	1	1	1	1
	2	2	2	2	2
	3	3	3	3	3
	4	4	4	4	4
	5	5	5	5	5
	6	6	6	6	6
	7	7	7	7	7
	8	8	8	8	8
	9	9	9	9	9

146. How many potassium ions can be transported into the cell with the hydrolysis of 200 ATP molecules?

SHORT FREE-RESPONSE QUESTION

147. The average cell expends from 20%–40% of energy on maintaining concentration differences of sodium and potassium ions across cell membranes. Intracellular potassium ion (K^+) concentrations are maintained at approximately 140 mM. Extracellular sodium ion (Na^+) concentrations are maintained between 5 and 15 mM.

Identify a cell type that you think expends a *greater fraction* (than 20%–40%) of its energy on sodium/potassium transport. **Explain** why the energy requirement to maintain K^+/Na^+ concentrations are higher in this cell type and **describe** the conditions under which the greatest fraction of energy expenditure on Na^+/K^+ transport is expected.

Questions 148–150 refer to ouabain, an inhibitor of the Na^+, K^+-ATPase transporter, and the following diagram and table.

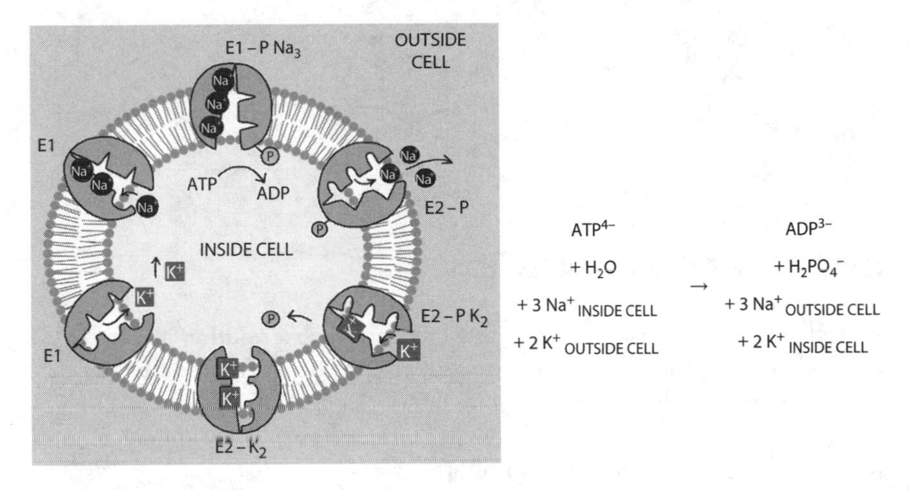

SHORT FREE-RESPONSE QUESTION

148. Certain plant and animal steroids such as ouabain inhibit the Na^+, K^+-ATPase ion transporter by binding to the transporter when it is in the E_2—P state, increasing its stability in this state. **State** the effect this could have on the transport of Na^+ and K^+. **Justify** your answer. **Propose** a potentially useful application of this molecule and **briefly describe** how or why it could be beneficial.

SHORT FREE-RESPONSE QUESTION

149. In people with hypertension (high blood pressure) there is an accumulation of sodium (and calcium) in the cells that line the blood vessels that results in the narrowing of the blood. The accumulation of intracellular sodium ion concentrations is *directly* caused by an inhibition of the Na^+, K^+-ATPase transporter in the cells that line the blood vessel.

The *chemical responsible* for the inhibition of the Na^+, K^+-ATPase transporter is present at very low concentrations in the plasma of people with hypertension. Mass spectrometric analysis has revealed the molecule is very close or identical to **ouabain.**

The structures of three cardiac glycosides (also called cardiotonic steroids)—ouabain, strophanthidin, and digitoxigenin—are shown as follows. These molecules all bind to the Na^+, K^+-ATPase transporter.

- **Indicate** the part of the **ouabain molecule** that is most likely the part that acts on the Na^+, K^+-ATPase.
- Briefly **state** your reasoning.

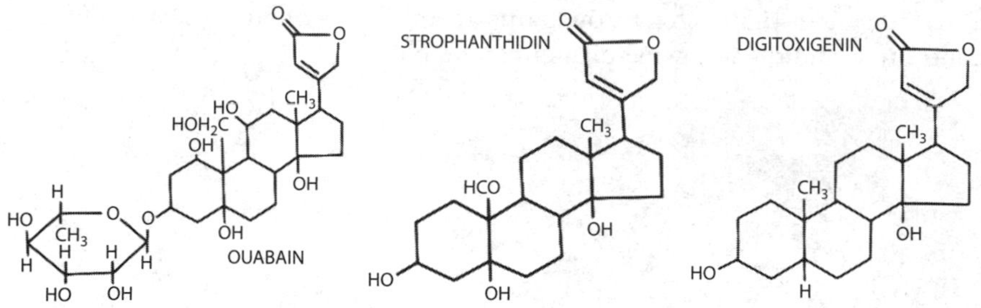

OUABAIN

STROPHANTHIDIN

DIGITOXIGENIN

SHORT FREE-RESPONSE QUESTION

150. Isolated animal cells maintained at a low density in a solution identical to plasma swell and burst when treated with ouabain. Propose a **brief explanation** of this observation.

SHORT FREE-RESPONSE QUESTION

151. State two reasons why simple diffusion alone cannot maintain cellular life.

Question 152 refers to the following table of nucleoside triphosphates (NTPs). NTPs are important in metabolism because the bond that joins the phosphate to the ribose sugar is an indispensable source of chemical energy to do biological work. Transfer of the phosphate group "transfers" the energy to the receiving molecule, activating it for participation in a reaction or process.

Nucleoside triphosphate (NTP)	Main function
ATP	Central energy pathways of metabolism
GTP	Drives protein synthesis
CTP	Phospholipid synthesis
UTP	Activates sugars for synthesis of polysaccharides

152. Which of the following statements is a reasonable inference based on the information in the table?

(A) There is a division of labor of NTPs in the cell. The energy source may be channeled through the recognition of nitrogenous bases.

(B) Cells use ATP as their source of energy for cellular work, but other NTPs exist as energy reserves.

(C) Cells preferentially use ATP but can use GTP, CTP, and UTP to regenerate ATP by transfer of their phosphate groups.

(D) Most cells use ATP, but some cells, particularly prokaryotes, use GTP, CTP, or UTP.

Questions 153 and 154 refer to the following diagrams.

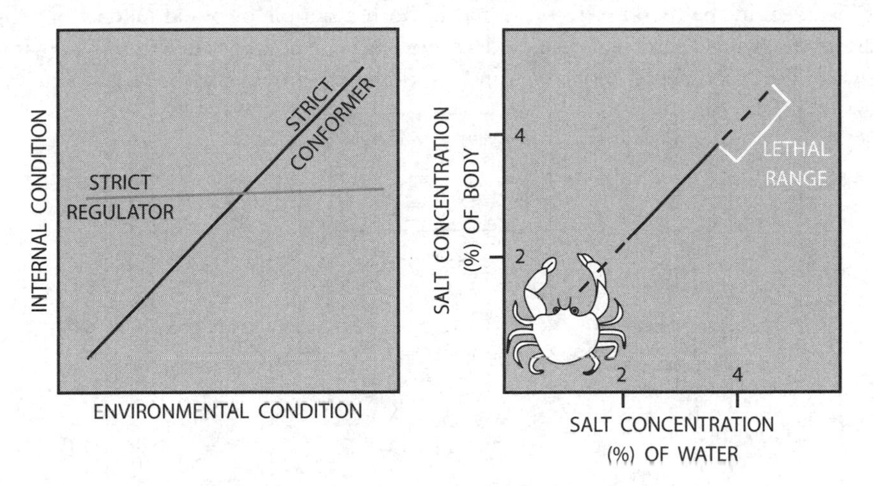

MULTIPLE-CHOICE QUESTIONS

153. Which of the following statements correctly identifies an animal that is a regulator along with the correct reason?

(A) a mouse, because it has a high surface-area-to-volume ratio

(B) a human, because it can change its environment to meet its needs

(C) a dog, because it can exert some level of selection within its environment

(D) a chimpanzee, because its body temperature typically stays within a narrow range of approximately 5°C

154. Which of the following is a correct conclusion regarding the crustacean in the diagram?

(A) The crustacean needs to match the salt concentration in its body to the salt concentration in its environment.

(B) As the salt concentration of the crustacean's body increases, the animal moves toward a saltier environment.

(C) The salt concentration of the crustacean's body tends to equilibrate with the environment.

(D) Crustaceans passively maintain salt concentrations in their tissues between approximately 2% and 4%, but at higher concentrations, they must use active transport.

Question 155 refers to an experiment in which the concentrations of several substances in the blood were measured in both a sample of blood taken from the artery supplying a particular tissue and a sample taken from the vein draining the same tissue. The experimental design and findings of the experiment are shown as follows.

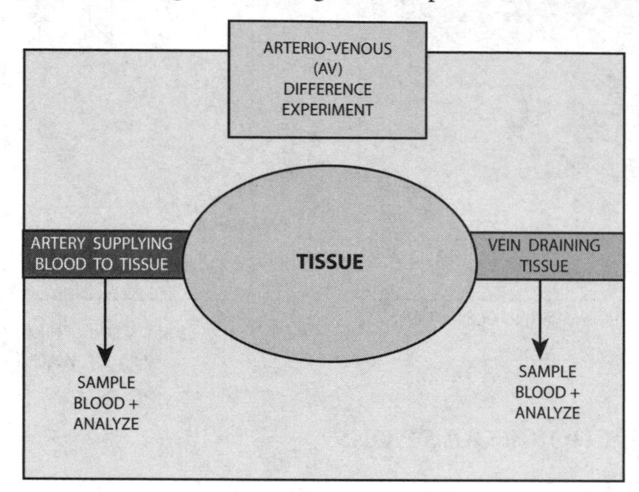

Substance	Artery	Vein
Glucose	Higher	Lower
Oxygen	Higher	Lower
Carbon dioxide	Lower	Higher
Lactic acid	Lower	Higher
Ammonia	Lower	Higher
Glycerol	Lower	Higher

155. Which of the following is an accurate conclusion based on the data?

(A) Oxygen is being converted to carbon dioxide in the tissue.
(B) Lactic acid is being used as an energy source by the tissue.
(C) Amino acid and triglyceride metabolism are occurring within the tissue.
(D) The tissue is working solely anaerobically.

156. Which of the following is true regarding the body temperature regulation in most endotherms (birds and mammals)?

(A) Body temperature is very tightly regulated at 37°C (98.6°F).
(B) Body temperature is constant over the entire body of organism throughout the day.
(C) Body temperature is regulated by the absorption of heat by the skin, sweat glands, respiratory surfaces, and muscles.
(D) The core of the body maintains the highest temperature, while losses of body heat at the surface are adjusted to maintain temperature.

Question 157 refers to the following diagram of body temperature regulation in humans.

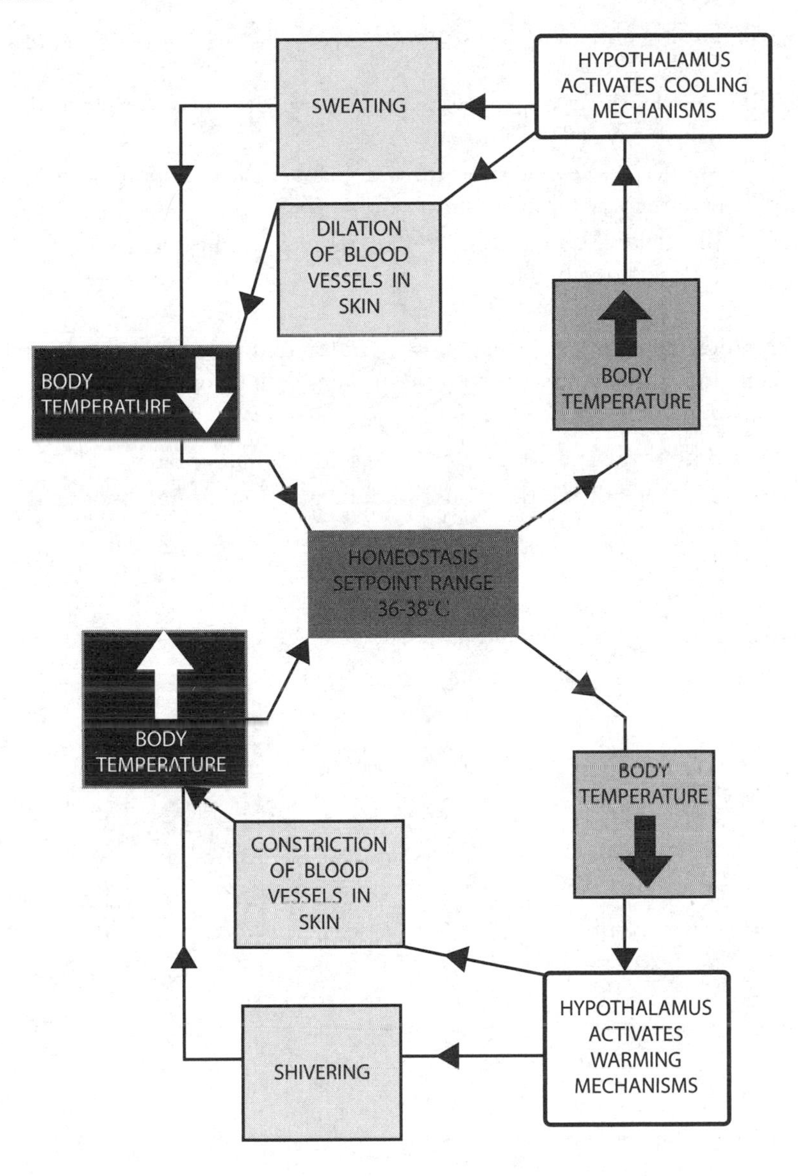

157. Which of the following best describes the mechanism of body temperature regulation shown in the diagram?

(A) Surface blood vessels are dilated to keep skin and muscles warm upon exposure to cold.
(B) Muscles shiver to generate heat, increasing the need for blood flow to deliver oxygen and nutrients.
(C) The act of sweating cools the skin because the excretion of sweat is endothermic (absorbs heat).
(D) Blood vessel constriction at the skin decreases heat losses at the surface of the body.

158. A mouse and a frog of equal mass are transferred from a room temperature container and placed at 5°C for 30 minutes. Which of the following outcomes is expected?

(A) Both organisms would have increased respiratory rates.
(B) The frog but not the mouse would have a significantly increased respiratory rate.
(C) The frog but not the mouse would have a significantly decreased respiratory rate.
(D) Both organisms would have a decreased respiratory rate.

159. Which of the following correctly compares the energy requirements of a rat and a snake of equal body mass living at room temperature?

(A) The rat and the snake have the same energy (caloric) requirements because they are the same mass and at the same temperature.
(B) The snake would require more energy because it is a carnivore.
(C) The snake would require more energy because it has no fur and little to no subcutaneous (under the skin) fat.
(D) The rat would require more energy because it is a mammal.

Questions 160–163 refer to the following diagrams of calcium homeostasis. Calcitonin and parathyroid hormone (PTH) are two hormones involved in the regulation of blood calcium concentrations.

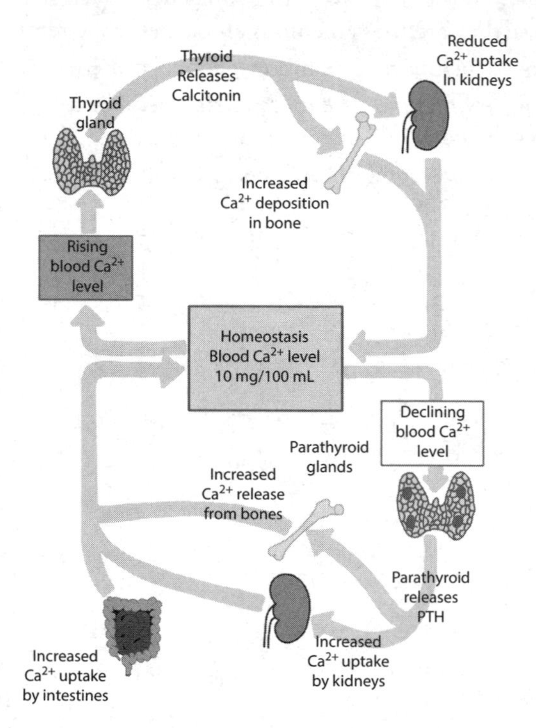

160. Which of the following statements most accurately describes the regulation of blood calcium concentrations based on the diagrams above?

(A) Maintaining calcium blood concentrations within strict limits requires no energy expenditure by the cell.

(B) Calcitonin causes calcium to be released by bone and PTH causes bones to take up calcium from the blood.

(C) Calcitonin and PTH are antagonist hormones that regulate blood calcium concentrations within the appropriate limits through negative feedback.

(D) Blood calcium levels must be regulated by positive feedback for homeostasis to be maintained.

161. Which of the following statements is a reasonable inference based on the diagrams above?

(A) Cells of the bone and kidney have receptors for calcitonin and PTH.
(B) The bones, kidney, and small intestine require calcium to maintain homeostasis.
(C) The small intestine absorbs PTH from the diet, which increases calcium absorption from the blood.
(D) PTH gets converted to calcium in the kidney and calcitonin is absorbed by the kidney, where it forms a complex with calcium ions that is excreted in the urine.

162. Which of the following is most consistent with the role of calcitonin in the body?

(A) The thyroid gland regulates blood calcium concentrations by releasing calcitonin.
(B) Calcitonin lowers blood calcium levels by promoting calcium uptake by bones and signaling the kidney to reduce calcium losses in the urine.
(C) Calcitonin works to counteract the effect of PTH and is released when PTH levels in the blood rise.
(D) Calcitonin is secreted by the thyroid gland when calcium levels fall below the set point, which triggers the secretion of PTH by negative feedback.

163. An experiment was performed in which a dilute solution of calcium ions was slowly administered directly into a rat's bloodstream for 1 hour. Calcium concentrations as well as levels of PTH and calcitonin were measured. Which of the following results is most likely to occur during the hour?

(A) Blood calcium levels would quickly increase and trigger the release of calcitonin, which lowers the level of calcium, which would trigger the release of PTH to increase blood calcium levels.
(B) Blood calcium levels would rise slowly at first, but then be offset by the effects of increasing calcitonin levels.
(C) The level of calcium would not change at all, because the combined effects of calcitonin and PTH would prevent any deviation from the calcium concentration set point.
(D) Increasing calcium concentrations in the blood would result in the secreting of PTH, which would promote calcium excretion in the urine by the kidney and uptake of calcium by bone.

Questions 164–168 refer to the regulation of blood osmolarity and the following diagram.

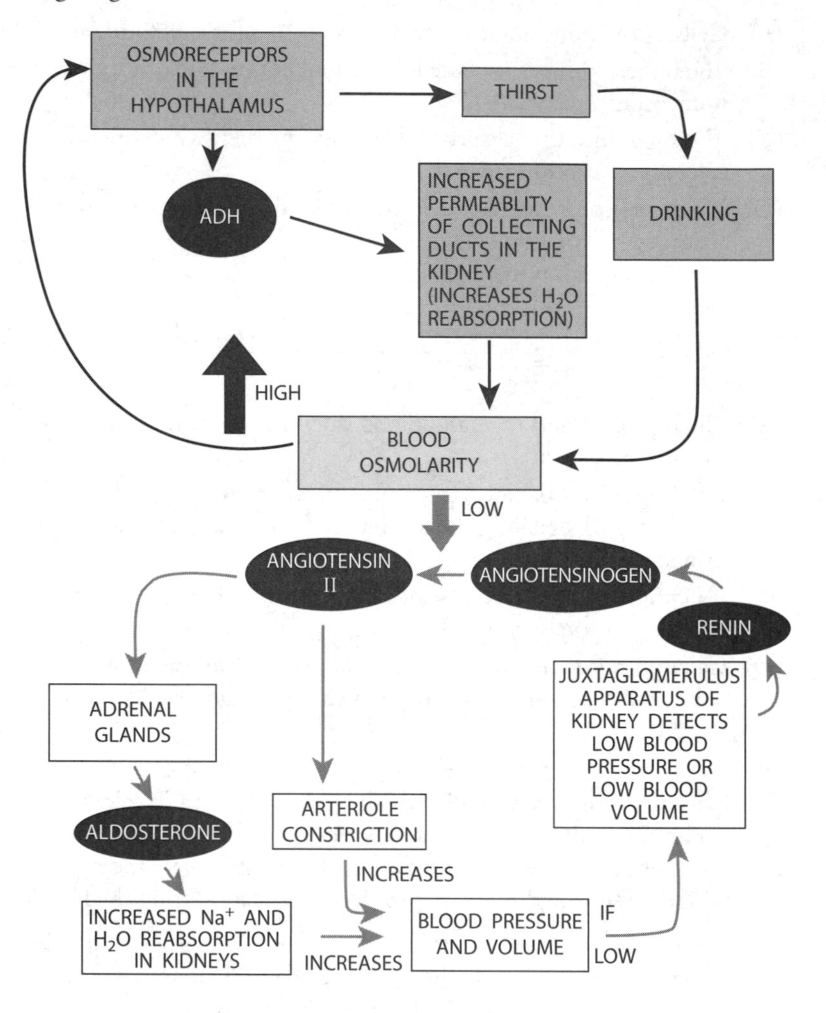

164. Which of the following statements most accurately summarizes the feedback cycle?

(A) Regulation of body fluid osmolarity occurs by a combination of physiological and behavioral mechanisms.

(B) Thirst encourages the consumption of water, which reduces blood osmolarity by diluting the body fluids.

(C) Body fluid osmolarity can be lowered by dilution (adding more water to the body through drinking), which promotes the conversion of angiotensinogen to angiotensin II.

(D) Body fluids can become more dilute by pumping Na^+ into the filtrate, concentrating the urine.

165. Which of the following statements accurately describes the relationship among ADH, renin, angiotensinogen, angiotensin II, and aldosterone?

(A) ADH and renin are secreted in response to increased body fluid osmolarity.
(B) ADH functions to decrease the osmolarity of the blood and renin functions to increase blood pressure.
(C) ADH functions to increase blood osmolarity and renin functions to decrease blood osmolarity.
(D) Aldosterone and renin are antagonistic hormones with opposite effects on blood osmolarity.

166. Which of the following is *not* consistent with the mechanism of osmolarity regulation shown in the diagram above?

(A) The distal tubule detects high body fluid osmolarity, and the pituitary detects low body fluid osmolarity.
(B) High osmolarity stimulates the dilution of body fluids through drinking and reducing water output in the urine.
(C) Low osmolarity situations stimulate changes in vasculature volume and retaining ions in body fluids.
(D) When blood osmolarity is too low, water is excreted into the urine, and when it is too high, excess Na^+ ions are secreted into the urine.

167. Which of the following statements is true regarding the regulation of blood flow in the body?

(A) All blood vessels either dilate or constrict depending on the chemical signals present.
(B) The volume of the circulatory system remains constant by homeostatic control.
(C) Changes in heart rate and contractility affect total blood pressure that can redistribute blood flow in the body.
(D) The same hormone can dilate some blood vessels and constrict others depending on the receptors present on the vessel.

168. Which of the following statements describes a reasonable mechanism by which the osmoreceptors in the hypothalamus detect changes in the osmolarity of extracellular fluids?

(A) Osmolarity changes in the extracellular fluid alter the volume of osmoreceptor cells of the hypothalamus, triggering a cellular response.
(B) Water receptors on osmoreceptor cells in the hypothalamus measure the ratio of water and solute.
(C) The pH of the extracellular fluid is used as an indirect measure of blood osmolarity.
(D) Sodium receptors detect an increased sodium ion concentration, which opens sodium channels and triggers an action potential in receptor cells.

Question 169 refers to the following graph of flower temperature versus environmental temperature for three different plants.

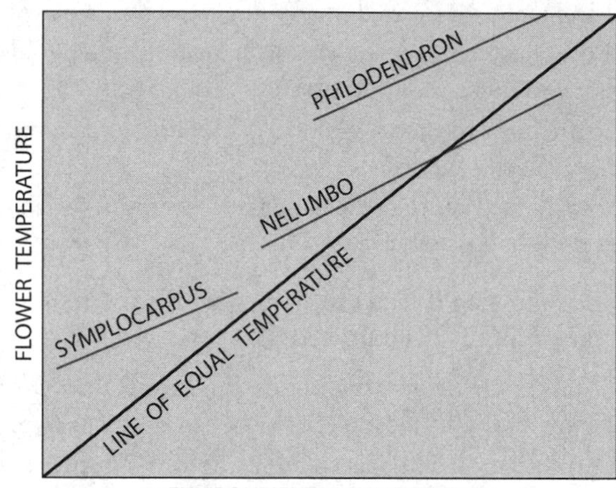

169. Which of the following statements most accurately explains the data above?

(A) Nelumbo regulates its flower temperature to equal the environmental temperature.

(B) In all plants studied, flower temperature is higher than environmental temperature for all temperatures measured.

(C) Philodendron maintains a temperature above the environmental temperature for all temperatures measured.

(D) Symplocarpus maintains a temperature slightly below environmental temperatures for all temperatures measured.

170. Which of the following is the most convincing evidence that a particular plant regulates its flower temperature?

(A) The number of mitochondria increase in response to lower temperatures.

(B) The rate of photosynthesis decreases in response to lower temperatures.

(C) The flower temperature directly correlates to the environmental temperature.

(D) The plant maintains a relatively constant flower temperature despite fluctuations in ambient temperature.

171. Which of the following is *not* an environmental pressure that would select for the maintenance of higher floral temperature?

(A) Thermogenic flowers attract and protect pollinators in cold environments.
(B) Higher temperatures increase the vaporization of odor attractants.
(C) Metabolic heat generation releases excess energy to maintain or reduce body mass.
(D) Flowers that maintain higher temperatures than the environment are less likely to be covered by snow.

SHORT FREE-RESPONSE QUESTION

172. Briefly explain why thermogenic flowers are typically large.

Question 173 refers to the following data in which the fatty acid composition of membrane phospholipids was analyzed in plants growing at two different temperatures.

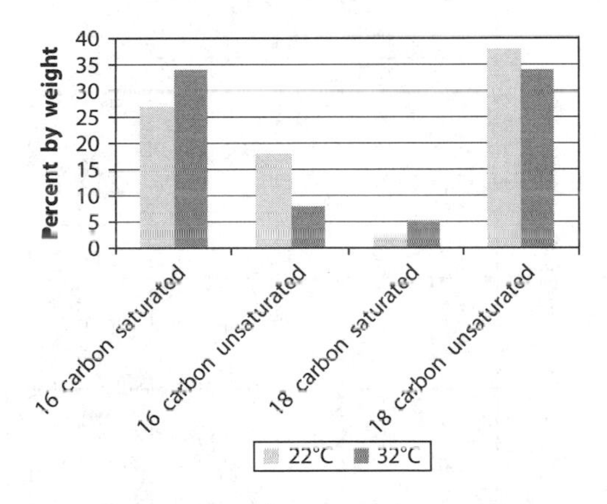

MULTIPLE-CHOICE QUESTIONS

173. Which of the following is true according to the data above?

(A) The saturation of fatty acids in phospholipids increases and decreases at increased temperatures.
(B) At higher temperatures, plants increase membrane fluidity by decreasing the saturation and increasing the chain length of the fatty acids in their phospholipids.
(C) The proportion of unsaturated fatty acids in membrane phospholipids decreases at higher temperatures.
(D) Increasing fatty acid saturation and chain length occur at both higher and lower temperatures.

174. Lipopolysaccharide (LPS) is a lipid-containing polysaccharide that is found *only* in the cell walls of eubacteria. The molecule itself is not harmful, but it is detected by the immune system. Which of the following observations is evidence that a fever is adaptive (*not* a failure of homeostasis)?

(A) A mouse injected with LPS develops a low-grade fever.

(B) A lizard injected with LPS will spend more time in the sun to reach and maintain a higher-than-usual body temperature.

(C) A human injected with LPS reports feelings of discomfort.

(D) Reducing a moderate fever does not significantly reduce the duration of an illness.

Questions 175 and 176 refer to the following diagram of an experiment in which an animal was injected with antigen A on day 0. Samples of blood were taken every day and the concentrations of both anti-A and anti-B antibodies were measured. On day 28, equal amounts of antigens A and B were given to the organism (the amount of antigen A given on day 28 was the same as had been given on day 0).

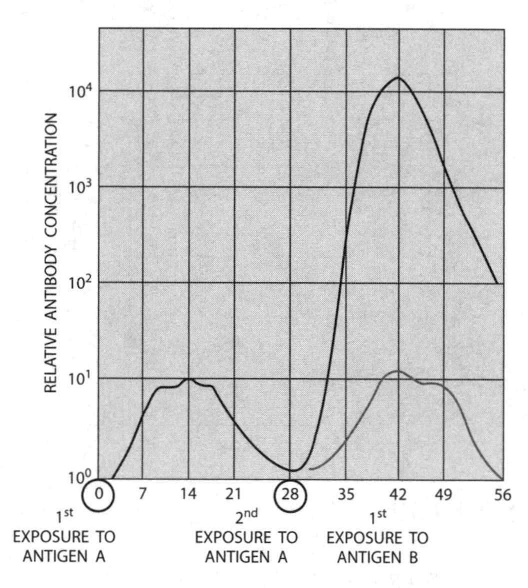

GRID IN

175. The rate of antibody production is much greater between days 28 and 42 than days 0–14. Calculate the approximate fold-increase in antibody production between these two periods of time. Report your answer to the nearest tens place. Do *not* use scientific notation.

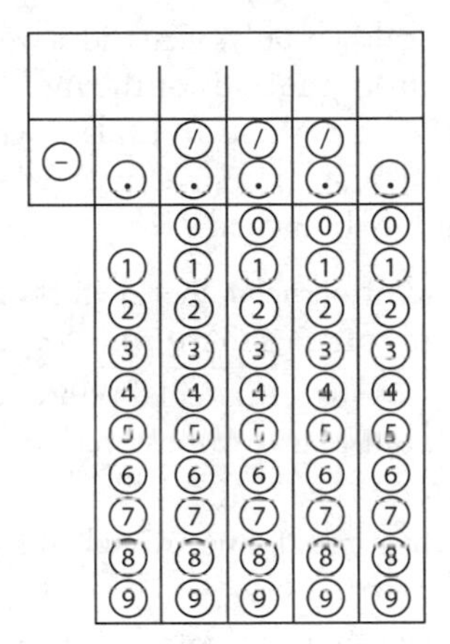

SHORT FREE-RESPONSE QUESTION

176. Explain why, for antigen A, the graph following first exposure differs from the graph following second exposure. **Propose** a purpose of the exposure to antigen B on day 28.

MULTIPLE-CHOICE QUESTIONS

177. Freshwater fish excrete nitrogenous waste primarily as ammonia, which is both toxic and water soluble. Terrestrial mammals convert ammonia to urea, which is also water soluble but much less toxic than ammonia. Which of the following statements provides a reasonable explanation for the fact that freshwater fish excrete ammonia rather than converting it to urea first?

(A) Ammonia is toxic only when it is present in body tissues.

(B) Freshwater fish make only small amounts of ammonia from amino acid metabolism.

(C) The conversion of ammonia to urea would increase urea concentrations that would slow the diffusion of ammonia out of the fish.

(D) Ammonia is diluted upon excretion by fish, and the conversion of ammonia to urea requires energy.

178. A nutrient is considered essential if:

(A) the body requires it to function properly.
(B) all cells require it to function properly.
(C) the body can synthesize it only in small amounts.
(D) it must be obtained in the diet.

179. The enzymes that break down polysaccharides can remove monosaccharides only from the ends of the molecule. Which of the following is a reasonable hypothesis to explain why animals store glucose as glycogen (a large, highly branched structure), whereas plants store glucose as starch (linear or slightly branched)?

(A) Animals move and often need energy in short bursts. A highly branched structure contains more ends for releasing glucose molecules.
(B) Animals store glucose as a branched polysaccharide to reduce osmotic pressure, but plants store starch to increase osmotic pressure, increasing turgor.
(C) Plants store starch in roots, but animals store glycogen in muscle and liver.
(D) Enzymes that break down polysaccharides are present in animal cells but not plant cells.

180. Fenestrations are protein-lined pores between endothelial cells in capillaries. They make the capillary leaky by permitting the diffusion of small molecules across the capillary wall. Fenestrated capillaries have larger holes in their walls than continuous capillaries. Different types of capillaries are present in different types of tissues. Which of the following types of capillaries would be *most* suitable for the cells to which they supply blood?

(A) fenestrated capillaries in the absorptive lining of the small intestine
(B) fenestrated capillaries in the brain
(C) continuous (unfenestrated) capillaries in the endocrine glands
(D) continuous capillaries in the glomerulus of the kidney

181. Which of the following is evidence that the number of cells in multicellular organisms is tightly regulated?

(A) Many cells have a limited number of times they can divide.
(B) Cells inherit a program for cell death (apoptosis) and can activate it when a serious stress is encountered (DNA breaks).
(C) In most adult multicellular organisms, the number of cells remains fairly constant.
(D) Cells in the embryo undergo apoptosis to sculpt individual digits in the hands of mammals.

182. In the process of metamorphosis when a tadpole becomes a frog, which of the following best describes the mechanism by which the frog's tail disappears?

(A) The tail cells are deprived of nutrients until the cells die and then the tail is shed.

(B) Blood flow to the tail is restricted until the cells die and the tail is shed.

(C) The developing frog eats its tail, which contains chemicals that help complete the metamorphosis.

(D) The cells of the tail undergo apoptosis, and the nutrients from the cells are recycled.

183. If a part of the liver is removed in an adult rat, the liver cells left will reproduce until the part removed is replaced. If a rat with a normal-sized liver is treated with phenobarbital, a chemical that stimulates liver cell division, the mass and cell number of the liver will increase. When phenobarbital is stopped, the liver returns to its original size and cell number. Which of the following statements is best supported by these observations?

(A) Organ and tissue size is strictly regulated by cell division.

(B) Adult tissues regulate their size by regulating cell division and programmed cell death (apoptosis).

(C) Chemicals (like phenobarbital) regulate tissue size by promoting cell growth.

(D) The number of cells that make up an organism can change with the needs of the organism.

184. Caspases are a family of intracellular proteases important in the activation of apoptosis and the amplification of apoptosis. The inhibitor of apoptosis (IAP) proteins are a family of proteins that inhibit apoptosis by inhibiting the action of caspases. They were first discovered in insect viruses. Which of the following most accurately expresses the advantage this protein would confer to the insect virus that carries it?

(A) It allows the cell to undergo cell division, replicating the virus with each cell division.

(B) It prevents the infected cell from dying before the virus has time to get replicated.

(C) It inhibits the action of caspases on viral proteins.

(D) It prevent the caspases from inducing apoptosis in newly formed viruses.

SHORT FREE-RESPONSE QUESTION

185. Caspases and other proteases (as well as clotting factors in the blood, hormones in the blood, and zymogens produced by the stomach and pancreas) are synthesized as inactive precursors that must be activated before they can function. **Explain** why it is necessary to synthesize these proteins in an inactive form.

Questions 186 and 187 refer to the following diagram.

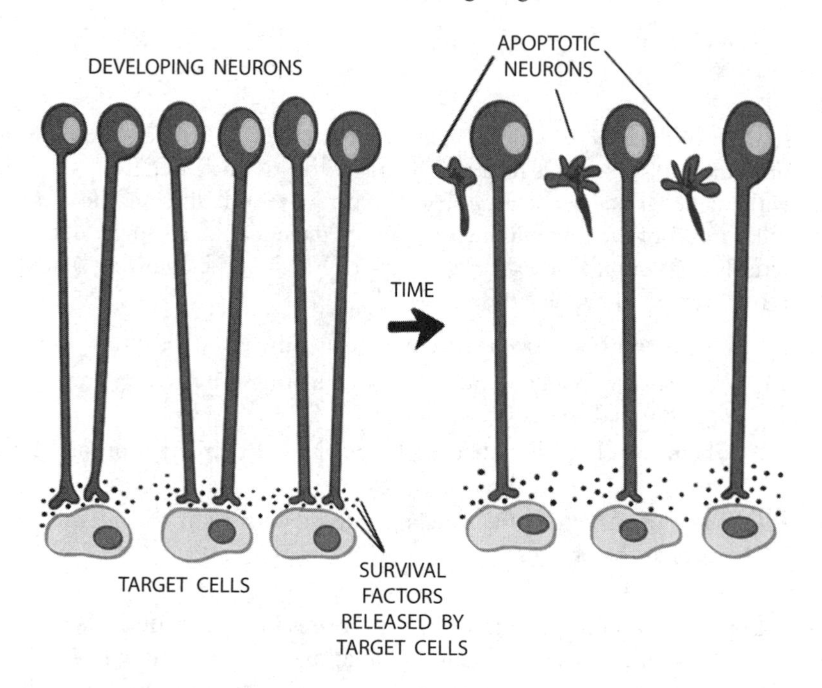

MULTIPLE-CHOICE QUESTIONS

186. More nerve cells are produced than can be supported by the limited amount of survival factors released by target cells. Survival factors inhibit apoptosis. Which of the following statements accurately describes a beneficial consequence of this strategy?

(A) It increases the energy expenditure of development.
(B) Neurons can choose their target cells, allowing greater cooperativity and communication in the nervous system.
(C) It ensures cells survive only when and where they are needed.
(D) The nervous system does not need to regulate cell division during development, because unnecessary cells will be eliminated.

187. Which of the following would be the direct result of a mutation that causes target cells to produce excess survival factors?

(A) Neurons would die.
(B) Neurons would make connections to the multiple target cells.
(C) Target cells would die.
(D) Multiple neurons would establish connections to each target cell.

188. *Caenorhabditis elegans* is a nematode worm whose development is well understood. Adult *C. elegans* hermaphrodites have exactly 1,031 somatic cells. Over the course of normal development, 131 cells die in a predictable pattern. Three genes required for the death of these cells have been identified. Which of the following answer choices includes a product of one or more of these genes?

(A) p53, a protein involved in tumor suppression
(B) caspases, proteases involved in activating apoptosis
(C) cyclins, proteins involved in regulation of the cell cycle
(D) PDGF, a growth factor necessary for initiating cell division

189. Which of the following statements *incorrectly* describes the pyramid of biomass?

(A) The mass of the pyramid's base is equal to the combined mass of all the trophic levels above it.
(B) The amount of biomass at any trophic level is dependent on the trophic level below it.
(C) Each trophic level is defined by the nutritional mode of the organisms within it.
(D) The biomass of a trophic level is the dry mass of all the organisms present in that trophic level.

190. Which of the following organisms can feed at more than one trophic level?

(A) omnivores, only
(B) omnivores and decomposers, only
(C) secondary consumers, only
(D) producers, omnivores, and decomposers

191. Which of the following statements accurately describes the flow of energy and nutrients in an ecosystem?

(A) Energy and nutrients are both recycled.
(B) Energy is recycled, but only 10% of the nutrients from one trophic level are transferred to the one above it.
(C) Nutrients are recycled within the biosphere, but energy is lost from every trophic level.
(D) Nutrients accumulate at the highest trophic levels, and energy flows into, through, and out of the biosphere.

192. Suppose a certain strain of plant produced albino offspring that are unable to grow or produce seeds unless they grow in close proximity to green plants of the same species. Which of the following statements correctly describes the relationship between these two plants?

(A) The albino plant required pollen from the green plant to produce seeds.
(B) The green plant is a producer, and the albino plant is a consumer.
(C) The albino plant is dependent upon the green plant for organic compounds.
(D) The albino and green plants are heterozygous.

193. Which of the following results when plants close their stomata on dry, sunny days?

(A) increased water loss
(B) increased water uptake
(C) decreased CO_2 loss
(D) decreased CO_2 uptake

194. Which of the following is common to *all* gas exchange systems in animals?

(A) Gases are actively transported across membranes.
(B) The countercurrent exchange of carbon dioxide and water occurs.
(C) The diffusion of gases occurs across moist membranes.
(D) Gas exchange requires a closed circulatory system.

195. A person's mean arterial blood pressure is 90 mm Hg with a systolic blood pressure of 120 and a diastolic blood pressure of 80. Which of the following most accurately explains why the mean arterial blood pressure is not simply the average of the systolic and diastolic pressures (100 mm Hg)?

(A) The mean arterial pressure is the average of many readings of systolic pressure only.
(B) The systole of the heart does not last as long as the diastole.
(C) The systolic pressure is measured in arteries, and diastolic pressure is measured in veins.
(D) The average is weighted because the diastolic pressure is more important than the systolic pressure.

196. A fever is sometimes erroneously considered a failure of homeostasis. An alternative hypothesis is that a fever occurs as the result of a higher temperature set point and the new, temporary temperature is homeostatically maintained. Which of the following observations does *not* support the hypothesis that a fever is the temporary maintenance of a new temperature set point?

(A) Chills occur at the onset of a fever indicating the body temperature is lower than the new, fever set point for temperature.
(B) Certain exogenous chemicals, like acetaminophen, can reduce body temperature during a fever even while the pathogen is still present.
(C) If a reptile is injected with an antigen, it will maintain a higher than normal body temperature until the antigen is cleared.
(D) Sweating occurs as the body returns to normal temperature, indicating that the temperature set point has reverted back to 37°C but the body is still at the fever temperature.

197. Under which of the following conditions would the pancreas most likely secrete insulin, a hormone that lowers blood glucose levels?

(A) after a carbohydrate-rich meal, regulated by positive feedback
(B) after a carbohydrate-rich meal, regulated by negative feedback
(C) before a protein-rich meal, regulated by positive feedback
(D) before a protein-rich meal, regulated by negative feedback

198. A bird and a snake are kept in a cold room (5°C, 41°F) for 6 hours. Which of the following changes in oxygen consumption and body temperature are expected?

	O_2 consumption Bird	O_2 consumption Snake	Body temperature Bird	Body temperature Snake
(A)	No change	No change	No change	No change
(B)	Increase	Decrease	No change	Decrease
(C)	No change	Increase	Increase	Decrease
(D)	Decrease	Decrease	Decrease	Decrease

199. A participant in an experiment is given air to breathe in which the carbon dioxide concentration has been increased. Which of the following results is expected? The reaction of carbon dioxide in the blood follows:

$$CO_2 + H_2O \rightarrow H_2CO_3 \rightarrow H^+ + HCO_3^-$$

(A) increased blood pH and respiration
(B) decreased blood pH and respiration
(C) increased blood pH and decreased respiration
(D) decreased blood pH and increased respiration

Questions 200–202 use the following answer choices.

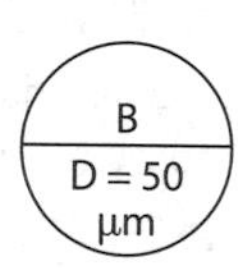

Surface area of a sphere $A = 4\pi r^2$.

Volume of a sphere $V = 4/3\pi r^3$.

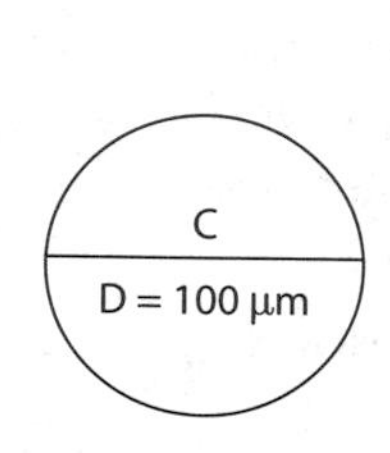

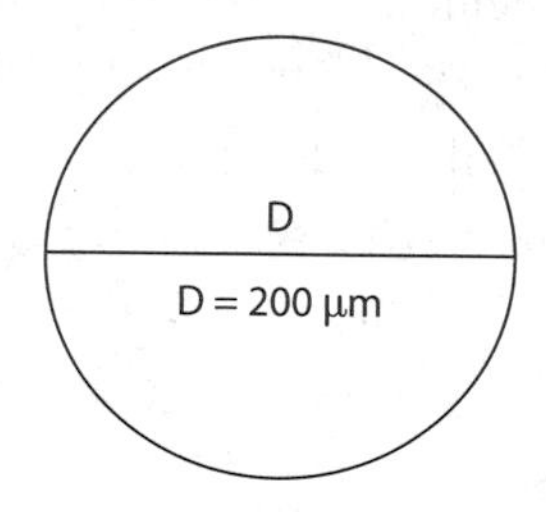

200. Which of the cells would be most efficient at exchanging nutrients and wastes with the environment?

(A) cell A
(B) cell B
(C) cell C
(D) cell D

201. In which of the cells would the diffusion of a molecule across the center of the cell take the longest amount of time?

(A) cell A
(B) cell B
(C) cell C
(D) cell D

202. Suppose each of the previous spheres were made of ice and left out at room temperature. Which of the following statements accurately predicts which would melt the fastest along with the correct reason?

(A) Sphere A because it has the smallest volume and therefore the smallest mass.
(B) Sphere B because it has a larger surface area than sphere A to gain heat from the environment but a smaller volume than spheres C and D, therefore a lower mass of ice to melt.
(C) Sphere D because it has the smallest surface-area-to-volume ratio to minimize heat loss to the environment.
(D) Sphere D because it has the largest surface area to maximize heat gain from the environment.

203. Membrane-bound organelles are present in practically all eukaryotic cells. Which of the following statements most accurately summarizes the advantage of internal membranes in eukaryotic cells?

(A) Membrane-bound organelles like the mitochondria and chloroplast are capable of semiautonomous replication.
(B) Internal membranes form partitions that can isolate specific reactions, increasing metabolic efficiency and the range of metabolic activities the cell can perform.
(C) Internal membranes allow prokaryotic cells to reside within eukaryotic cells to form cooperative cells.
(D) Compartmentalization allows cells to reproduce faster and with greater efficiency.

LONG FREE-RESPONSE QUESTION

204. The properties of surfaces are critical in performing their exchange functions. **List three structural properties** of exchange surfaces in organisms. **Choose** an example of a cellular or organismal structure associated with exchange and **briefly describe** the structural features that support its function.

Questions 205 and 206 refer to the following diagram.

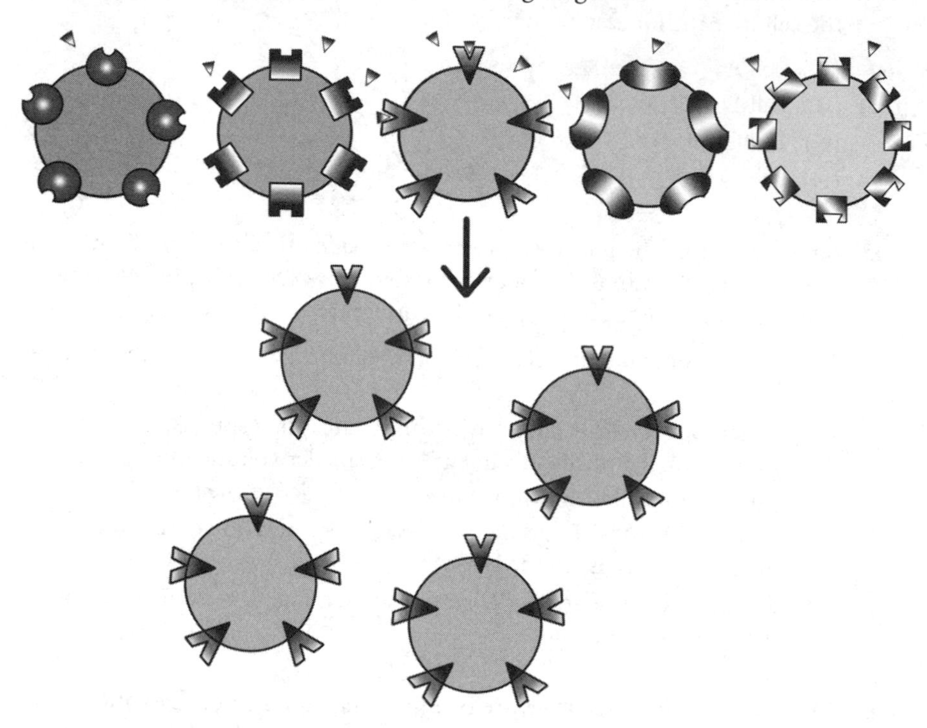

MULTIPLE-CHOICE QUESTIONS

205. The immune system is able to recognize specific foreign invaders it has not previously encountered and quickly build up a defense against them. Which of the following analogies most accurately illustrates the mechanism by which this is accomplished by the immune system?

(A) Choose a lottery ticket out of a bag with your eyes closed.
(B) Try on several pairs of shoes until you find one that fits comfortably and then wear them regularly.
(C) Make several different kinds of cookies for a friend, note which kind they like most, and make more of that type of cookie.
(D) Ask someone for something every day until he or she gives it to you.

206. Which of the following most accurately summarizes the function of adaptive immunity?

(A) It is an inherited adaptation that allows cells like macrophages to nonspecifically recognize and kill foreign invaders.
(B) It allows the immune system of an individual to learn about the environment in which it lives.
(C) It allows parasites to adapt to their host's environment.
(D) It is necessary for cells within a multicellular organism to undergo natural selection during development.

207. Which of the following diagrams best illustrates the process by which a macrophage (a type of phagocytic cell) engulfs a pathogenic bacterium as part of the innate (nonspecific) immune response?

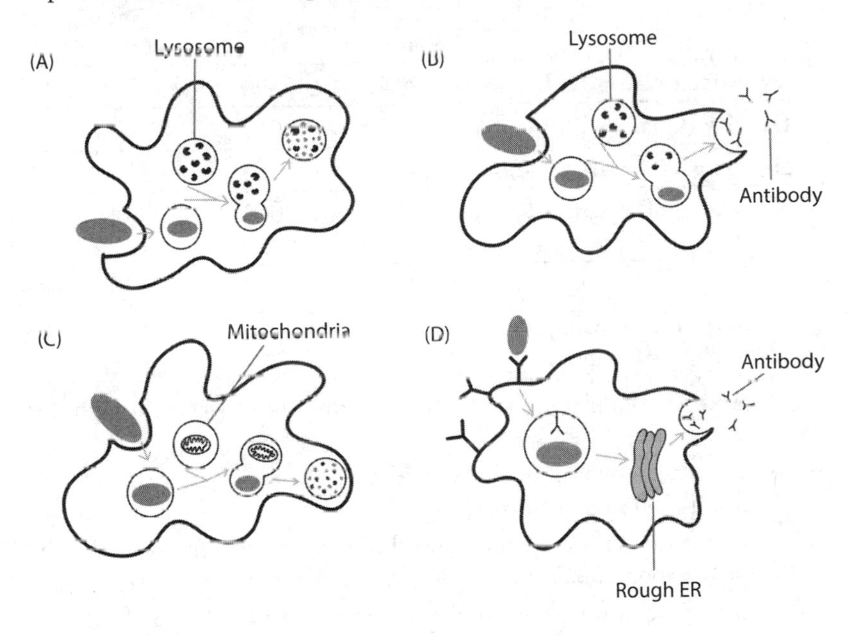

Question 208–209 refer to the following diagram and data table.

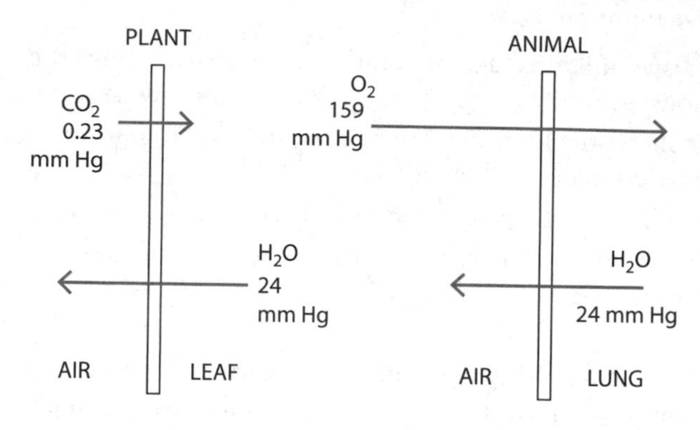

	Water	Air	Ratio Water:Air
O_2 concentration (L/L)	0.007	0.2090	~1:30
Density (kg/L)	1.000	0.0013	~800:1
Heat capacity (cal/L • °C)	1,000	0.3100	~3,000:1
Kilograms of medium per liter O_2	143	0.0062	~23,000:1

208. Which of the following statements is a reasonable inference based on the diagram and data above?

(A) Animals and plants absorb the same amount of water through their surfaces.

(B) Cellular respiration in animals requires more oxygen than cellular respiration in plants.

(C) Plants use carbon dioxide instead of oxygen for respiration, but both plants and animals create water as a product.

(D) It is difficult for plants to obtain enough carbon dioxide from the atmosphere by diffusion without incurring great water losses.

209. Which of the following statements is a logical inference from the data?

(A) The oxygen concentration is greater in air than in water, but the higher density of water makes the oxygen more easily absorbed.

(B) A fish must move at least 23,000 kilograms of water over its gills to absorb 1 liter of oxygen.

(C) The high heat capacity of water decreases its ability to hold oxygen.

(D) A greater mass of respiratory medium must be moved over gills compared to the mass of air ventilated in the lungs.

210. Which of the following statements describes an adaptation in fish that maximizes the absorption of oxygen from water?

(A) Movement of water over gills is almost exclusively one-way and in the opposite direction of blood flow in the gills to maximize diffusion.
(B) The large fraction of the energy requirements of fish are the result of accelerating large masses of water back and forth over the gills.
(C) Because the concentration of oxygen is low in water, blood entering the gills must have as low of an oxygen concentration as possible to increase the rate of diffusion from water.
(D) Fish swim in the opposite direction of their blood flow to maximize the passage of water over their gills.

211. The extensive and intricate respiratory system of birds is very different from the respiratory system of mammals. Which of the following true statements best supports the hypothesis that the respiratory adaptations of birds are *not a necessary prerequisite* for flight but *may confer considerable advantages* for flight?

(A) Bats have typical mammalian lungs, are good fliers, and are capable of migrating over long distances.
(B) Resting oxygen consumption is similar for birds and mammals.
(C) During flight, 8- to 10-fold increases in oxygen consumption are required for both birds and bats.
(D) The lung volume of a typical bird is only a little more than half that of a similarly sized mammal.

Questions 212 and 213 refer to the following curve of population size over time in a population of snowshoe hares.

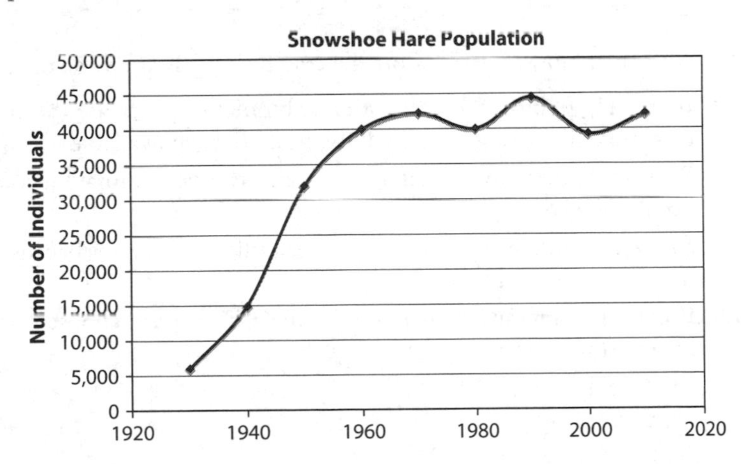

212. The graph indicates that the population of snowshoe hares is most likely:

(A) stabilized by density-dependent factors after 1960.
(B) regulated by density-independent factors between 1960 and 2010.
(C) growing at an exponential rate from 1930–1950, 1980–1990, and 2000–2005 because there are few limiting factors.
(D) headed for extinction.

213. Which of the following statements is *not* a reasonable assumption based on this data?

(A) A few years after 1990, there was a fairly large increase in the population of lynx in the area (the predator of the snowshoe hare).
(B) A decline in the population of lynx occurred in 1982.
(C) A sustained increase in the amount of primary productivity in the environment occurred in 1930.
(D) The average life span of the snowshoe hares decreased in 1960.

Question 214 is based on the following data.

Temperature (°C)	Freshwater O_2 content ($mL\ O_2/L\ H_2O$)	Seawater O_2 content ($mL\ O_2/L\ H_2O$)
0	10.3	8.0
10	8.0	6.4
15	7.2	5.8
20	6.6	5.3
30	5.6	4.5

214. Which of the following is a reasonable conclusion based on the data?

(A) Increasing photosynthesis in aquatic biomes can increase the amount of animal life because there will be more oxygen available.
(B) Warmer bodies of water cannot support as much animal life as cooler bodies of water.
(C) Saltwater contains less oxygen gas per milliliter of water because it does not support plant life.
(D) If placed in a temperature gradient, most fish would choose water of lower temperatures.

Questions 215 and 216 are based on the following information.

Vibrio fischeri is a species of bioluminescent marine bacteria. When grown in liquid culture, the bacteria produce light only when large numbers of them are present (a high population density). It was observed that whenever bacteria from a luminescing culture seeded a new culture, they would produce light at low population densities, but only for a short time and then they stopped producing light completely. They would produce light again only once the population in that culture grew to a high enough density.

215. Which of the following is a reasonable mechanism for the changes in bioluminescence with bacteria density?

(A) Low-density bacterial cultures contain an inhibitor of luminescence whose concentration increases as their population grows.
(B) Bacterial metabolism releases excited electrons that emit photons in solution that are visible when large numbers of bacteria are present.
(C) The bacteria carried bioluminescent particles with them during transfer into the new culture that were either diluted or inactivated by exposure to a new, low-density culture.
(D) Light is produced only when bacteria make physical contact with other bacteria.

216. It was eventually discovered that luminescence is activated only after the concentration of a particular molecule has sufficiently accumulated. Which of the following is the most likely purpose of this for the bacteria?

(A) to sense the density of their population based on the concentration of the molecule
(B) to attract a particular mating type based on the flashes of light that are produced
(C) to provide photons (energy) for photosynthesis
(D) to signal a low-nutrient density in the culture

Questions 217 and 218 refer to an experiment in which one person was monitored in a calorimetry chamber over a 24-hour period. Measurements of energy input and output were taken once every hour for the 24-hour period.

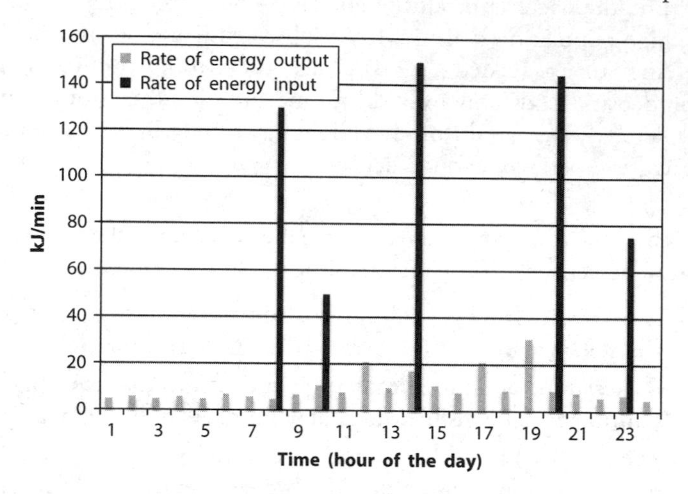

217. Which of the following statements is true according to the data above?

(A) Energy intake is constant throughout the day.
(B) Energy output is constant throughout the day.
(C) Energy expenditure is fairly constant with occasional bursts of work.
(D) Energy output is usually equal to energy input.

218. Which of the following most accurately explains why the person in the chamber never has an energy output of zero?

(A) At a cellular level, work is always being done by a living organism.
(B) The person did not sleep for the 24-hour period of the study.
(C) Even the most minute body movements require energy.
(D) The person in the study was fairly young, and resting metabolic rate approaches zero only at an advanced age.

219. Which of the following most accurately explains how multicellular animals are able to accommodate long periods with no energy input?

(A) Every cell has its own store of energy-rich molecules.
(B) Some cells are specialized to store energy-rich molecules in times of excess energy input and release energy-rich molecules in times of little or no energy input.
(C) Energy restriction tends to extend maximum life span.
(D) Energy balance occurs over long periods of time so the day-to-day input of energy is not physiologically critical.

220. Which of the following is *not* a component of energy regulation in animals?

(A) a physiological mechanism for detecting high-energy and low-energy situations in the body
(B) structures for energy storage and release
(C) behavioral strategies for obtaining food
(D) the ability to synthesize essential nutrients according to available food sources

221. Which of the following is an example of local enzyme regulation?

(A) allosteric regulation of enzymes by negative feedback
(B) phosphorylation of enzymes resulting from hormone induced activation of kinase (phosphorylating) enzymes
(C) tissue-specific expression of enzyme isoforms
(D) behavioral responses to cues such as hunger, thirst, or cold

SHORT FREE-RESPONSE QUESTION

Ecosystem	Net primary production (kcal/m²/year)	Ecological efficiency (%)	Number of trophic levels (n)
Open ocean	500	25	7
Coastal marine	8,000	20	5
Temperate grassland	2,000	10	4
Tropical forest	8,000	5	3

222. Choose two of the communities represented in the table above and **explain** why they differ in the number of trophic levels they can support.

Questions 223 and 224 refer to an experiment done in the late 1960s by George Cahill, Jr., in which several obese men were starved for 40 days. They received water, vitamins, and minerals but no source of food energy. Measurements of several physiological variables were taken after a last, small meal containing carbohydrates, fats, and protein was ingested and then throughout the 40 days of starvation.

Glucagon is a hormone that raises blood glucose levels, and *insulin* lowers blood glucose levels. *Gluconeogenesis* is the process by which the liver synthesizes glucose from nonglucose precursors such as certain amino acids (but not fatty acids).

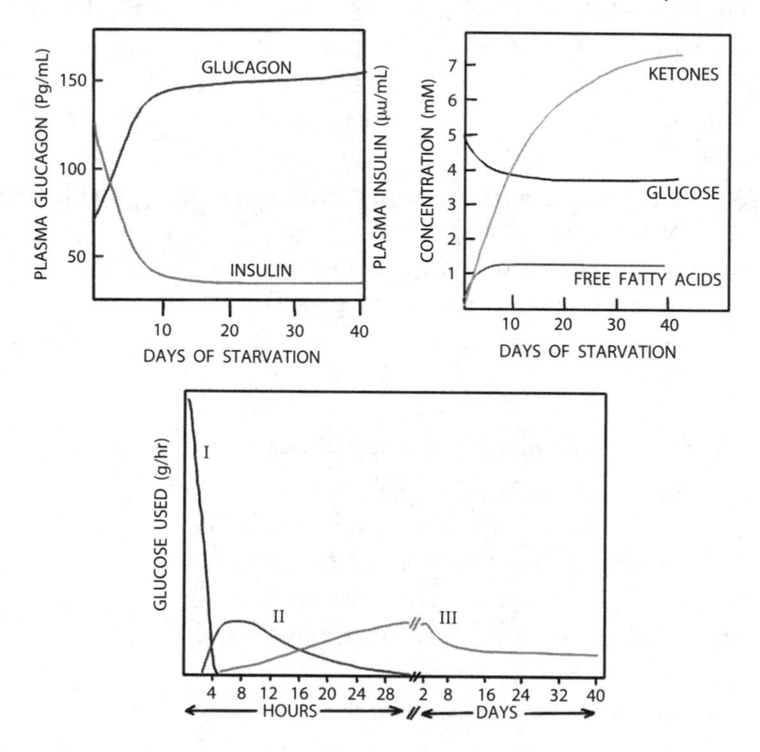

MULTIPLE-CHOICE QUESTIONS

223. Which of the following statements accurately explains the concentrations of glucagon, insulin, and glucose in the blood throughout the experiment (40 days of starvation) and the rate of glucose usage?

(A) The initial drop in blood glucose concentration causes glucagon levels to rise, which causes insulin levels to fall, resulting in an overall decrease in blood glucose concentration in the first 10 days of starvation despite decreased glucose usage.

(B) Increased glucagon decreases glucose levels by decreasing glucose usage.

(C) Low insulin and high glucagon levels help keep glucose in the blood while cells can use fatty acids and ketone bodies for energy.

(D) Glucose concentrations and usage drop to make insulin levels fall, so no further decrease in glucose concentration occurs.

224. The third graph of glucose used versus time contains three separate curves. Each curve represents a different source of glucose. Which of the following answer choices appropriately identifies the source of glucose for each curve?

	Curve I	Curve II	Curve III
(A)	Dietary glucose	Gluconeogenesis	Glycogen breakdown
(B)	Dietary glucose	Glycogen breakdown	Gluconeogenesis
(C)	Dietary glucose	Gluconeogenesis	Glycogen breakdown
(D)	Glycogen breakdown	Dietary glucose	Gluconeogenesis

GRID IN

225. What is the water potential of a plant cell bathed in pure water until there is no net movement of water into or out of the cell? Include a negative sign if necessary.

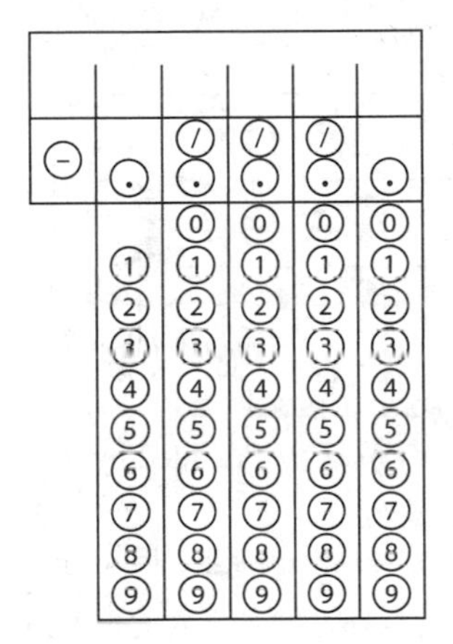

$\Psi_{WATER} = 0$

$\Psi_{S,\,CELL} = -1$ MPa

Questions 226–230 are based on the following data.

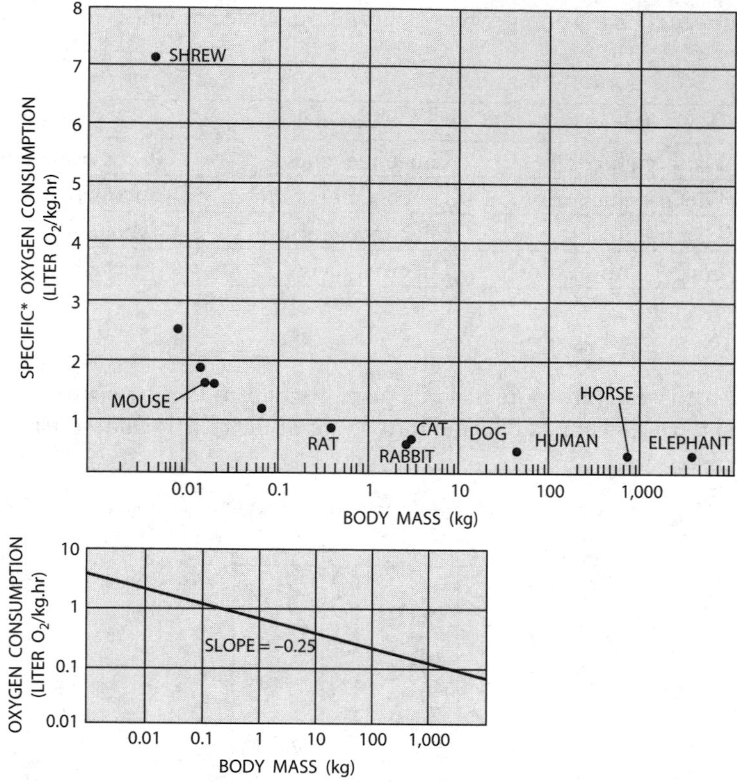

*Specific means "per unit mass" or "divided by mass."

226. Which of the following statements is a logical conclusion based on the data?

(A) Large mammals consume less oxygen than smaller mammals.

(B) A horse and a human would consume oxygen at the same rate if they were the same size.

(C) Shrews are smaller than elephants because they have a higher rate of oxygen consumption.

(D) Diving animals tend to be large because the rate of oxygen consumption relative to body size is lower in larger mammals.

SHORT FREE-RESPONSE QUESTION

227. Smaller mammals tend to have shorter life spans than larger mammals. **Choose one** physiological reason that could account for the shorter life span and **explain** how it could affect an organism's life span.

GRID INS

228. Calculate the volume of oxygen consumed by a 500-gram rat (0.5 kg) in a 24-hour period. Round to the nearest whole number.

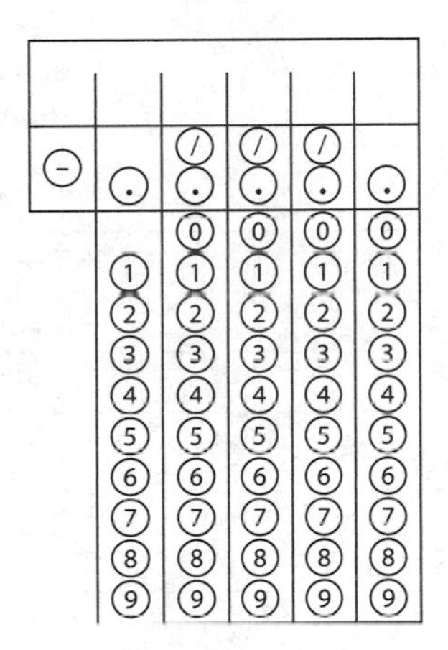

229. Calculate the volume of oxygen consumed by a 60-kilogram human in 1 hour. Round to the nearest whole number.

SHORT FREE-RESPONSE QUESTION

230. The first graph was constructed from the same data as the second graph. **Briefly explain** the purpose of using a regression line to express data.

Questions 231–233 refer to the following table.

D = dark
FR = brief exposure to light in the far-red part of the visible spectrum
R = brief exposure to light in the red part of the visible spectrum

Trial #		↓	Effect on flowering in long-night (short-day) plants	Effect on flowering in short-night (long-day) plants
1	LIGHT	DARK	No flowering	Flower
2	LIGHT	DARK	Flower	No flowering
3	LIGHT	DARK R DARK	No flowering	Flower
4	LIGHT	DARK R D FR DARK	Flower	No flowering
5	LIGHT	DARK R D FR D R DARK	No flowering	Flower
6	LIGHT	DARK R D FR D R D FR	Flower	No flowering

MULTIPLE-CHOICE QUESTIONS

231. Which of the following statements most accurately summarizes the data?

(A) The total amount of time in the dark determines whether flowering will occur in long-night (short-day) plants.
(B) Red light shortens the dark period in long-night (short-day) plants.
(C) Exposure to either red light or far-red light inhibits flowering.
(D) The total amount of time in the light, including exposures to red and far-red light, ultimately determines whether flowering will occur in short-night (long-day) plants.

232. Which of the following trial comparisons best supports the hypothesis that photoperiod regulation depends on night length and *not* day length?

(A) trials 1 and 2
(B) trials 1 and 3
(C) trials 2 and 3
(D) trials 2 and 4

233. Which of the following true statements is the strongest evidence that plants detect night length precisely?

(A) Some long-night (short-day) plants will not flower if night length is one minute shorter than the critical night length.

(B) Tobacco plants can grow tall but won't flower in the summer even when temperature, moisture, and mineral conditions are optimized for flower development.

(C) Many long-night (short-day) plants can be stimulated to flower by manipulating day length.

(D) Tomato plants grown in controlled conditions will flower, regardless of day length, once they've reached a certain stage of maturity.

Questions 234 and 235 refer to a spectacular act of cooperation that occurs in the social amoeba *Dictyostelium discoideum* (also known as slime mold). *D. discoideum* are normally solitary, but during starvation conditions *D. discoideum* cells aggregate to form a temporary, migratory slug that later develops into a fruiting body. Some of the *D. discoideum* cells will become part of the stalk and the others will form the fruiting body (spore-producing cells). After reproduction, the cells of the aggregate die.

The *D. discoideum* cells that die to produce a stalk hold the spore-producing cells aloft. Those cells will produce reproductive spores (regions IV through VI in the diagram of the life cycle of *D. discoideum* below) and are the only the cells that will pass their genes to the next generation of *D. discoideum.*

The **dim A gene** is required to receive the chemical signal (DIF-1) that causes differentiation into prestalk cells. Cells that express the dim A gene and receive the DIF-1 signal will differentiate into stalk cells and will not produce reproductive spores. Cells that *do not* possess the dim A gene *do not* respond to the DIF-1 signal released by neighboring cells and *do not* differentiate into prestalk cells. These cells form the fruiting body that produces reproductive spores.

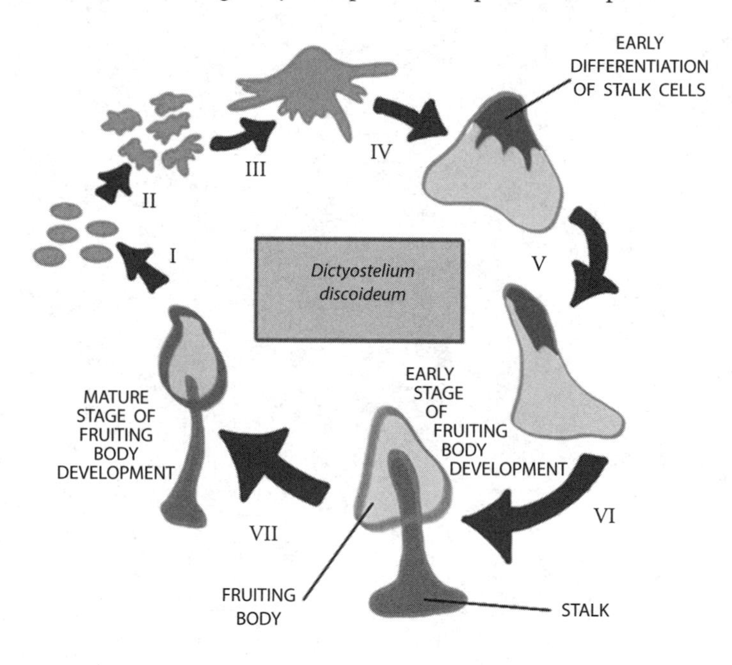

234. Which of the following statements explains why acts of cooperation such as stalk formation are a challenge for evolutionary biologists to understand?

(A) *D. discoideum* can't be both a solitary organism and a multicellular cooperative.
(B) The presence or absence of one gene should not affect the behavior of a cell so dramatically.
(C) Reproductive cells of *D. discoideum* have no advantage over stalk producers but are still selected for in the population.
(D) The *D. discoideum* that form the stalk never pass on their genes, and over many generations, stalk producers are expected to be eliminated from the population.

Question 235 refers to an experiment in which the dim A gene was knocked out to examine the effects of ignoring the DIF-1 signal. The dim A gene encodes a protein necessary for the DIF-1 response pathway. The knock-out mutants produce prespore cells instead of prestalk cells. Dim A$^-$ cells coaggregated normally and in equal numbers with AX4 (wild-type, spore-forming) cells. The data are shown as follows (adapted from Foster, K. R., and Thompson, C. R. L., 2004).

The graph on the left shows the proportion of spore cells in the fruiting body of dim A$^-$/AX4 aggregates. The table on the right shows the number of spores per plate when comparing AX4/AX4 and dim A$^-$/AX4 aggregates.

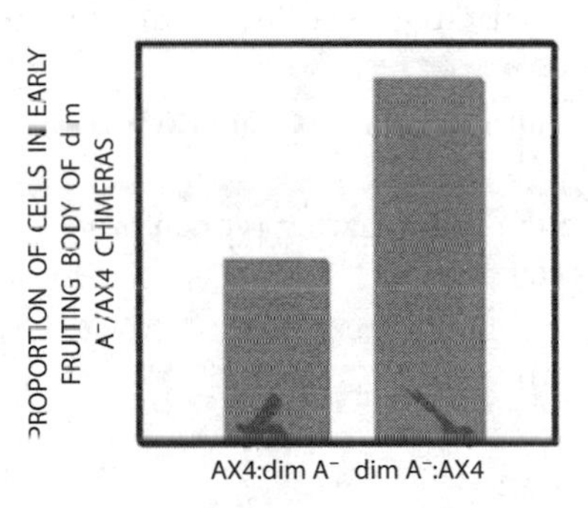

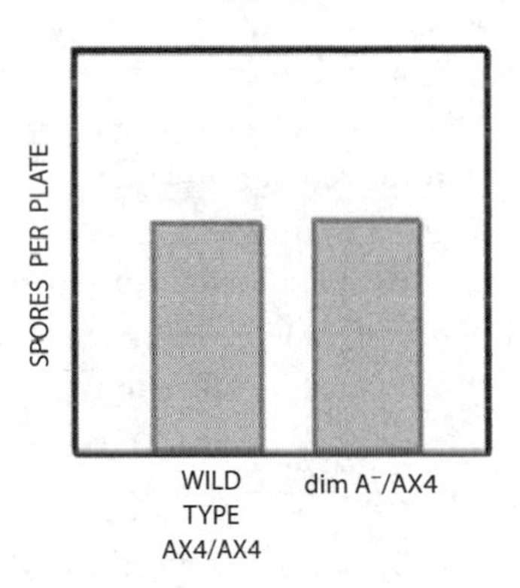

235. The combined results of these two analyses demonstrate that:

(A) The number of dim A$^-$ spores in the fruiting body of the dim A$^-$/AX4 aggregates is reduced but heterozygote spores can carry the dim A$^-$ gene into future generations through carriers.
(B) AX4/AX4 aggregates and dim A$^-$/AX4 aggregates produce equal numbers of spores because a mutation in the dim A gene causes it to proliferate.
(C) Dim A$^-$ cells are excluded from the fruiting body of the dim A$^-$/AX4 aggregates, but they end up producing an equal number of spores as the AX4 cells.
(D) The dim A$^-$ cells are outnumbered by AX4 cells in the dim A$^-$/AX4 aggregates, but they inhibit spore formation of the AX4 cells.

Question 236 and 237 refer to the diagram of the unicellular budding yeast, *Saccharomyces cerevisiae.* Besides their use as a model organism in research, *S. cerevisiae* have been a useful species for humans, important in alcoholic fermentation and baking, for thousands of years. *S. cerevisiae* can differentiate into three distinct cell types: haploid α mating type, haploid a mating type, and diploid α/a, which cannot mate but can form spores.

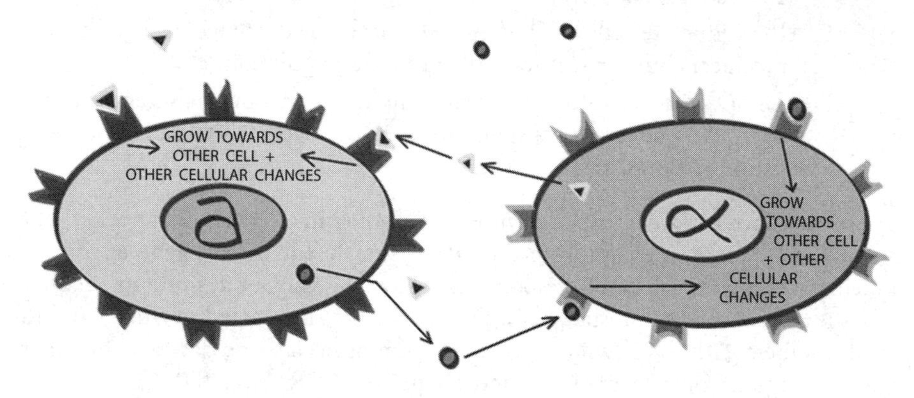

236. Which of the following statements most accurately summarizes the processes illustrated?

(A) The two mating types can recognize the opposite mating type and respond to the chemical factors it secretes.

(B) The α and a mating types communicate chemically to form a multicellular organism.

(C) Each mating type must locate the same mating type in order to form the diploid, spore-producing cell type.

(D) Changes in cell physiology occur only when a cell is activated by chemical factors from the environment.

237. The preceding process is evidence that:

(A) cells can only reproduce when they are triggered by external factors from the environment.

(B) diploid yeast are less adapted to their environment than haploid yeast.

(C) all cells are capable of sexual reproduction.

(D) cell signaling is evolved in unicellular organisms.

Question 238 refers to the following answer choices.

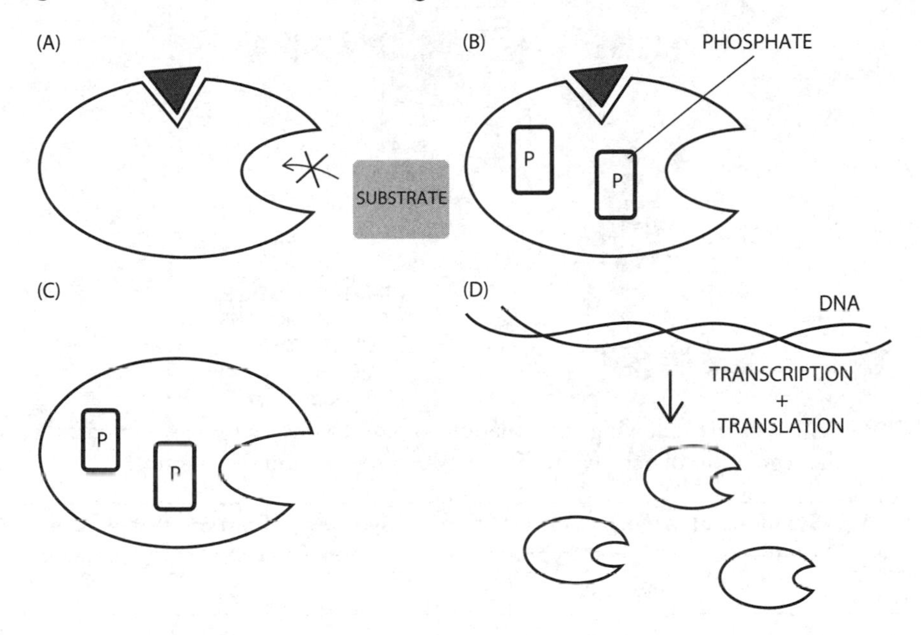

238. Which of the preceding choices illustrates a regulatory process that acts over the longest period of time?

Questions 239–242 refer to the following diagrams of the *trp* and *lac* operons in *E. coli*.

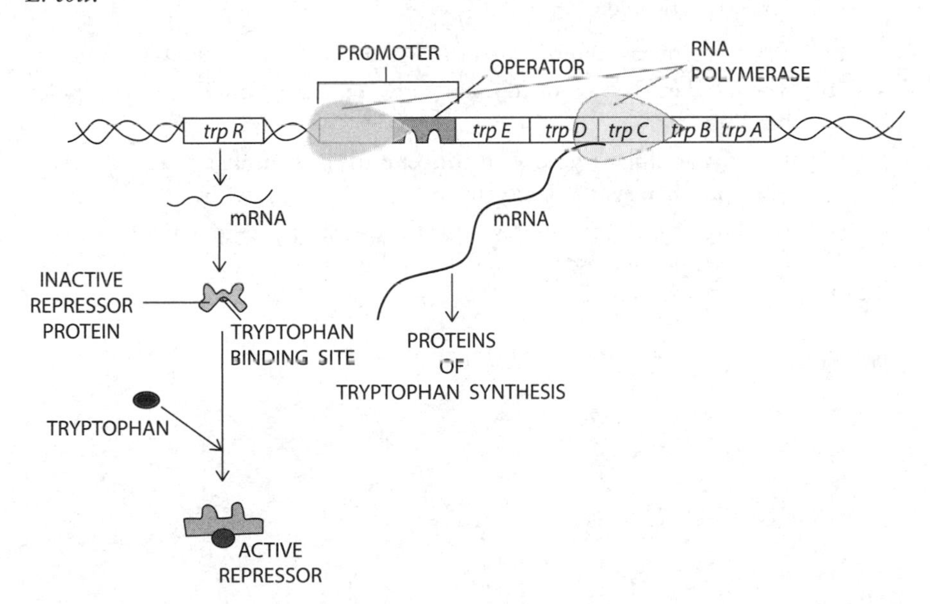

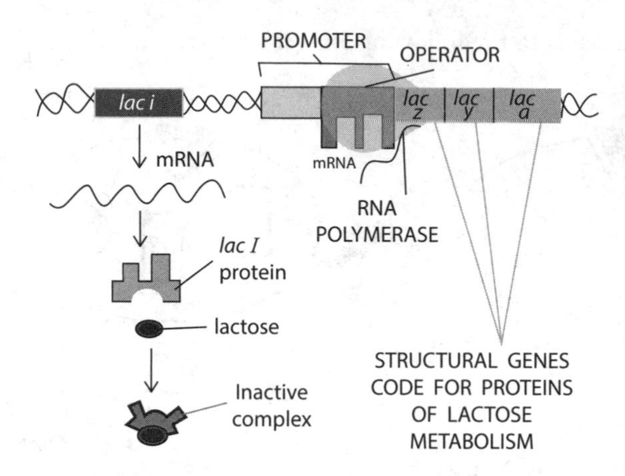

239. Which of the following combinations of conditions accurately compares the regulation of gene expression between the *trp* and *lac* operons?

	Synthesis of enzymes of tryptophan synthesis is ON when tryptophan is	**Synthesis of enzymes of lactose metabolism is ON when lactose is**
(A)	Present	Present
(B)	Present	Absent
(C)	Absent	Present
(D)	Absent	Absent

240. Which of the following is the mechanism by which the transcription of the *trp* operon is activated?

(A) RNA polymerase directly interacts with mRNA to activate transcription.
(B) The *trp R* mRNA binds to the inactive repressor to induce tryptophan binding.
(C) Low intracellular tryptophan concentrations stimulate transcription of tryptophan synthesizing enzymes.
(D) Binding of the active repressor to the operator provides a docking site for RNA polymerase to initiate transcription.

241. The *trp* repressor protein detects levels of tryptophan in the cell. If an experiment used radiolabeled tryptophan in the media used to grow bacteria and later found the radiolabeled tryptophan bound to the *trp* repressor proteins, which of the following must be true?

(A) The presence of radioactivity in the cell induces changes in gene transcription.
(B) A transport protein for the molecule of interest is present in the cell membrane.
(C) Bacteria can synthesize molecules using radioactive atoms.
(D) Tryptophan and lactose have the same mechanism of action on the operons they regulate.

242. Proteins are in a constant state of flux in cells. They are continually being synthesized and degraded; however, the synthesis of proteins is energetically expensive. Which of the following *does not* explain why cells would degrade proteins only to resynthesize the same kind of protein later?

(A) Protein degradation releases energy, which allows the degradation of old proteins to fuel the synthesis of new proteins.
(B) The relative rates of synthesis and degradation allow cells to adjust the levels of each protein to changing cellular requirements.
(C) Protein degradation prevents the accumulation of proteins that are not useful to the cell.
(D) Cells can use the available amino acids to make the specific proteins needed by the cell.

243. The Nile perch is a species of fish that was introduced into Lake Victoria in the 1950s. It has caused the extinction or near-extinction of several hundred native species in the lake. Which of the following is an accurate statement regarding the introduction of the perch to Lake Victoria?

(A) Eutrophication of a lake can decrease its carrying capacity.
(B) Humans can negatively impact the balance of an ecosystem.
(C) The Nile perch is a successful predator.
(D) The diversity in Lake Victoria is likely to increase because most extinction events are followed by a large increase in diversification.

Questions 244 and 245 refer to the mechanism of red and far-red light detection in plants illustrated below.

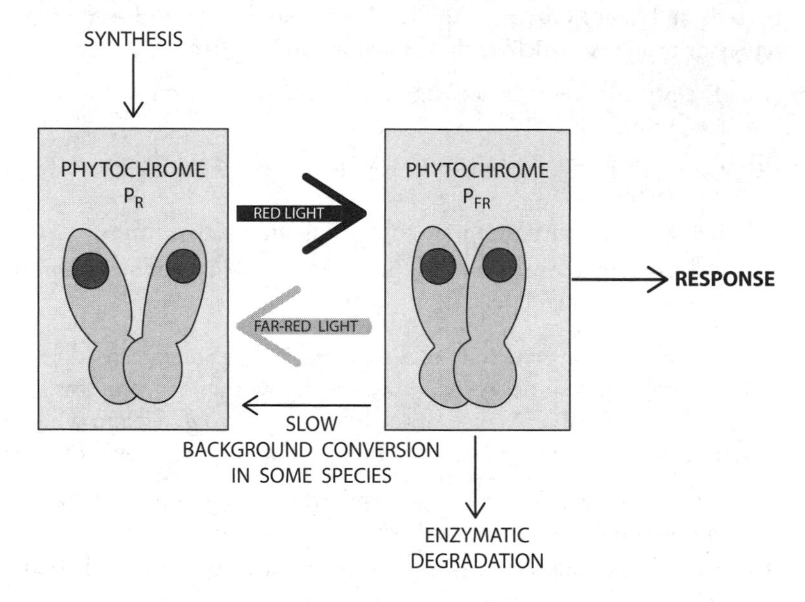

244. Phytochrome P_{FR} is a red-light photoreceptor in plant cells. Which of the following processes is the *least* likely to be regulated by phytochrome?

(A) flowering
(B) seed germination
(C) phototropism
(D) apical dominance

245. At sundown, P_{FR} is most abundant form of phytochrome. It slowly converts back to the P_R form. At dawn, more phytochrome is in the P_R form. The daylight rapidly converts P_R to P_{FR}. The switch to the P_{FR} isoform resets the plant's biological clock. Which of the following statements is a likely consequence of the daylight-mediated conversion of P_R to P_{FR} on the plant's biological clock?

(A) The plant's circadian rhythm is innate so the plant cannot learn about the environmental light conditions through phytochrome.
(B) The shorter the day, the more P_{FR} accumulates, resetting the clock according to day-length.
(C) The longer the night, the more P_R form builds up and the longer it takes to convert to P_{FR}.
(D) The plant's clock gets reset every morning regardless of length of the night.

LONG FREE-RESPONSE QUESTIONS

246. Organisms use both negative and positive feedback. **Compare** and **contrast** their *functions* and *mechanisms,* using one example for each.

247. **Choose one example** of positive feedback in an organism (you may choose from the following list or use your own example) and **describe** the mechanism by which positive feedback is able to amplify the physiological variable.

- Lactation in mammals
- Onset of labor in childbirth
- Ripening of fruit
- Ovulation in mammals
- Blood clotting

SHORT FREE-RESPONSE QUESTION

248. **Identify** which of the following graphs illustrates negative feedback and which illustrates positive feedback. **Briefly explain** your reasoning.

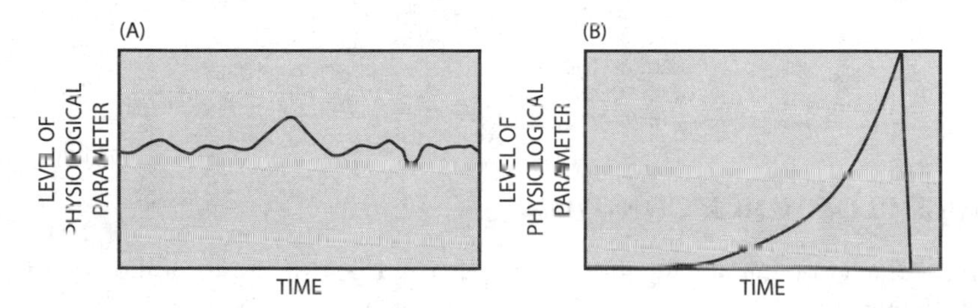

Question 249 refers to the following diagram, which illustrates a transplantation experiment performed by Spemann and Mangold in 1924. In the experiment they removed a portion of the gastrula, an early embryo, from a newt (the donor) whose cells had been stained with a pigment. The portion they removed was transplanted into the gastrula of a second newt who had the same portion removed (the dorsal lip). The cells of the second newt, the recipient, were not pigmented.

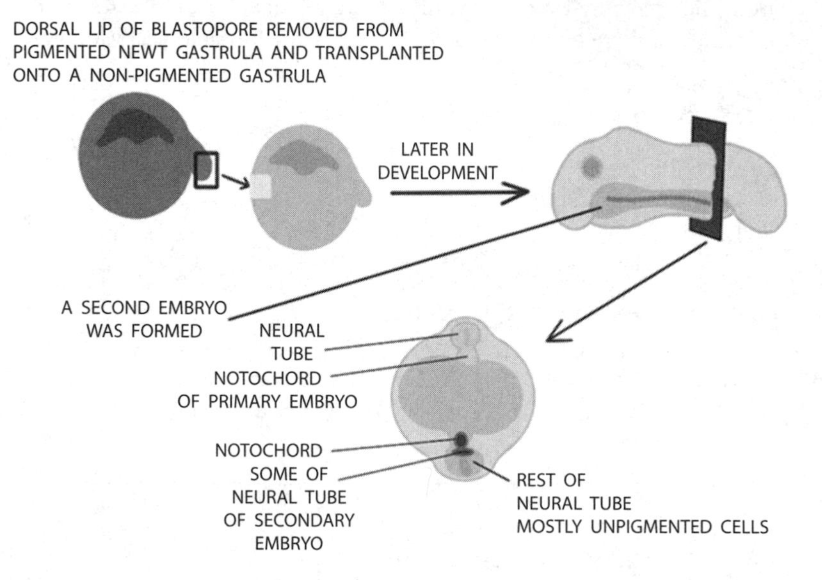

MULTIPLE-CHOICE QUESTIONS

249. Which of the following most accurately describes the results of the experiment?

(A) Fusion of two embryos produces a viable embryo with two nervous systems.

(B) Pigmented blastopore cells induce mutations in embryos, causing the ectopic (abnormal place/position) development of certain body structures.

(C) Pigmented cells differentiate during development, but their daughter cells do not contain pigment.

(D) The dorsal lip of the blastopore induced the development of a second notochord and neural tube in the newt it was transplanted into.

Information

Big Idea 3: Living systems store, retrieve, transmit, and respond to information.

MULTIPLE-CHOICE QUESTIONS

250. Which of the following statements most accurately describes the roles of DNA and RNA in the cell?

(A) DNA is in eukaryotic cells, and RNA is in prokaryotic cells.
(B) DNA is used to control cellular processes, and RNA is used to make proteins.
(C) DNA is the genetic information in cells, and RNA is the genetic information in viruses.
(D) DNA is the repository of information, and RNA serves in the expression of the information.

Questions 251 and 252 refer to the following diagram.

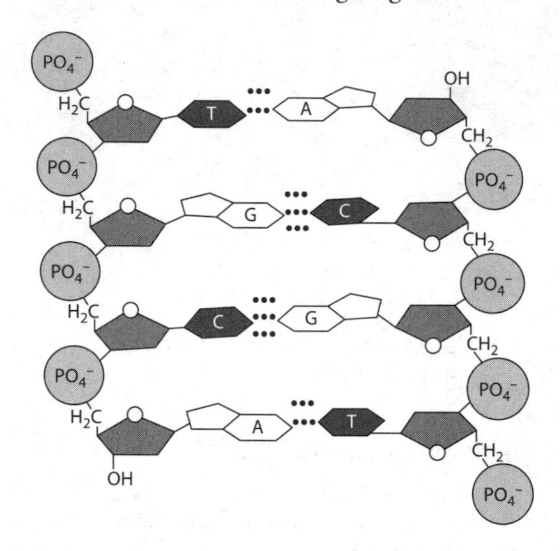

251. Which of the following most accurately summarizes the type of information contained in the molecule?

(A) Structural only—the base sequences encode the amino acid sequences of proteins.
(B) Regulatory only—the DNA provides the instructions for cell operation.
(C) Structural and regulatory—genes encode the amino acid sequence of proteins, and regulatory sequences regulate the functioning of the proteins.
(D) Structural and regulatory—genes encode proteins and functional RNA sequences, and regulatory regions control the production of these molecules.

252. Which of the following structural features of the molecule is most important for the high fidelity replication of this molecule?

(A) The four bases of DNA contain both purines and pyrimidines.
(B) Two hydrogen bonds connect adenine-thymine base pairs, and three hydrogen bonds connect guanine-cytosine base pairs.
(C) Each single strand of DNA contains the information required to synthesize the opposite strand.
(D) The sugar-phosphate backbones are positioned anti-parallel to each other.

253. Retroviruses, like HIV, use RNA as genetic material and therefore tend to mutate rapidly to produce many genetically different populations of retroviruses relatively quickly. Which of the following statements most accurately explains why RNA viruses have such high mutation rates?

(A) Reverse transcription is more prone to errors than DNA replication.
(B) Twice as much RNA is needed to store the same amount of information as DNA, so the RNA contained in the virus must be fragmented in order to fit into the capsid.
(C) The virus depends on the RNA polymerase of its cellular host to replicate its genome.
(D) Accidental base pairing between complementary bases produces sections of double-stranded RNA molecules, increasing the probability that RNA polymerase will introduce mutations.

254. Which of the following statements is *incorrect* concerning gene expression in multicellular organisms?

(A) The DNA sequences are the same in each cell, but the expression pattern is different for each kind of cell.
(B) Different cell types (e.g., liver and muscle) contain different proteins.
(C) All the cells of a multicellular organism contain the same genes, but each cell type differs in the regulatory sequences that control the expression of the genes.
(D) Promoters are regions of DNA that help regulate gene expression.

255. Which of the following statements most accurately explains why a yeast cell can be used to express an animal gene?

(A) Yeast and animals are both heterotrophs.
(B) The genetic code is universal.
(C) DNA replication is similar in yeast and animals.
(D) Yeast and animals share a recent common ancestor.

Questions 256 and 257 refer to an experiment in which bacteriophages were grown in culture to contain radioactive phosphorus *or* radioactive sulfur. Bacteria were infected with one of the two labeled phages and then interrupted by agitation in a blender, which separated the bacteriophage capsid from the bacteria cells. The mixture was then centrifuged and analyzed for radioactivity. The experimental procedure and results are shown as follows. (Based on Hershey and Chase, 1952.)

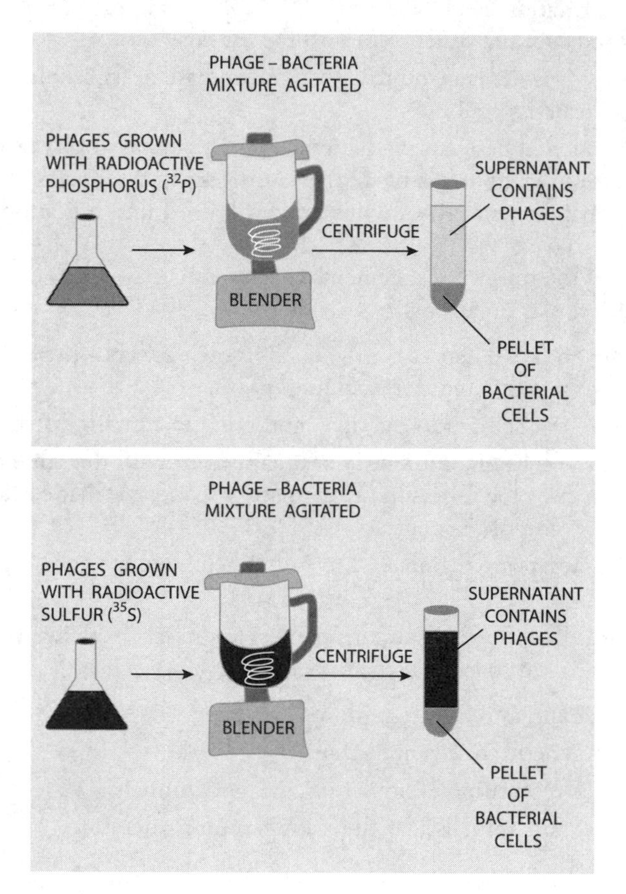

256. Which of the following statements accurately summarizes the results of the experiment?

(A) Viruses use protein while bacteria use DNA as infectious agents.
(B) Viruses are obligate cellular parasites that require living cells in order to reproduce.
(C) Viral proteins are needed to synthesize virus capsids in bacteria.
(D) DNA transfer into the bacteria cell is required for phage infection.

257. Which of the following statements accurately summarizes the purpose of using radioactive sulfur and phosphorus?

(A) Radioactive phages and medium can determine the mechanism of DNA translocation from viruses into bacteria cells.

(B) Elements with different modes of radioactive decay emit different wavelengths of light, allowing the contrasting visualization of proteins and nucleic acids.

(C) Differences in the elemental composition of molecules can be exploited experimentally to trace their location during the course of the experiment.

(D) Bacteria and viruses have different modes of reproduction and inheritance that can be exploited by scientists to engineer bacteria capable of producing viruses.

Question 258 refers to the following diagram.

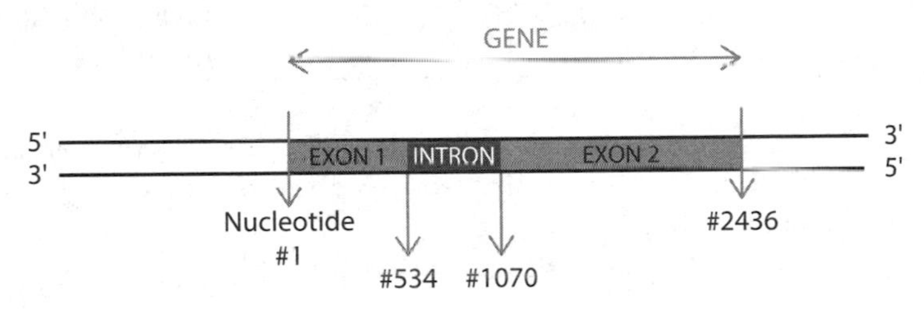

258. Which of the following diagrams of a polypeptide chain show the correct spatial relationship between the sequence of DNA and the polypeptide that results from its expression? Assume the top strand (5' end on the left) is transcribed.

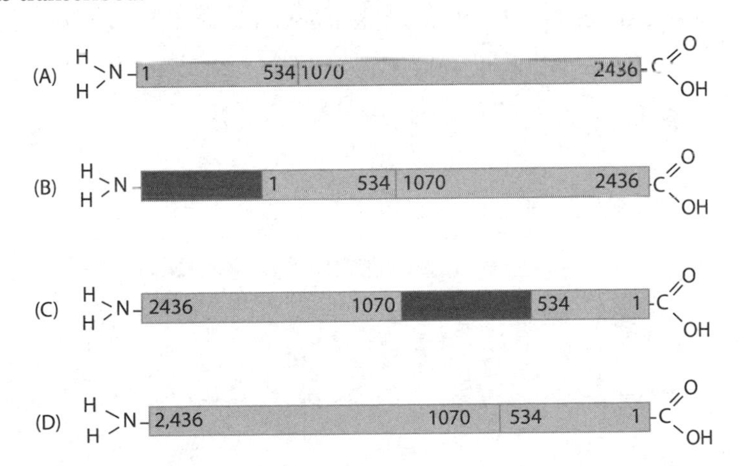

Questions 259–263 refer to a series of experiments involving the bacteria *Streptococcus pneumoniae* (pneumococcus). There are two strains of *S. pneumoniae*, the pathogenic S-strain and the harmless R-strain. The S-cells have a smooth appearance due to the presence of a polysaccharide capsule surrounding the cells. The R-cells appear rough because they lack the polysaccharide capsule.

In the first experiment, mice were injected with one of four different solutions containing living R-cells (group 1), living S-cells (group 2), heat-killed S-cells (group 3), or heat-killed S-cells mixed with living R-cells (group 4). The results are given in the following table.

Group of mice	Mouse gets injected with	Fate of mouse
1	Living R-cells only	All live
2	Living S-cells only	All die
3	Heat-killed S-cells	All live
4	Heat-killed S-cells plus living R-cells	All die Live S-cells isolated from dead mice

Figure 1 illustrates the results of many generations of cultivation of *S. pneumoniae* in vitro. Some S-cells lose the ability to make the polysaccharide capsule, transforming them into R-cells (Figure 1).

In a separate observation, R-cells grown with S-cells were transformed into S-cells (Figure 2).

The procedure and results of an experiment to identify the class of biomolecule responsible for the transformation of R-cells into S-cells is shown as follows. S-cell homogenates were fractionated into RNA, protein, DNA, lipid, and carbohydrate components. R-cells were incubated in *one* of the five fractions.

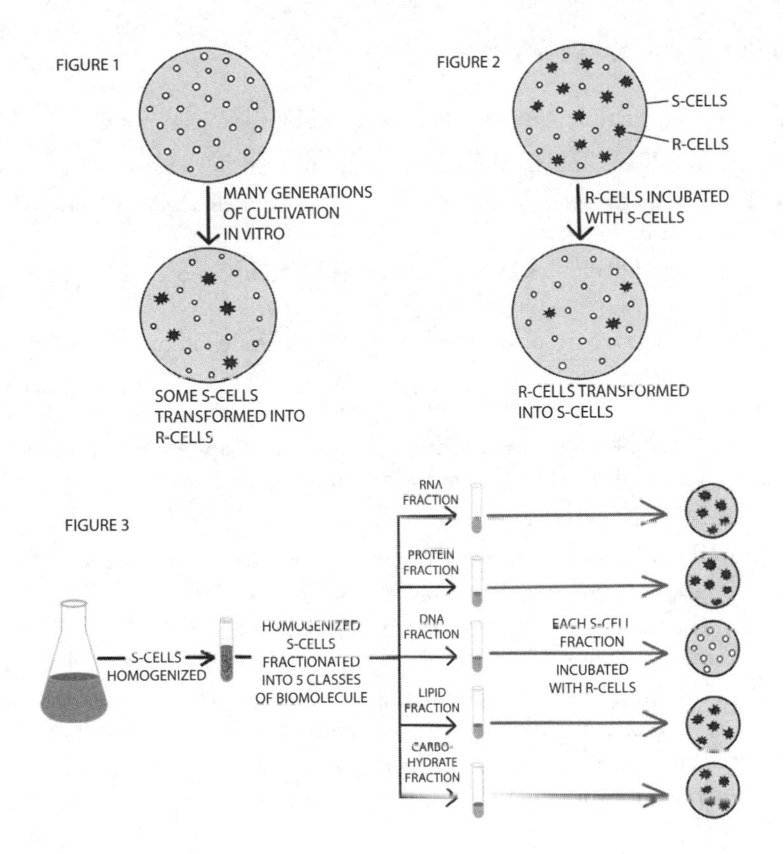

259. Which of the following is the most accurate conclusion based on the *initial transformation* of the nonpathogenic R-cells into S-cells (Figure 2)?

(A) DNA is the genetic material.
(B) Mutations can be beneficial to bacteria.
(C) S-cells contain molecules that carry heritable information.
(D) R-cells incubated with S-cell extracts become resistant to antibiotics.

SHORT FREE-RESPONSE QUESTIONS

260. **Describe** one possible mechanism by which an R-cell could be transformed into an S-cell after it takes up exogenous (foreign) DNA

261. **Explain** why R-cells incubated in the RNA, protein, lipid, and carbohydrate fractions did not transform into S-cells.

MULTIPLE-CHOICE QUESTIONS

262. Which of the following was demonstrated by the experiment?

(A) S-cells are only harmful when administered in the absence of R-cells.
(B) R-cells compete with the S-cells, limiting their ability to harm the mouse.
(C) Heat-killed S-cells were transformed into living S-cells by some factor that was released by live R-cells.
(D) R-cells were transformed into S-cells by some factor released by dead S-cells.

263. If the S-cell homogenate was divided into two samples, one treated with a protease and the other treated with a nuclease, which of the following results is expected?

(A) The mice receiving either the mixture containing protease or the mixture containing nuclease would live.
(B) The mice receiving the mixture containing protease would die, but the mice containing the mixture containing nuclease would live.
(C) The mice receiving the mixture containing nuclease would die, but the mice containing the mixture containing protease would live.
(D) The mice receiving either the mixture containing protease or the mixture containing nuclease would die.

Question 264 refers to the following data.

Source of DNA	Adenine: Guanine ratio	Thymine: Cytosine ratio	Adenine: Thymine ratio	Guanine: Cytosine ratio
Hen	1.45	1.29	1.06	0.91
Yeast	1.67	1.92	1.03	1.20
Haemophilus influenzae	1.74	1.54	1.07	0.91
E. coli K-12	1.05	0.95	1.09	0.99
Avian tubercle bacillus	0.40	0.40	1.09	1.08

264. Which of the following statements is true according to the data?

(A) The genetic code is the same for animals, fungus, bacteria, and viruses.

(B) % cytosine bases = $\left[\dfrac{1-(2\times\%\text{ adenine bases})}{2}\right]\times 100.$

(C) Base-pairing can occur between any two bases in *Escherichia coli.*

(D) Adenine:guanine base pairs make up 40% of the bases in Avian tubercle bacillus.

265. DNA analysis of adult and developing frogs reveals that the same genes are present in all the cells of the frog throughout its development. However, certain proteins found in developing frogs are absent from adult frogs. Which of the following statements correctly explains this observation?

(A) Developmental genes are not expressed in adults.

(B) No genes expressed in adults are expressed in developing organisms.

(C) Proteins are greatly modified as an organism develops.

(D) Gene expression is induced during development and repressed in adults.

Questions 266–268 refer to the following experiment.

The nucleus was removed from a fibroblast cell of an adult mouse (mouse 1). The nucleus from mouse 1 was then inserted into an egg cell from which the nucleus was removed (from mouse 2). The resulting diploid cell began to divide and the young embryo was implanted into the uterus of mouse 3 and allowed to develop. Mouse 3 eventually gave birth to a healthy mouse (4, not shown).

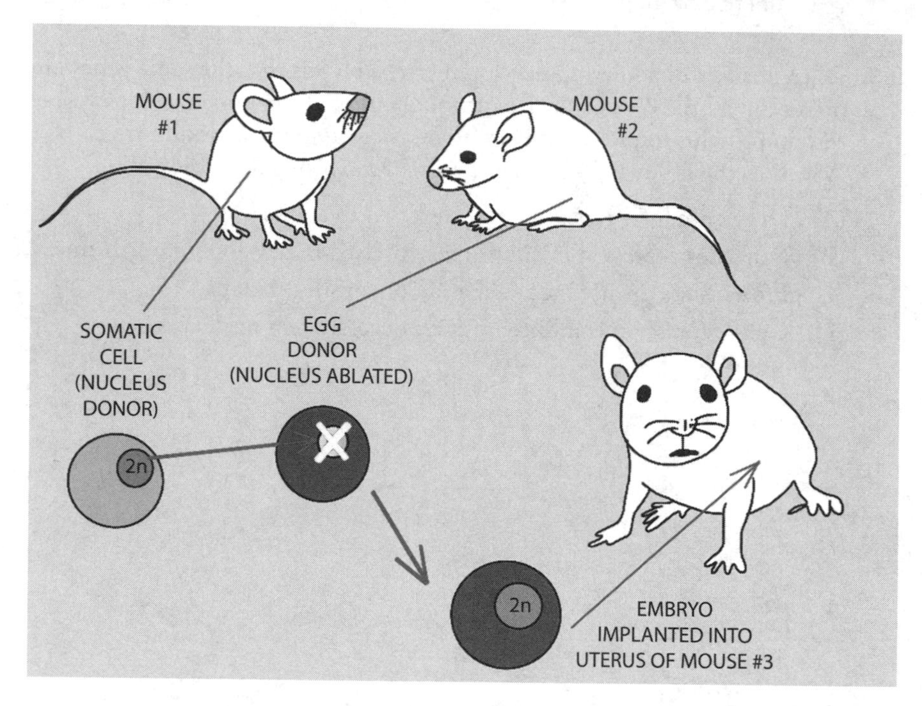

266. Which of the following correctly identifies the procedures described previously?

(A) DNA sequencing
(B) nuclear transplantation (cloning)
(C) embryology
(D) nuclear fission

267. The phenotype of the young mouse (#4) is expected to be

(A) practically identical to the adult mouse from which the nucleus was taken (#1).
(B) practically identical to the mouse who donated the egg (#2).
(C) similar to the mouse who donated the egg (#2).
(D) a combination of the nuclear and egg donor mice (#1 and #2).

268. A cloned animal may exhibit traits that are not present in the animal from which it was cloned (the nucleus donor). Which of the following most accurately explains this observation?

(A) The egg donor, not the nucleus donor, determines the phenotype of the clone.
(B) The nucleus donor usually, but not always, determines the phenotype of the offspring.
(C) The egg (of mouse 2) into which the nucleus of mouse 1 was transferred contained extra-nuclear genes.
(D) The genes of the nucleus donor were not properly expressed in the offspring.

LONG FREE-RESPONSE QUESTION

269. Describe two mechanisms of horizontal gene transfer in prokaryotes.

State one possible benefit of horizontal gene transfer and **explain** how it is beneficial. **Propose** a hypothesis to explain why horizontal gene transfer does not occur in most multicellular organisms.

Questions 270–274 are all related to β-gal production in *Escherichia coli.*

Questions 270–273 refer to an experiment in which four identical groups of *E. coli* were grown in a lactose-free, glucose medium. After 5 minutes, a solution of lactose was added to each of the culture mediums.

β-galactosidase is an enzyme that breaks down lactose. The levels of β-galactosidase were measured every minute throughout the experiment. The data are shown as follows. Group 1 mutants never made β-galactosidase, so they are not represented in the graph.

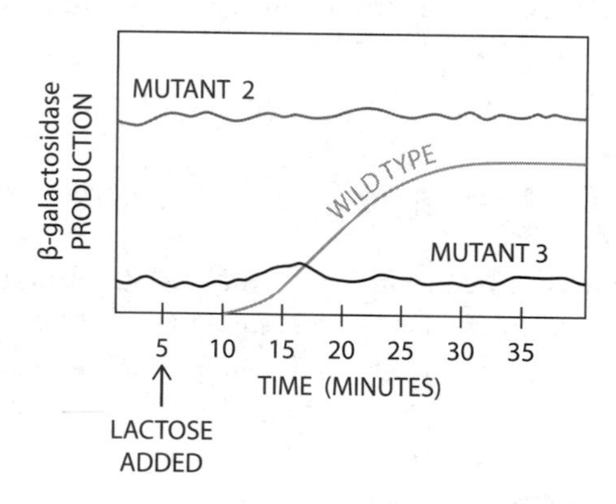

MULTIPLE-CHOICE QUESTIONS

270. Which of the following is a reasonable hypothesis to explain why mutant 1 did *not* produce β-galactosidase in the presence or absence of lactose?

(A) The β-galactosidase gene had a mutation that produced an early stop codon in the mRNA.
(B) Mutant 1 is unable to metabolize glucose and galactose.
(C) Mutant 1 is unable to break down lactose.
(D) Mutant 1 prefers glucose, but will use lactose, when no other substrate is available.

271. Which of the following procedures could be used to determine if transcription of the β-galactosidase (β-gal) gene is activated by the addition of lactose?

(A) Measure β-galactosidase mRNA before and after the addition of lactose.
(B) Isolate the β-gal gene before and after the addition of lactose.
(C) Add an inhibitor of RNA synthesis just before adding lactose; then measure β-gal protein.
(D) Sequence the β-gal gene.

272. *E. coli* is one of many bacterial species present in the human digestive tract. Why does *E. coli* have a system of activating β-galactosidase production?

(A) Most people don't drink milk all the time.
(B) Lactose secretion by the human intestine is inducible.
(C) Lactose is toxic to intestinal cells.
(D) The products of lactose hydrolysis can be toxic to *E. coli.*

273. Which of the following is the best explanation for the constant high levels of β-galactosidase seen in mutant 2?

(A) Mutant 2 is highly sensitive to the presence of even minute amounts of lactose.
(B) The amino acid sequence is incorrect.
(C) There is a mutation in the regulatory portion of the β-galactosidase gene.
(D) Mutant 2 cannot metabolize glucose.

Question 274 refers to the following data that plots the rate of β-gal production in *E. coli* cells *relative to the wild-type E. coli* as a function of time. Wild-type *E. coli* do not express the β-gal gene in the absence of lactose. Lactose is added after 5 minutes.

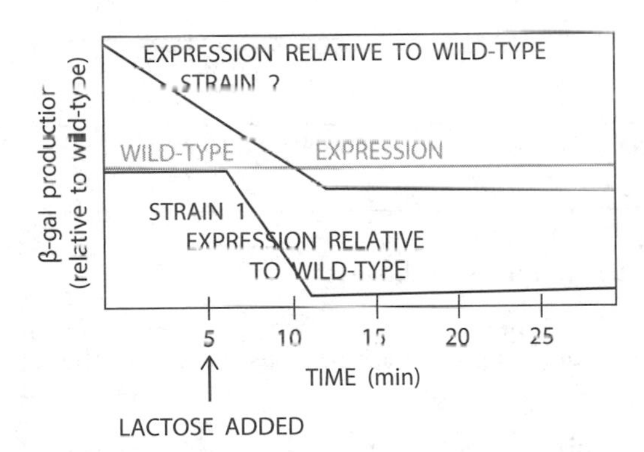

274. Which of the following statements correctly matched the bacterial strain(s) to the β-gal expression pattern?

(A) The wild-type is insensitive to lactose.
(B) Strains 1 and 2 decrease the rate of β-gal production in response to lactose.
(C) Strain 2 constantly overproduces β-gal and is not sensitive to lactose.
(D) Strain 1 does not produce β-gal at a significant rate at any time during the experiment.

Organism	Genome size (thousands of nucleotide pairs per haploid genome)	Number of genes
Mycoplasma genitalium a eubacterium that lives in the human genital tract	580	468
E. coli eubacteria that live in the human gut and many other places	4,639	4,289
Aeropyrum pernix an aerobic archae that lives at the hot-steam vents	669	2,620
Saccharomyces cerevisiae budding yeast, a eukaryote	12,069	~6,300
Arabidopsis thaliana a flowering plant	~142,000	~26,000
Caenorhabditis elegans nematode (roundworm)	~97,000	~19,000
Drosophila melanogaster fruit fly	~137,000	~14,000
Homo sapiens human	~3,200,000	~20,000–25,000
Protopterus aethiopicus lungfish	130,000,000	unknown

Question 275 refers to the table above.

275. Which of the following most accurately summarizes the relationships in the table above?

(A) Genome size increases with the size of the organism.
(B) The number of genes is directly proportional to the complexity of the organism.
(C) The number of genes is always greater than the number of kilobase-pairs (1,000 base pairs).
(D) There is no simple relationship among the number of genes, the size of the genome, and the size and complexity of the organism.

Question 276 refers to the following graph of the number of amino acid changes (per 100 amino acids) in four different proteins over time.

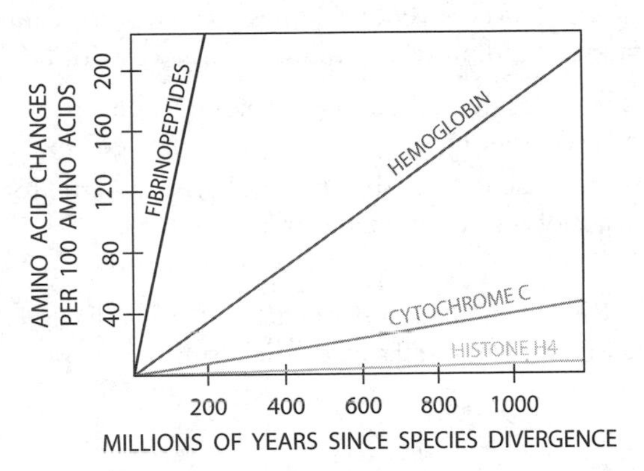

276. Which of the following statements is the most logical conclusion from the data above?

(A) Fibrinopeptides are more important than hemoglobin, cytochrome c, or histone H4.

(B) There was strong selection against organisms whose histone H4 protein had variations in their amino acid sequence.

(C) Hemoglobin is the most ideal amino acid sequence to use to date the time of divergence from a common ancestor.

(D) Fibrinopeptide sequences are more affected by genetic drift than hemoglobin, cytochrome c, or histone H4.

277. Eukaryotic DNA replication results in one mistake per 10^9 base pairs, yet the actual number of incorrect nucleotides incorporated by DNA polymerase is much greater. In addition, nucleotides can be chemically modified in a variety of ways, including the absorption of UV radiation. How can the process of DNA replication occur with such high fidelity despite mutations introduced in the initial polymerization or by mutations that occur between replications (cell divisions)?

(A) Most mistakes are made in noncoding regions, so they make little to no difference to the cells.

(B) Most mistakes are made after the parent cell or zygote has already divided hundreds of times, so the mistakes that are introduced affect few cells and do not affect zygotes.

(C) The translated product of mutant genes acts as a signaling molecule that binds to the promoter of the mutated gene and marks it for excision by DNA polymerase.

(D) DNA polymerase detects and repairs all mutations during replication regardless of the source of the mutation.

SHORT FREE-RESPONSE QUESTION

278. Introducing errors occurs 100,000 times greater in the combined processes of RNA synthesis and protein synthesis compared with DNA synthesis.

- **Explain** why this does not have a greater effect on phenotype (compared to a similar error in DNA).
- **Explain** why selection did not act as strongly for high fidelity mechanisms of transcription and translation.

Questions 279–282 refer to the transformation of *E. coli* bacteria with a plasmid carrying the gene for ampicillin resistance. The results of the procedure are shown as follows.

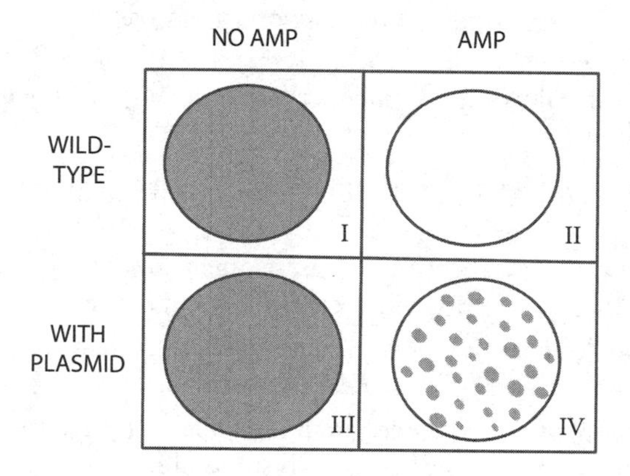

MULTIPLE-CHOICE QUESTION

279. Which plates have *only* ampicillin-resistant bacteria growing on them?

(A) I and III only
(B) II only
(C) III only
(D) IV only

SHORT FREE-RESPONSE QUESTION

280. For each plate, state the reason for the pattern of bacterial growth.

1) Plate I ______________________________
2) Plate II ______________________________
3) Plate III ______________________________
4) Plate IV ______________________________

MULTIPLE-CHOICE QUESTIONS

281. If the plasmid containing the ampicillin-resistance gene also contained a gene for the green fluorescent protein (GFP), which of the following plates would be expected to have the highest percentage of bacteria that could produce GFP?

(A) I and III only
(B) II only
(C) III only
(D) IV only

282. Which of the following statements correctly describes the purpose of using ampicillin resistance in the plasmid?

(A) so that ampicillin could be used to sterilize the nutrient agar plates
(B) to maintain a sterile environment to grow the bacteria
(C) to prevent viral infection of the bacteria
(D) to kill the bacteria that were not transformed

LONG FREE-RESPONSE QUESTION

283. **Explain** the differences in the regulation of gene expression between prokaryotes and eukaryotes.

Questions 284–289 refer to the following diagram of the *trp* operon, a well-studied example of the regulation gene expression in *E. coli.* The synthesis of tryptophan requires 5 enzymes. The genes encoding these enzymes are part of the *trp* operon. Their expression is coordinated by one promoter that includes an operator. When tryptophan is present, it binds to a repressor protein that binds to the operator and blocks transcription by RNA polymerase. The repressor protein can bind to the operator only when tryptophan is bound to it.

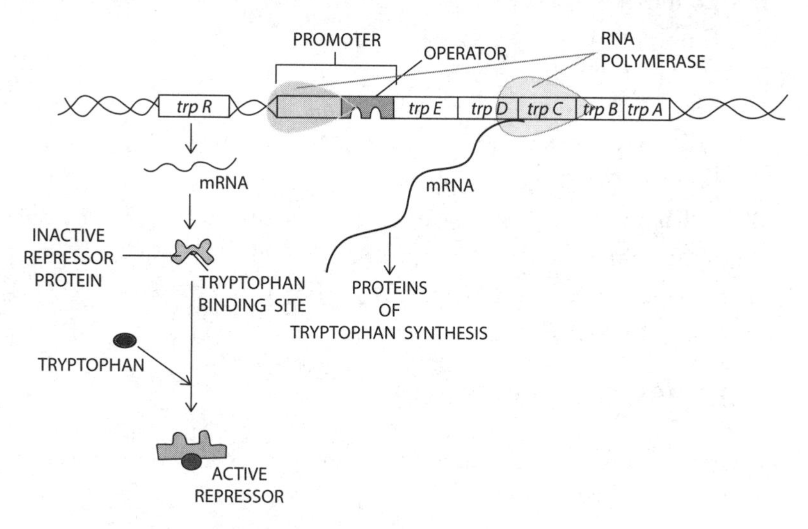

MULTIPLE-CHOICE QUESTIONS

284. What is the purpose of *trp* operon repression in *E. coli*?

(A) It allows bacteria to secrete tryptophan in the absence of tryptophan.
(B) It regulates the uptake of tryptophan from the environment.
(C) It conserves energy when tryptophan is available.
(D) It prevents protein hydrolysis when tryptophan is not available.

285. Which of the following statements most accurately describes the mechanism by which the repressor protein functions?

(A) The repressor protein is produced only when tryptophan is present and then binds to the operator to prevent transcription of the genes that code for the enzymes of tryptophan synthesis.
(B) The repressor protein is produced constitutively (all the time) and is turned off by the binding of tryptophan.
(C) The repressor undergoes a shape change when bound to tryptophan.
(D) The repressor is an enzyme that degrades mRNA unless tryptophan binds to inhibit the enzyme.

286. Operons can be inducible or repressible. The *trp* operon is repressible. Which of the following statements correctly connects the type of operon with its regulatory advantage?

(A) *Inducible:* genes are expressed until something inhibits their expression.
(B) *Repressible:* it produces a particular product unless it becomes available.
(C) *Inducible:* it inhibits the production of the required molecules unless they are available.
(D) *Repressible:* it inhibits the production of the required molecules unless they are available.

287. Several strains of *E. coli* bacteria can synthesize tryptophan while some others cannot. Which of the following is *least* likely to be a source of this genetic variation?

(A) virus transmission of genes between bacteria (transduction)
(B) horizontal gene transfer (gene exchange between cells that are not parent-progeny)
(C) mutation
(D) random (independent) assortment of chromosomes

Questions 288 and 289 also refer to the following information and data table.

Suppose a plasmid was engineered so that the regulatory sequences (the promoter/operator) of the *trp* operon controlled the expression of the green fluorescent protein gene (GFP). The GFP protein glows green under UV light. The plasmid also contains an *amp resistance* gene that is constitutively (always) expressed. The *amp resistance* gene allows the bacteria that express it to grow in the presence of ampicillin, an antibiotic. Bacteria were transformed with the plasmid and immediately plated on one of four plates. The plates were incubated at 37°C for 24 hours and then analyzed. The results are shown as follows.

	Type of culture media		Color of bacteria	Color of colonies
Plate	**Ampicillin**	**Tryptophan**	**under white light**	**under UV light**
1	–	–	White lawn	White lawn with tiny, faint spots of green when viewed under a microscope
2	–	+	White lawn	All white
3	+	+	All white	All white
4	+	–	All white	All green

+ indicates the presence of the substance in the culture media
– indicates the absence of the substance in the culture media

288. Which of the following statements correctly explains why GFP is expressed in *E. coli* cells after transformation with the plasmid?

(A) The absence of tryptophan in the media results in expression of the GFP gene.

(B) Ampicillin activated the expression of the GFP gene.

(C) The presence of tryptophan in the media activates GFP expression, but ampicillin makes the colonies appear white.

(D) The combination of ampicillin and tryptophan activate the expression of the GFP gene.

289. Which of the following is most likely the reason that tiny, faint spots appear on plate 1 under UV light?

(A) The absence of ampicillin allows the colonies without the plasmid to grow.

(B) The absence of ampicillin activates the expression of GFP on the plasmid.

(C) Some of the plated cells had been transformed.

(D) There is a mutation in the DNA.

Question 290 refers to the process of bacteriophage transduction illustrated in the following diagram.

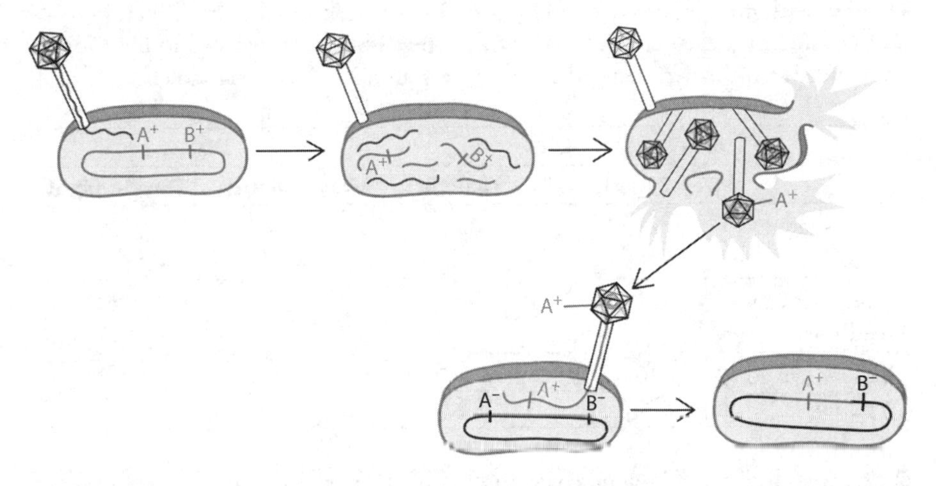

290. What is the effect of the process on the genetic variation in a bacterial population?

(A) Genetic diversity decreases because the bacteriophages kill the bacteria cells they infect.
(B) Genetic diversity decreases because the recipient bacteria can incorporate the phage DNA into their chromosome without ever expressing them.
(C) Genetic diversity increases because the donor bacterium's genes undergo recombination with the recipient bacterium's genome.
(D) Genetic diversity increases because the DNA introduced by the phage causes mutations in the recipient genome, creating new alleles with high frequency.

291. Which of the following statements best explains why a person who is heterozygous for Tay-Sachs disease *does not* show signs of the disease?

(A) The disease has a dominant inheritance pattern.
(B) A heterozygote makes enough of the normal protein.
(C) Heterozygosity results in hybrid vigor.
(D) Heterozygotes for Tay-Sachs have an advantage (heterozygote advantage).

Question 292 refers to an experiment with pea seeds. Pea seeds of the same species were germinated and allowed to grow for 14 days in one of two dishes. Dish A was covered with an opaque cover and placed in the dark for the first 7 days. Dish 2 was covered by a clear cover. On day 8, all the dishes were exposed to light for the remainder of the experiment. The data are given in the following table.

	Dish A		Dish B	
	Day 7 (dark)	**Day 14 (light)**	**Day 7 (light)**	**Day 14 (light)**
Germinated seeds	21	28	28	28
Green-leaved seedlings	1	21	20	20
Yellow-leaved seedings	20	7	8	8

292. Which of the following comparisons, in the absence of other data, best supports the hypothesis that leaf color is genetically controlled?

(A) dish A day 7 and dish B day 7
(B) dish A day 7 and dish A day 14
(C) dish A day 14 and dish B day 14
(D) dish A day 7 and dish B day 14

Question 293 refers to *Thalassoma bifasciatum*, a Caribbean reef fish that is one of several fish species whose sex is determined by the sex of the other fish it encounters at the reef.

- If an immature *T. bifasciatum* arrives to the reef and finds a single male defending a territory with multiple females, the immature *T. bifasciatum* will become a female.
- If an immature *T. bifasciatum* arrives to an undefended reef with multiple females, the immature *T. bifasciatum* will become a male.
- If the territory male dies, one of the mature females will become a male (the ovaries shrink and die and testes develop).

293. These observations suggest that:

(A) the environment can act as an agent of natural selection on a genotype regardless of the phenotype.
(B) environmental cues can result in different phenotypes in the same individual.
(C) the effect of sexual selection on males is greater than for females.
(D) males and females will always be present in equal numbers in populations of *T. bifasciatum*.

294. Which of the following is a true statement regarding the diversity of genetic information on Earth through time?

(A) The diversity of genetic information has increased because the number of genes and alleles has generally increased over time.
(B) The diversity of genetic information has remained fairly constant because the genes and alleles of lower fitness are removed and replaced by genes and alleles of higher fitness.
(C) The diversity of genetic information has remained fairly constant because organisms of greater complexity make up a large percentage of all the genes in the gene pool of the biosphere.
(D) The diversity of genetic information has decreased because natural selection has eliminated many genes and alleles over the billions of years.

295. Animals have two modes of long-distance communication between cells: endocrine and synaptic (nervous). Which of the following pairs most accurately describes the structural features of each?

	Endocrine	Nervous
(A)	Many diverse cell types that move throughout the body, secreting chemicals that target other cell types	Many diverse cell types, some with very long projections that extend through long distances in the body, chemically communicating with cells that are linked with them
(B)	Many diverse cell types, some with long projections that extend through long distances in the body secreting, chemicals to target cells	Many diverse cell types that move throughout the body, secreting chemicals that target other cell types
(C)	Many diverse cell types mostly contained in glandular structures that each secrete various chemicals into the extracellular fluids	Many diverse cell types, some with very long projections that extend through long distances in the body, communicating with cells that are linked with them
(D)	Many diverse cell types that move throughout the body, secreting chemicals that target other cell types	Many diverse cell types mostly contained in glandular structures that each secrete various chemicals into the extracellular fluids

296. Which of the following is an example of intracellular regulation?

(A) An enzyme is inhibited by an elevated concentration of a particular molecule.

(B) Phosphorylation of enzymes due to a cyclic AMP activation of protein kinase A.

(C) Lactic acid secretion by muscles lowers local blood pH, resulting in blood vessel dilation, increasing blood flow to the area.

(D) The vagus nerve decreases heart rate during times of rest.

SHORT FREE-RESPONSE QUESTIONS

297. One hormone can often bind to many receptor subtypes. (Receptor subtypes are receptors with the same ligand-binding site but different intracellular responses to ligand binding.) **Explain** one regulatory benefit of this type of system. Provide **one** example and use it to **support** your explanation.

298. According to the fossil record, prokaryotic cells were present on Earth for approximately 2.5 billion years before the first multicellular organisms appeared. **Explain** why it may have taken so long for multicellular life to appear on Earth.

299. Cellular communication involves transduction signals from other cells, organisms, or the environment. **List** the cellular structures required for cell-cell communication and provide a **brief description** of their function.

Question 300 refers to the following information and graphs.

The half-life of a substance is the time it takes for one-half of the number of molecules of the substance to be removed, degraded, or modified. All proteins and most molecules in the body have a specific half-life.

The following graphs show the relative concentrations of intracellular molecules after the synthesis of that molecule has increased or decreased by 10-fold according to its specific half-life. (The data were adapted from Alberts, Johnson, Lewis, Raff, Roberts, and Walter. *Molecular Biology of the Cell*, 4th edition, 2002.)

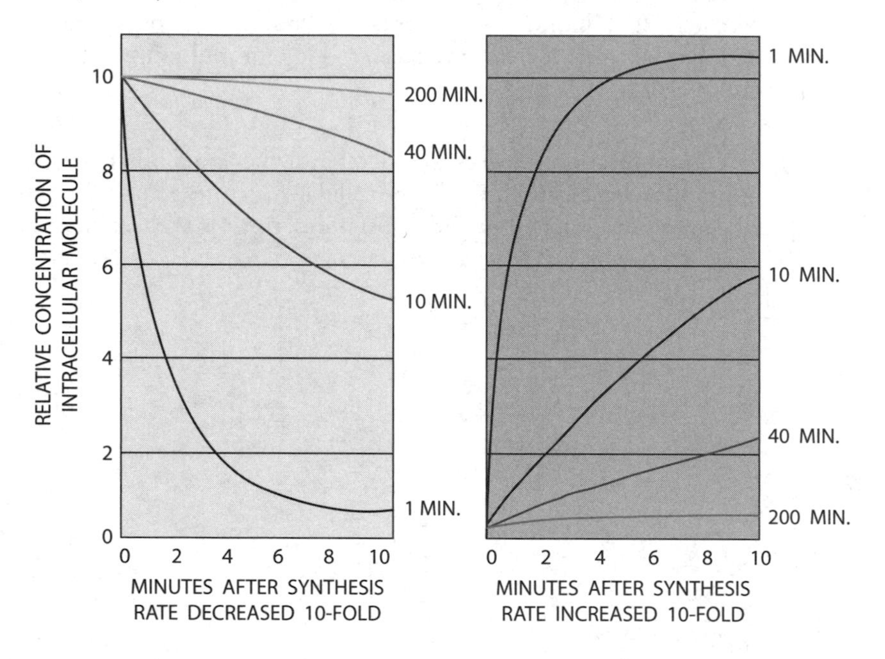

MULTIPLE-CHOICE QUESTIONS

300. Which of the following half-lives would be expected for a regulatory molecule and why?

(A) short half-life because its concentration rapidly changes with synthesis and rapidly disappears when synthesis is slowed or stopped

(B) long half-life because its concentrations remain relatively stable regardless of synthesis rates

(C) moderate half-life because its concentrations remain relatively stable but can change when needed

(D) moderate half-life because the maximum amount of the molecule will be present regardless of the synthesis rate

Questions 301–303 refer to information in the following table that compares the mechanisms of chemical communication in the endocrine and nervous system (synaptic signaling).

Type of chemical signaling	Ligand (signaling molecule) concentration	Receptor affinity	Functional characteristics
Endocrine signaling	Low concentrations of ligand. Concentration achieved slowly over minutes, hours, or days.	High affinity receptors. Most ligands bound to receptors.	Slow transport of signaling molecule through the blood. Changes in enzymes already present–fast. Changes in gene expression and protein synthesis–slow.
Synaptic signaling (Nervous system)	High concentration of ligand. Concentration increase very rapid (less than 1 second).	Low affinity receptors. Ligand can dissociate rapidly.	Precise signaling from one cell to another, specific cell. Long-distance transport achieved very quickly.

301. The endocrine and nervous systems both coordinate the functioning of the body. Which of the following answer choices *least correctly* relates the structure to its function in *either* the endocrine and nervous system?

(A) The long cells of the nervous system allow specific sets of cells to communicate very rapidly.

(B) The fat solubility of some hormones allows them to cross cell membranes and alter gene expression by directly binding to intracellular receptors that act as transcription factors.

(C) The limitations of the time for diffusion is minimized through tiny synaptic spaces between cells of the nervous system increasing the speed at which they can communicate.

(D) Glands are located at specific locations throughout the body according to the extracellular composition that best supports their growth and maintenance.

302. Which of the following is *not* an essential feature of chemical signaling?

(A) cells that secrete signals based on the composition of the extracellular fluid
(B) diverse chemical receptors to provide a variety of responses by target cells
(C) diverse chemical signals to provide a specificity of responses by target cells
(D) a variety of glands that regulate each homeostatic variable independently of the others

303. Which statement correctly describes transport across axonal membranes in the nervous system?

(A) Simple diffusion directly through the membrane bilayer balance sodium and potassium concentrations in neurons at rest, but during an action potential, active transport allows their relative concentrations to change quickly.
(B) Channels allow diffusion of specific ions to occur rapidly.
(C) Active transport via the sodium-potassium pump is linked to chemiosmosis in neurons.
(D) Neurotransmitters are secreted through the axon via active transport but pass into the synapse via simple diffusion.

Questions 304–308 refer to the following diagram, which shows different modes of communication between cells that are not physically linked through gap junctions or plasmodesmata. The structures are *not* drawn to scale.

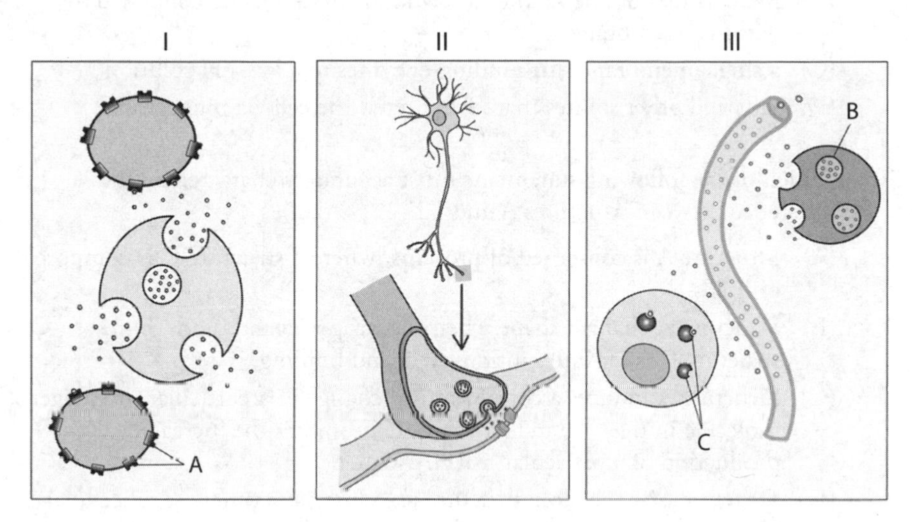

304. Which of the following statements most accurately explains why endocrine signaling is a critical part of coordination in a multicellular organism?

(A) to coordinate the movements of a relatively large body size and large distances

(B) to provide many different cell types in diverse locations in the body with information about the state of the organism

(C) to allow the organism to rapidly respond to changes in the environment

(D) to provide a mechanism of positive feedback for the maintenance of homeostasis through the exchange of hormones between cells

305. What is the specific function of structure A in Figure I?

(A) the secretion of chemical signals

(B) the coordination of the circadian rhythm

(C) the exocytosis of neurotransmitters

(D) the activation of a cascade of intracellular events in response to a specific signal

306. What is structure B in Figure III?

(A) a double-membrane surrounding DNA and proteins
(B) many copies of one kind or a few kinds of molecules contained within a membrane
(C) a single membrane surrounding enzymes in a low pH solution
(D) salts and other solutes that just entered the cell via pinocytosis

307. Which of the following statements most accurately characterizes the difference between structures A and C?

(A) Structure A is composed of proteins, whereas structure C is composed of lipids.
(B) Structure A changes shape when the proper ligand binds, but structure C is set into motion by ligand binding.
(C) Structure A induces a concentration change in a particular messenger molecule in the cell, and structure C promotes the increased production of a particular mRNA sequence.
(D) Structure C is the same structure as A but has a different subcellular location to perform a similar function closer to the nucleus.

308. Which of the following describes the events illustrated in Figure II?

(A) All nearby cells form close associations for the transmission of chemical information.
(B) Close associations between some cells allow them to communicate rapidly.
(C) Two-way communication between cells of the nervous system is necessary for the coordination of a multicellular organism.
(D) Cells of the nervous system communicate electrically.

Question 309 refers to the following experiment done in 1921 by Otto Loewi.

The heart of a frog was isolated *with* the vagus nerve intact. Loewi bathed the heart in solution and electrically stimulated the nerve (shown below, left). He observed that the stimulation caused the heart to beat more slowly. Loewi removed the heart and nerve from the solution. He then isolated a second frog heart *without* the vagus nerve and placed the heart in the same solution that bathed the heart *with* the vagus nerve (from the first part of the experiment). After immersion, the beating of the heart *without* the vagus nerve slowed (below, right). (The heart *with* the vagus nerve and the heart *without* the vagus nerve were not in the solution together at any time during the experiment.)

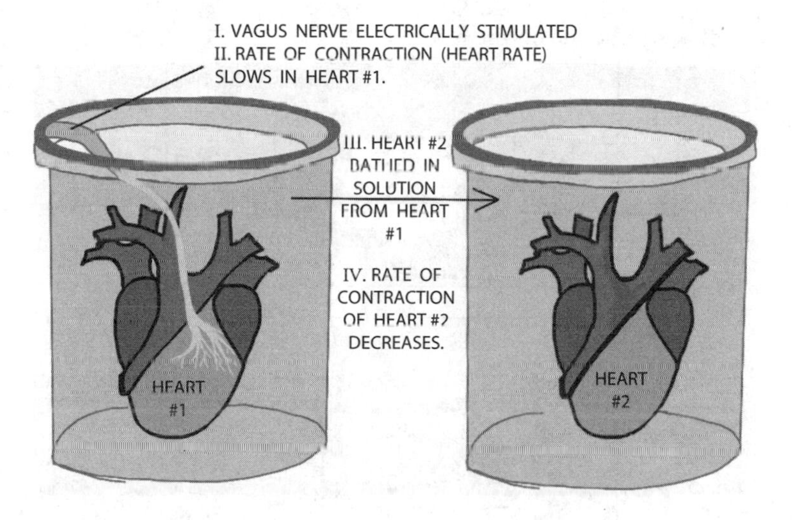

309. Which of the following hypotheses was this experiment designed to test?

(A) Nervous stimulation is required for the heart rate to slow.

(B) The vagus nerve slows the beat rate of the heart by electrical conduction.

(C) The heart rate was slowed by a chemical factor released by the vagus nerve.

(D) The heart can beat in the absence of nervous stimulation.

Questions 310–312 refer to the information contained in the illustrations of cell surfaces, the diagram of the action potential, and the table of intra- and extracellular ion concentrations that follow.

Ion	Intracellular Concentration (mM)	Extracellular Concentration (mM)
Na^+	5–15	145
K^+	140	5
Ca^{2+}	10^{-4}	1–2
Cl^-	5–15	110

The effect of acetylcholine on the heart

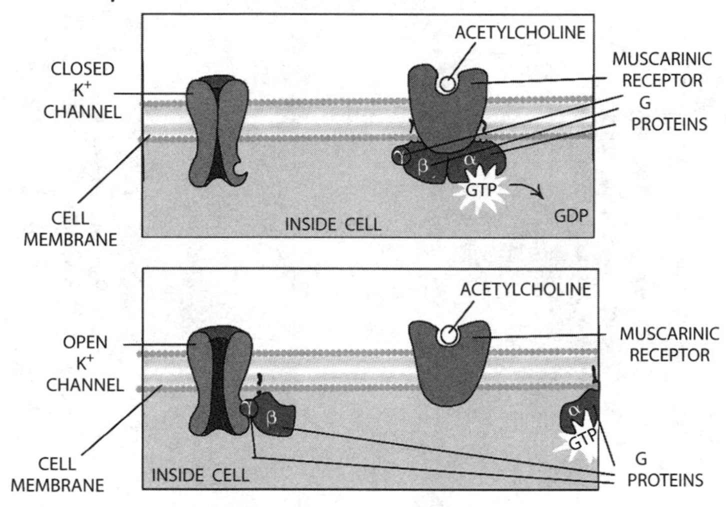

The effect of acetylcholine at the neuromuscular junction of skeletal muscle

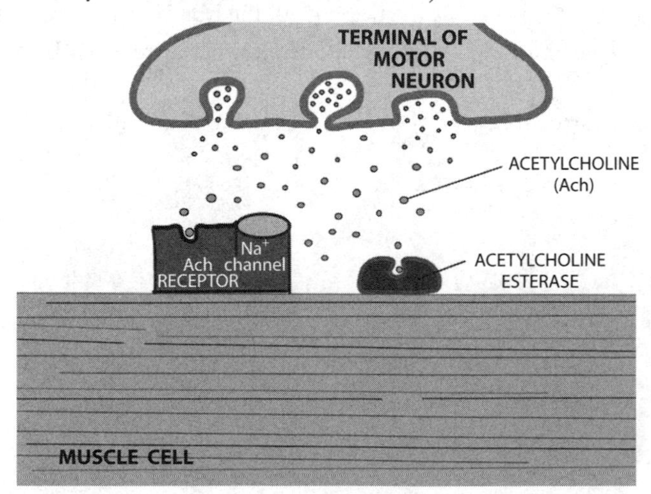

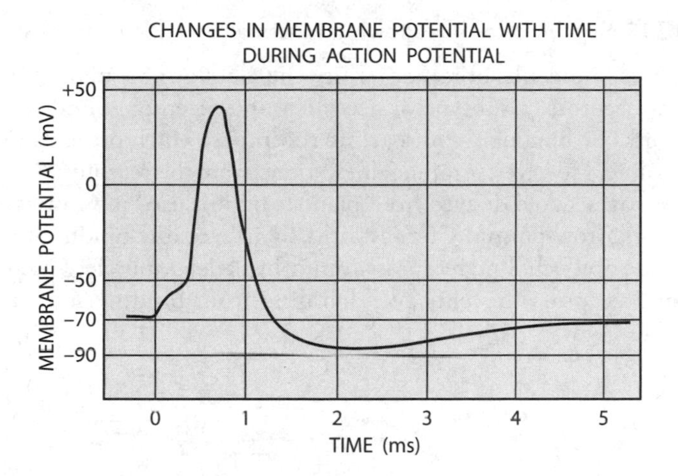

LONG FREE-RESPONSE QUESTION

310. **Explain** how the binding of acetylcholine causes a response in the heart ***or*** skeletal muscle. Include a **brief explanation** of the mechanism by which the same molecule (acetylcholine) can trigger opposite responses in two different but similar cell types (they're both muscle).

MULTIPLE-CHOICE QUESTIONS

311. Several neurotoxins exert their effect by increasing the rate of acetylcholine breakdown in the synapse. Which of the following would be the direct result of exposing skeletal muscle to this kind of toxin?

(A) the inability of the muscle to initiate another contraction
(B) the inhibition of acetylcholine release by the neuron
(C) the onset of tetanus, a prolonged muscle contraction
(D) down-regulation of the acetylcholine receptor

Question 312 also refers to the following information and chemical structures.

Atropine is a molecule obtained from the belladonna plant. It binds to muscarinic receptors, a subtype of acetylcholine receptor found in the heart. Atropine works by binding to **muscarinic receptors,** which prevents the binding by acetylcholine. However, atropine does not activate the receptor.

Curare is an alkaloid derived from plants, and it is used in Central and South America as an arrow-poison. Curare works by irreversibly binding to **nicotinic receptors**, the acetylcholine receptors found in skeletal muscle. Curare binding to nicotinic receptors prevents acetylcholine from binding, but it does not activate the receptor.

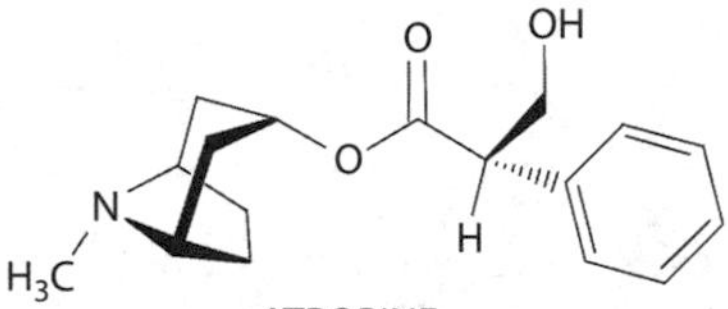

ATROPINE

Curare

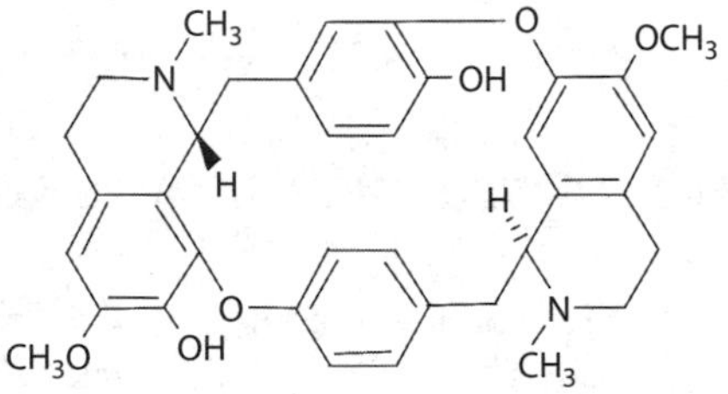

312. Which of the following is a logical expectation based on the information provided?

(A) Delivery of atropine into the bloodstream directly increases heart rate.
(B) Atropine induces muscular paralysis.
(C) Delivery of curare into the bloodstream directly decreases heart rate.
(D) Curare induces tetanus, a prolonged muscular contraction.

Question 313 refers to the following diagram.

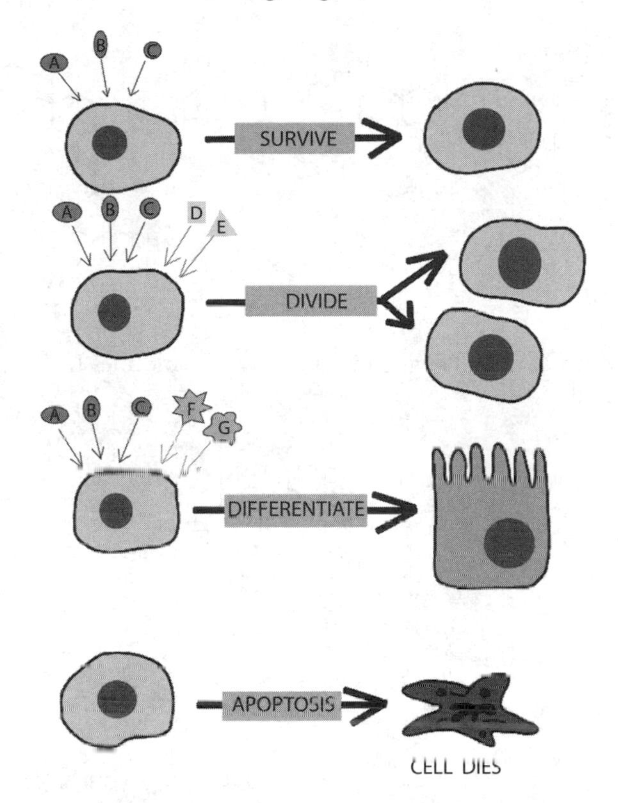

313. Which of the following statements is *not* a logical, biological interpretation of the diagram?

(A) Cells need signals from their environment to survive.

(B) Specific molecules are required to coordinate gene expression during development.

(C) Of the signals shown, no specific signals are required to induce cell death.

(D) Survival, division, and differentiation but not apoptosis are required to maintain homeostasis.

Question 314 refers to the following molecular structures.

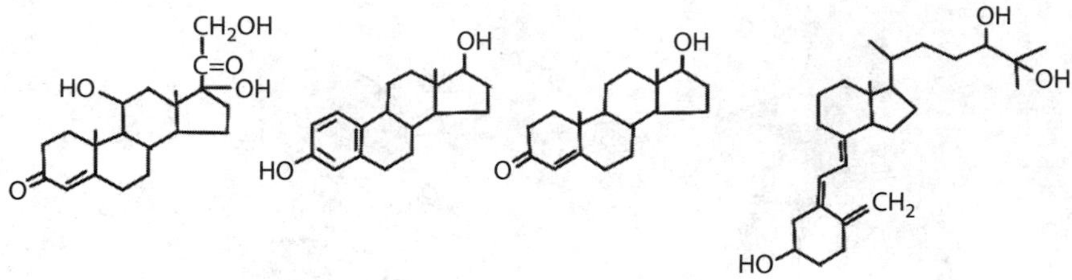

I II III IV

314. From which of the following molecules were molecules I, II, II, and IV synthesized?

(A)

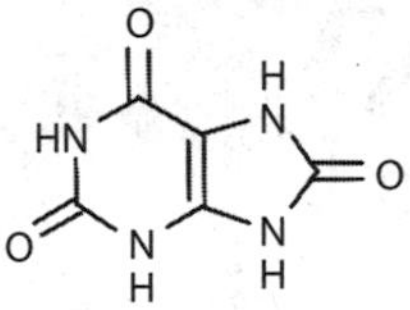

(B)

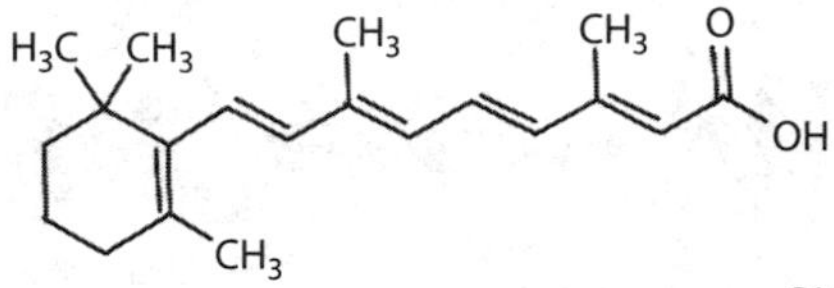

(C)

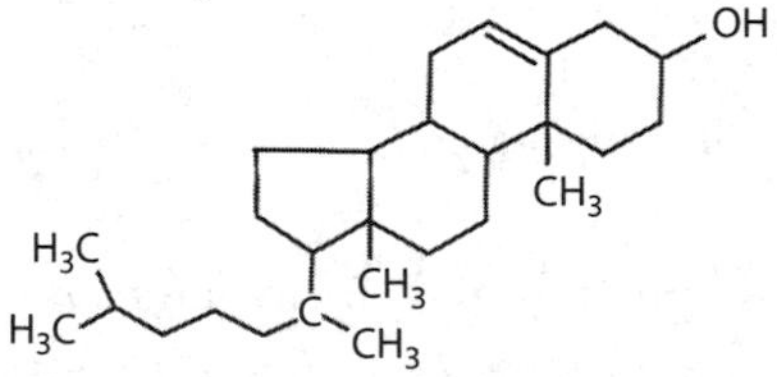

(D)

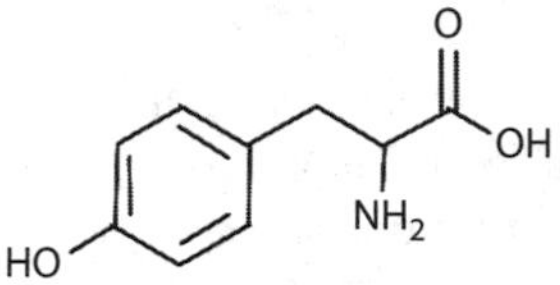

Question 315 refers to the following molecular structures.

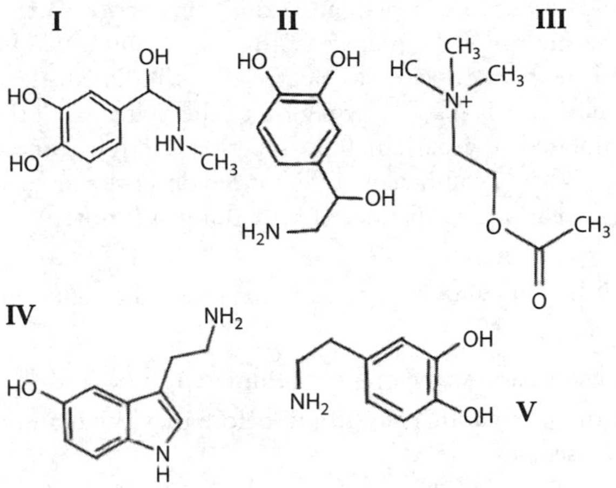

315. Which molecules likely bind to the same type of receptor?

(A) I and III

(B) I and IV

(C) IV and V

(D) I, II, and V

Questions 316 and 317 refer to the following situation.

During a medical procedure, a patient's skull was opened to reveal the surface of the brain (the surface of the brain has no pain receptors). When a section of the brain called the *motor cortex* was electrically stimulated, the patient's hand twitched in response. When a nearby region of the brain called the *somatosensory cortex* was stimulated, the patient reported that she felt something brushing against her face. When a completely different region of the brain was stimulated, the patient laughed as if she experienced something as funny.

316. Which of the following statements is *not* a logical conclusion based on these observations?

(A) Different parts of the brain have different functions.
(B) The brain can anticipate stimuli before they occur and initiate a response.
(C) Sensations felt in the body are generated by the brain.
(D) Processing of body sensations and movement generation occur in nearby parts of the brain.

317. The fact that the patient laughed when a particular part of her brain was stimulated:

(A) suggests that humans have pre-programmed responses to all stimuli.
(B) may help explain why people can't tickle themselves.
(C) proves that scientists have a sense of humor.
(D) provides convincing evidence that laughter is a response to a stimulus.

318. Which of the following is an accurate statement concerning genes?

(A) Genes exist in alternate forms called introns and exons.
(B) A gene can code for the amino acid sequence of a polypeptide.
(C) Recessive alleles are never expressed.
(D) Genes make up the majority of nucleotide sequences in the human genome.

SHORT FREE-RESPONSE QUESTION

319. **Describe two** structural features of the neuron that support its function in the body and **explain** how the structural feature supports the specific function of that structure in the cell.

MULTIPLE-CHOICE QUESTION

320. Many medications act to interfere with signaling. Which of the following is *least* likely to be the mechanism by which these medications exert their effect?

(A) The medication enters the cell and binds to a transcription factor, which upregulates the transcription of rRNA genes.

(B) The medication binds to a hormone receptor that prevents binding of the hormone but does not activate the receptor.

(C) The medication enters cell and inhibits the production of an intermediate (like a 2nd messenger) needed for signaling.

(D) The medication causes a gland to produce a factor that regulates another gland.

Questions 321–325 refer to glycogen metabolism in liver and skeletal muscle. A diagram of the intracellular processing of glycogen in liver and muscle in response to adrenaline (epinephrine) follows.

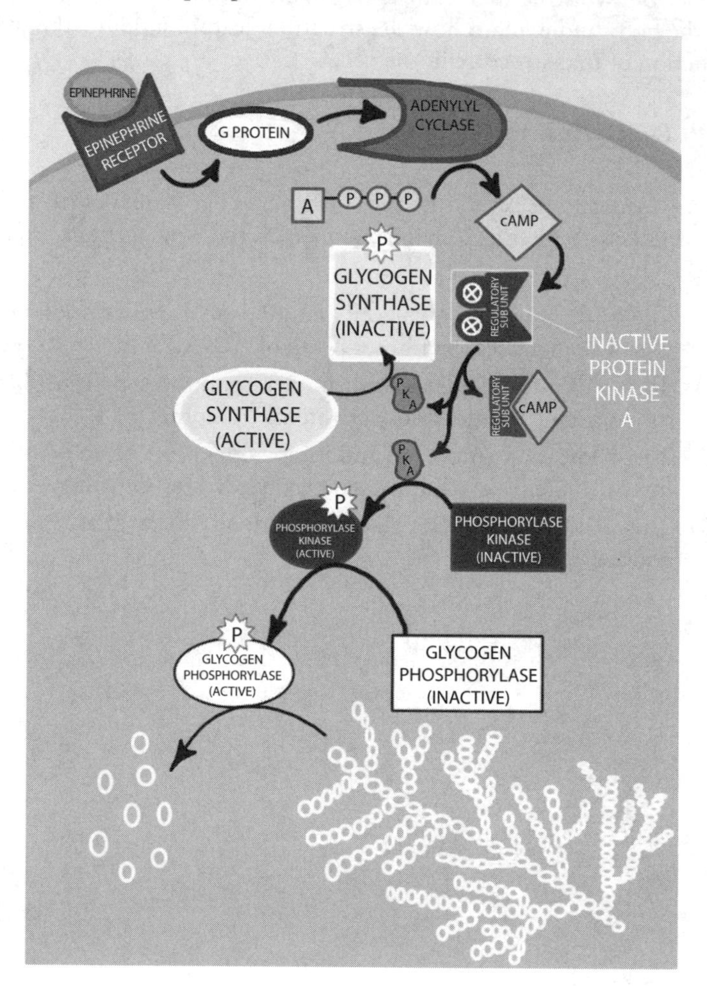

SHORT FREE-RESPONSE QUESTION

321. Liver and muscle both store glycogen in times of elevated blood glucose levels and can break it down later in order to increase blood glucose levels (the liver) or during exercise (muscle). Both tissue types have enzymes for glycogen storage and enzymes of glycogen synthesis present at the same time.

Explain how is possible for these enzymes of opposite function to coexist within the cell without performing a futile cycle of glycogen breakdown and synthesis.

MULTIPLE-CHOICE QUESTIONS

322. Epinephrine is a hormone secreted in times of stress and starvation in many animals. Which of the following correctly relates the action of epinephrine to its effects?

(A) Epinephrine activates all enzymes in the cell through phosphorylation.
(B) Epinephrine exerts its effects through mediating enzyme action.
(C) Epinephrine activates glycogen synthesis and breakdown.
(D) Epinephrine increases the rate of ATP synthesis in the cell by the activation of adenylyl cyclase.

323. Insulin binding to insulin receptors also causes a phosphorylation cascade. In liver and muscle, it results in the phosphorylation and inactivation of glycogen synthase kinase. When active, glycogen synthase kinase phosphorylates glycogen synthase, which is inactive when phosphorylated (see diagram on previous page). What effect will insulin signaling have on glycogen synthesis?

(A) It will increase glycogen synthesis by inactivating glycogen synthase kinase, the inhibitor of glycogen synthase.
(B) It will increase glycogen synthesis by activating glycogen synthase kinase, the activator of glycogen synthase.
(C) It will decrease glycogen synthesis by inactivating glycogen synthase kinase, the activator of glycogen synthase.
(D) It will decrease glycogen synthesis by inactivating glycogen synthase kinase, the inhibitor of glycogen synthase.

324. Which of the following is the most likely mechanism by which the presence of cAMP activates PKA (protein kinase A)?

(A) The binding of cAMP to a regulatory subunit of PKA causes a shape change.
(B) cAMP activates adenylyl cyclase to form ATP, which is needed as a phosphate donor.
(C) cAMP acts as a coenzyme and phosphate donor of PKA.
(D) The high-energy phosphate from cAMP phosphorylates PKA.

Question 325 refers to the following flowchart of the intracellular signaling amplification cascade of epinephrine.

For each (1) epinephrine molecule that binds to a G-coupled receptor...

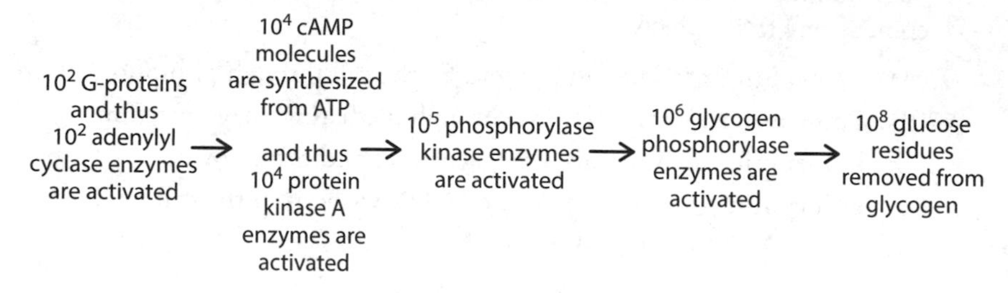

GRID IN

325. Calculate the number of glucose molecules released from glycogen by each activated glycogen phosphorylase enzyme. Report your answer to the hundreds place. Do not use scientific notation.

Start your answer in any column, just make sure you've got enough spaces.

You have symbols for fractions & a negative number.

Leave extra spaces blank.

Question 326 refers to the following experiment.

Cells in the S phase (synthesis, the 2nd stage of interphase when DNA replication occurs) of the cell cycle were fused to cells in the G_1 phase (gap 1, the 1st stage of interphase). A group of cells in the M phase (mitotic phase, during the separation of the duplicated chromosomes) were fused to cells in the G_1 phase. The results are shown as follows.

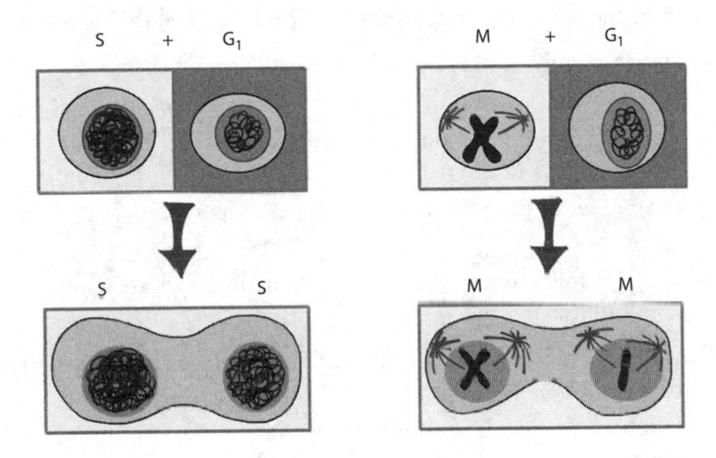

SHORT FREE-RESPONSE QUESTION

326. Explain the results of the experiment. **State** a hypothesis that was tested.

MULTIPLE-CHOICE QUESTIONS

327. Which of the following statements about cell size is *not* true?

(A) Cells are small because their surface area is larger than their volume.
(B) As cells grow, both their surface area and volume increase.
(C) As cells grow, their surface-area-to-volume ratio decreases.
(D) Large cells have a lower surface-area-to-volume ratio.

328. Which of the following statements about mitotic cell division is *not* true?

(A) Two genetic clones of the original cell are produced.
(B) Genetic recombination does not occur.
(C) Only diploid cells can undergo mitosis.
(D) Mitotic cell division functions in growth, healing, and normal cell turnover.

329. Which of the following best describes the significance of mitosis?

(A) symmetrical division of cytoplasm
(B) accurate segregation of organelles
(C) high fidelity DNA replication
(D) provide daughter cells with a complete copy of the genome

330. Which of the following statements about meiotic cell division is *not* true?

(A) It occurs only in haploid cells to create genetic variation.
(B) It creates cells with only one set of chromosomes.
(C) The chromosomes in daughter cells are different from the chromosomes in the parent cell.
(D) Homologous pairs of chromosomes are separated.

331. Which of the following statements correctly describes an advantage of being diploid (having two sets of chromosomes)?

(A) Individuals cannot be carriers of harmful alleles in a population of haploid organisms.
(B) Haploid organisms cannot create genetic variation through sexual reproduction.
(C) Diploid organisms have two copies of each gene, so harmful alleles are not always expressed.
(D) Diploid organisms can reproduce asexually through mitotic cell division.

332. Which of the following organisms is incapable of meiotic cell division?

(A) mushroom
(B) protist
(C) bacteria
(D) tree

333. How can a human cell in an early stage of meiosis I be distinguished from a human cell in an early stage of mitosis?

(A) The cell in meiosis will have half the number of chromosomes.
(B) A spindle forms only in cells undergoing meiosis.
(C) Twenty-three sets of tetrads will be observed in the cell undergoing meiosis.
(D) The chromosomes condense only in the cell undergoing mitosis.

Questions 334–338 refer to the following information.

Proteins of the Hedgehog (Hh) family are paracrine (local) signaling factors that are critical for the induction of particular cell types and the creation of cell boundaries during embryonic development.

In the absence of the Hedgehog protein, the Patched protein (Ptc) prevents the Smoothened protein (Smo) from functioning. When Smo is inactive, the Gli protein (Cubitis interruptus, Ci, is the homolog in invertebrates) enters the nucleus where it acts as a transcriptional repressor for Hedgehog responsive genes. When Smo is active, Gli cleavage is inhibited and the Gli protein enters the nucleus where it acts as a transcription factor, activating the transcription of Hedgehog responsive genes.

Vertebrates have at least three kinds of Hedgehog proteins: Sonic hedgehog (Shh), Desert hedgehog (Dhh), and Indian hedgehog (Ihh).

Sonic hedgehog (Shh) signaling is the primary determinant of anterior-posterior polarity in the vertebrate limb field.

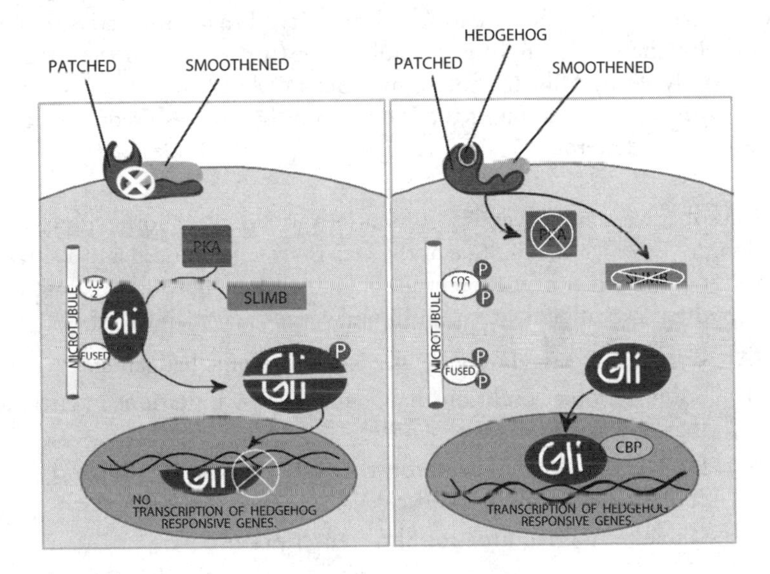

334. Mutations in Shh produce major limb and facial abnormalities. Mutations of Dhh produce gonadal dysgenesis. Mutations in Ihh produce brachydactylyl (anomalous digit formation). These specific observations suggest that the three types of Hedgehog proteins:

(A) are different alleles of the same gene.

(B) are expressed in different cell types during development.

(C) are active at the same time but different locations during development.

(D) result from gene duplications that occur during development.

335. Mutations that inactivate Hedgehog or the Smo/Gli signaling pathway cause major anatomical malformations. Mutations that activate the pathway ectopically (in an abnormal place or position) cause certain types of cancer. Which of the following situations would be the most effective in the treatment of a cancer that is caused by the activation of the Hedgehog signaling pathway (assume each could be specifically accomplished without side effects)?

(A) activation of Smo through the Ptc receptor
(B) increased CBP binding to Gli
(C) activation of Gli cleavage by PKA/SLIMB
(D) maintaining the phosphorylated state of Cos 2 and Fused proteins on microtubules

Question 336 also refers to the following information.

Sonic hedgehog (Shh) has an unusual activity. Final Shh activity requires activation by chemically linking to cholesterol (not shown in the diagram), a reaction catalyzed by Shh. If cholesterol does not bind to Shh, Shh cannot be secreted from cells. Activation of the Ptc receptor by Shh also requires Shh to be complexed to cholesterol.

336. Mice that were homozygous for mutations in Shh were born with *cyclopia*, a disorder in which the midline of the face is severely reduced and a single eye forms in the center of the forehead. Which of the following is a reasonable hypothesis regarding the relationship between *cyclopia* and Shh?

(A) Cyclopia is caused only by mutations in Sonic hedgehog.
(B) Some cyclopia conditions may be caused by mutations in enzymes that code for cholesterol synthesis.
(C) High cholesterol levels inhibit the reaction between Shh and cholesterol and can cause cyclopia.
(D) Cholesterol is required for Shh transcription.

337. How does the Gli protein recognize Hedgehog responsive genes?

(A) The promoters of responsive genes contain a nucleotide sequence whose shape is complementary to the DNA binding site of the Gli protein.
(B) The Gli protein recognizes the promoters of many genes, but only those that are Hedgehog responsive will get expressed as a result.
(C) The Gli protein tethers the DNA molecules and then translocates along the DNA until it finds Hedgehog responsive genes.
(D) The Gli protein recognizes the TATA box of the promoter.

338. Sonic hedgehog (Shh) is an example of a morphogen, a molecule that exerts its effects on cells during development as a function of its concentration gradient. Which of the following most accurately explains the effect of a morphogen during development?

(A) All cells exposed to a particular morphogen will differentiate into the same type of cell.
(B) Cells exposed to high concentrations of the morphogen differentiate more slowly than those exposed to low concentrations.
(C) Active transport allows all cells to accumulate morphogens, but those cells that are exposed to high concentrations expend less energy to do so.
(D) Cells exposed to a particular morphogen can differentiate into different subtypes depending on its concentration.

Questions 339–341 refer to the following data. The effect on plant growth is positive in the region of the graph *above* the line representing ungrazed plants and negative if *below* the line.

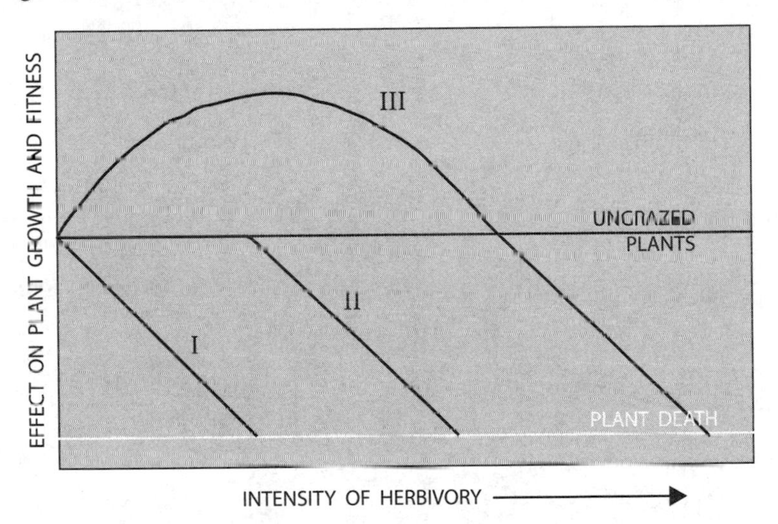

339. Which of the following statements most accurately states the difference in response of type I and type II plants to herbivory?

(A) They are both negatively affected by herbivory, but more type II plants die than type I.
(B) Type II plants are less resistant to herbivory than type I.
(C) At a relatively low herbivory intensity, more type I plants will die.
(D) Type II plants are not affected by herbivory until it reaches a particular level, and then they die very quickly.

340. Which of the following would *not* be responsible for the shape of the curve for type III plants?

(A) an increase in fruit production
(B) increased root growth
(C) production of a compound that is toxic or unpalatable for its predator
(D) an increased number of flowering stalks

341. Which of the following is a way to measure plant fitness?

(A) resistance to disease
(B) number of seeds produced
(C) biomass of plant
(D) responsiveness to environment

Questions 342 and 343 refer to the following diagram of the series of events that results in parasitoid wasp eggs hatching out of a dead caterpillar.

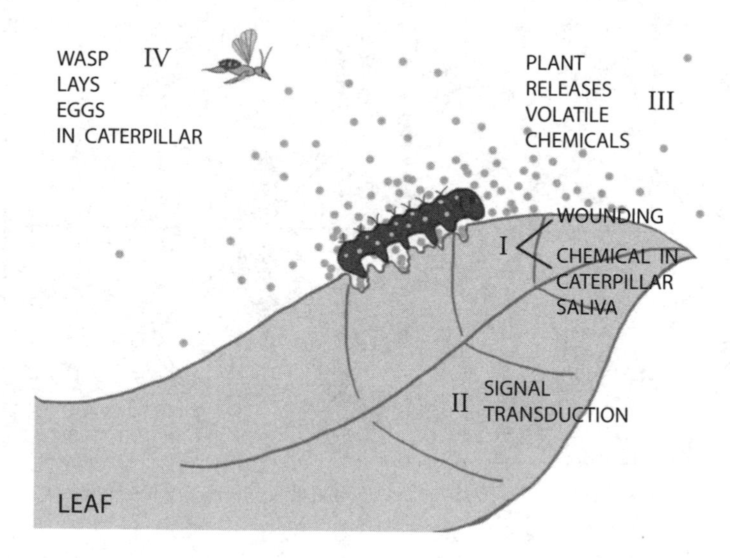

342. Which of the following hypotheses most accurately explains the process?

(A) When the plant senses damage (wounding), it secretes a substance that is toxic to both the caterpillar and the wasp.
(B) The wasp protects the caterpillar against the chemicals released by the plant so she can use the caterpillar as an incubator for her eggs.
(C) The chemical secreted by the plant is a food source for the wasp.
(D) The plant recruits a parasitoid of the caterpillar as a defense against herbivory.

343. Which of the following processes is required for this signal transduction pathway?

(A) Binding of the volatile chemical to a receptor activates a behavioral response in the wasp.

(B) Leaf damage triggers an action potential in a neuron, activating the exocytosis of the volatile chemical.

(C) A chemical in the caterpillar saliva enters the phloem and is transported to the roots for processing.

(D) Plant wounding increases water losses and causes the stomata to close.

Question 344 refers to the following chemical structures.

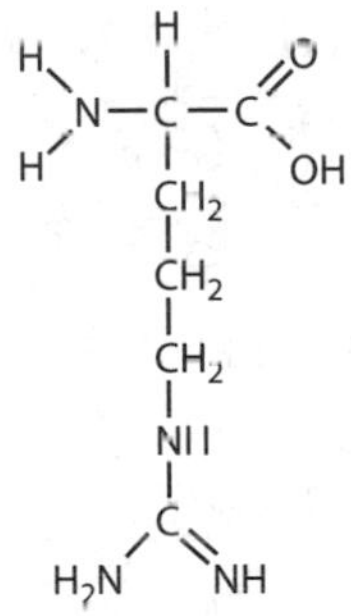

ARGININE

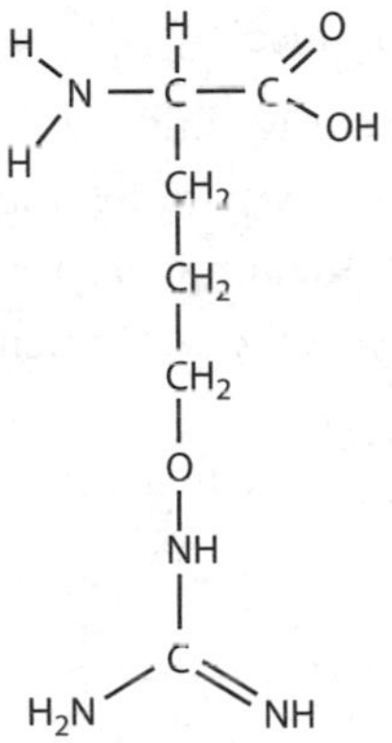

CANAVANINE

SHORT FREE-RESPONSE QUESTION

344. Some plants produce an unusual amino acid called canavanine. Canavanine resembles arginine, an amino acid all organisms use to make proteins. The plants that make canavanine do not use it to make proteins, but the insects that eat the plant incorporate it into their proteins in place of arginine. **Explain** how the incorporation of canavanine into a protein could be harmful to the insect. **Propose a hypothesis** to explain how the amino acid could have been incorporated into the protein when it is not one of the 20 amino acids used to make proteins.

Question 345 refers to the following information.

Pheremones are volatile compounds detected by the vomeronasal organ (VNO) in many animals. Pheremones are involved with chemical communication between organisms of the same species. Normally, male mice will attempt to mate with receptive, unrelated female mice.

When male mice have their VNO removed surgically, they attempt to mate with all mice they come into contact with regardless of age, gender, or relatedness and also attempted to mate with inanimate objects.

MULTIPLE-CHOICE QUESTION

345. What does this observation suggest about the direct function of pheromones in mice?

(A) reduce sex drive in mice
(B) confer specificity to mating preferences
(C) increase variation in a species
(D) regulate sexual development

Question 346 refers to the following hypothetical plasmid.

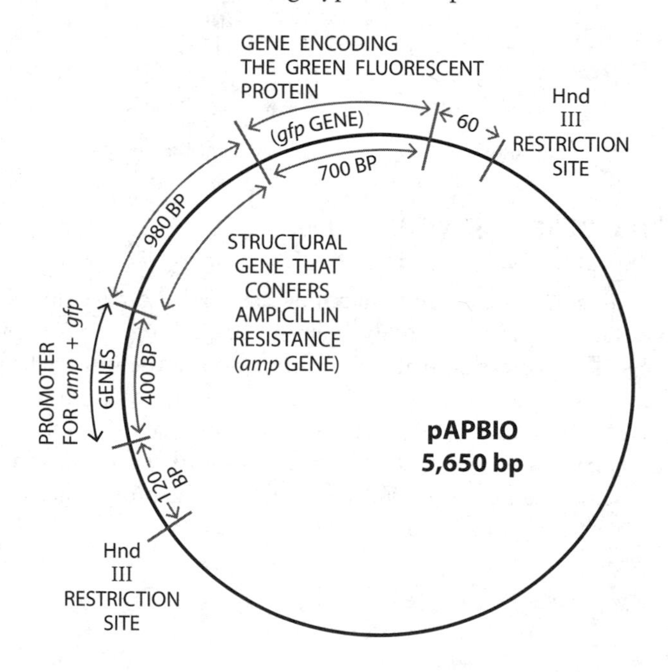

SHORT FREE-RESPONSE QUESTION

346. Sketch the results of the digestion of the plasmid with Hnd III on the following picture of an agarose gel. **Include a control** in which the plasmid is not digested.

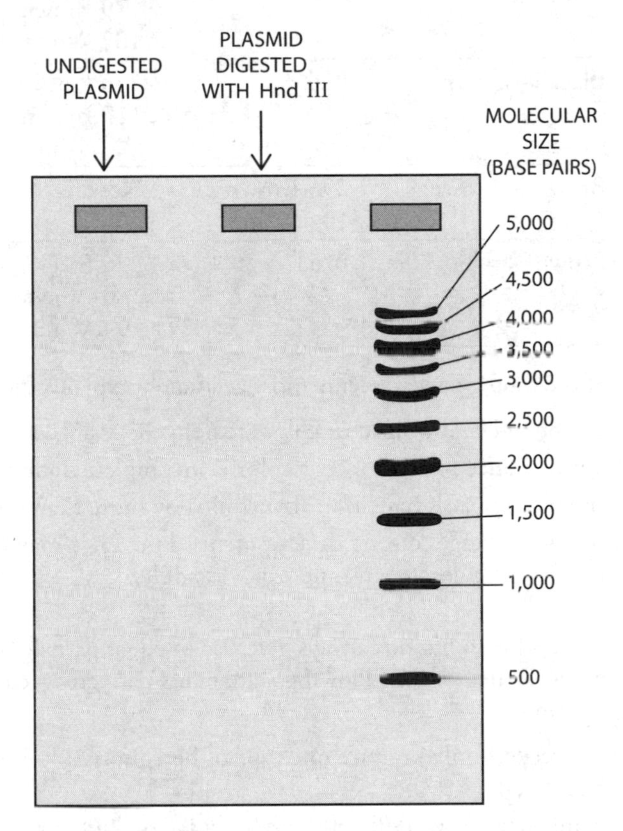

MULTIPLE-CHOICE QUESTIONS

347. When young male chicks of one species are exposed *only* to the songs sung by a different species, they are, as adults, unable to sing the song they were exposed to *and* the song of their own species. Which of the following hypotheses is the most likely explanation of this observation?

(A) Birds can sing only the song of their own species.
(B) Birds are unable to learn songs of another species.
(C) Birds learn to sing the specific song of their species during a critical period in their life.
(D) Birds cannot learn to sing any song unless it is practiced.

Question 348 refers to the following data of crosses between mice of different coat color.

Parents	F_1	F_2
Black × white	Black	219 Black 79 Brown 102 White
Black × brown	Black	253 Black 113 Brown 34 White
Brown × brown	296 Brown 104 White	No data
Brown × white	Black	226 Black 80 Brown 94 White

348. Which of the following statements most accurately explains the data?

(A) Black, brown, and white coat color are not genetically controlled traits.
(B) The inheritance pattern of coat color is incomplete dominance.
(C) Coat color is a polygenic trait (controlled by more than one gene).
(D) Coat color changes during development so the mice may have different coat colors as juveniles than as adults.

349. Hybrids crossed with hybrids always give rise to equal parts hybrids and true breeders. Which of the following statements correctly accounts for this observation?

(A) True breeders only produce one type of hereditary factor, whereas hybrids produce two.
(B) The presence of two different hereditary factors in hybrids give rise to a different phenotype.
(C) Hybrids can revert to true breeders when self-pollinated.
(D) The recessive phenotype is observed only in the homozygous (true breeding) condition.

350. The genotype of a purple-flowered *Pisum sativum* plant is unknown. Which of the following is the most feasible approach to determining its genotype?

(A) Sequence the plant's DNA.
(B) Sequence the DNA of the pollen.
(C) Allow the plant to self-pollinate and observe the phenotypes of its offspring.
(D) Analyze the chemical composition of the petal to determine the amount of pigment present.

Question 351 refers to the following crosses between red-, white-, and pink-flowered snapdragon plants.

P		Red × White	
F_1		100% Pink	
F_2	25% Red	50% Pink	25% White

351. Which of the following statements correctly explains the data?

(A) Heritable factors are not discrete (particulate); they are capable of blending to produce a greater number of phenotypes.
(B) There are at least three alleles that code for flower color in the snapdragon.
(C) Pink is the dominant phenotype for flower color in the snapdragon.
(D) There is segregation of the red and white flower alleles in the gametes produced by pink-flowered plants.

GRID IN

352. A cross of a red-flowered and a white-flowered snapdragon produced 95 pink plants and white plants. What is the chi-square value for this result if the probability that the result was due to chance is 1%.

	Degrees of freedom							
P	**1**	**2**	**3**	**4**	**5**	**6**	**7**	**8**
0.05	3.84	5.99	7.82	9.49	11.07	12.59	14.07	15.51
0.01	6.64	9.32	11.34	13.28	15.09	16.81	18.48	20.09

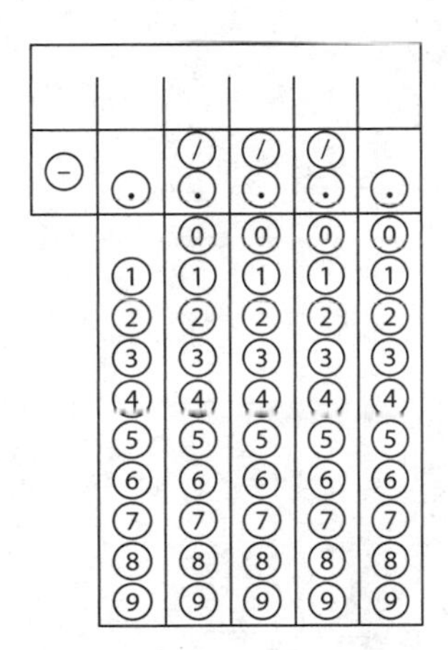

Questions 353–356 refer to the following crosses.

Cross 1:

Round, yellow seeds were grown to produce plants with purple flowers and long stems (tall). The plants were true breeding for yellow seed color and round seed shape.

Cross 2:

Green, wrinkled seeds were grown to produce short plants with white flowers. The plants were true breeding for green seed color and wrinkled seed shape.

Cross 3:

25 tall, purple-flowered plants (progeny from cross 1) were crossed with 25 short, white-flowered plants (progeny from cross 2) and were allowed to pollinate and go to seed.

1,000 seeds were collected. All were round and yellow. The harvested seeds were planted, and 100% of the plants were tall with purple flowers.

SHORT FREE-RESPONSE QUESTION

353. Suppose all 1,000 of the tall plants with purple flowers that grew from the seeds harvested in cross 3 plants were allowed to pollinate (each other). **Calculate** the expected ratios of seed phenotypes.

GRID IN

354. Calculate the number of different gametes that can be produced by a plant that is heterozygous for seed shape and color, flower color, and stem length.

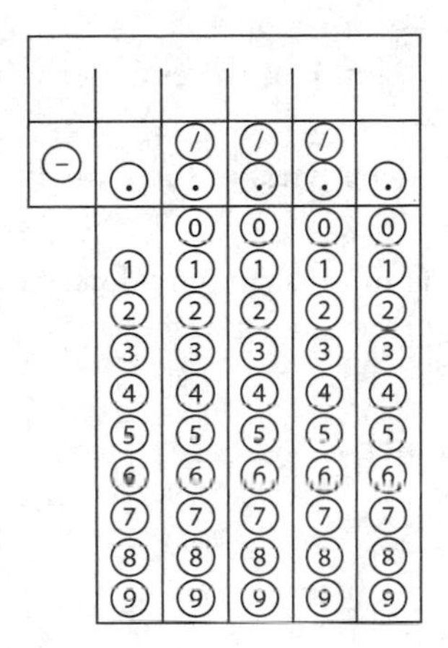

SHORT FREE-RESPONSE QUESTIONS

355. Suppose the tall, purple-flowered plants from cross 3 are crossed again and the seeds are then planted. **Calculate** the expected ratios for height and flower color. **Show your work** and **state** your assumptions.

356. Suppose out of 400 offspring, 200 are tall with purple flowers, 84 are tall with white flowers, 66 are short with purple flowers, and 30 are short with white flowers. **Calculate** the chi-square value for the data. **State** the null hypothesis and **explain** if the null hypothesis is supported by the data.

MULTIPLE-CHOICE QUESTIONS

357. Which of the following developmental processes is the best example of vertebrate embryonic induction?

(A) The mesoderm develops into the musculoskeletal system.
(B) Cartilage is replaced by bone.
(C) The lens of the eye forms as the endoderm interacts with the optic cup.
(D) The endoderm differentiates to form the lining of the digestive tract.

358. Which of the following most accurately describes the behavior used by mice to find their way through a maze?

(A) instinct
(B) classical conditioning
(C) insight
(D) trial and error

Question 359 refers to the following diagram, which represents the code of the honeybee waggle dance. Only bees returning from a highly profitable food source perform the waggle dance. The angle of the zig-zag line relative to gravity represents the angle of flight relative to the food source (F). The duration of the vertical line (called a run, since the bee is performing) represents the distance to the food source. (Adapted from an original illustration by Hölldobler-Forsyth, 1997.)

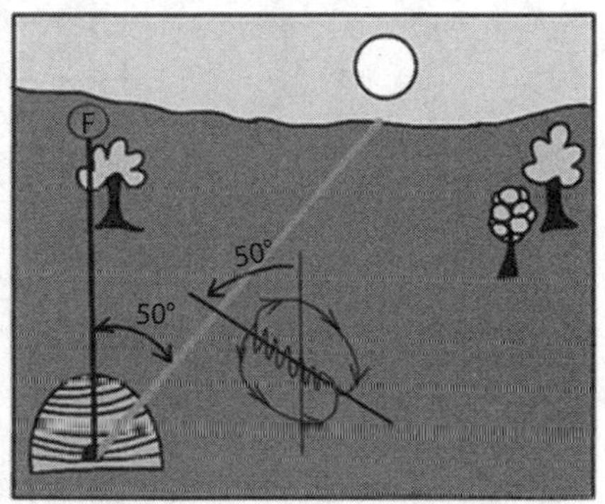

359. Which of the following statements is a reasonable inference about honeybees from the information?

(A) Bees are genetically programmed with information about the food source relative to their hive.

(B) Bees use visual cues in navigation and communication.

(C) The waggle dance is innate and cannot be learned.

(D) Food sources at large angles relative to the sun are unobtainable to bees.

Question 360 refers to the following diagram.

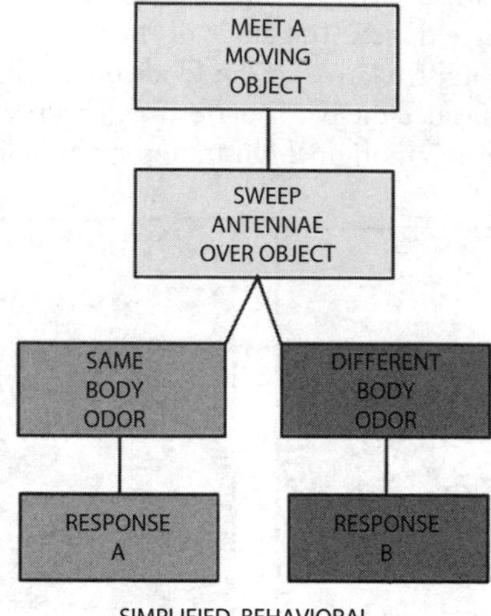

SIMPLIFIED BEHAVIORAL
ALGORITHM OF A SOCIAL ANT

360. Which of the following statements is *not* a logical inference based on the diagram on the previous page?

(A) Ant antennae contain chemoreceptors.
(B) Ants are able to recognize members of their colony.
(C) Ants have pre-programmed responses to specific stimuli.
(D) Ants learn to distinguish between members of their own colony and members of other colonies using tactile (touch) information.

Questions 361 and 362 are based on the following diagram.

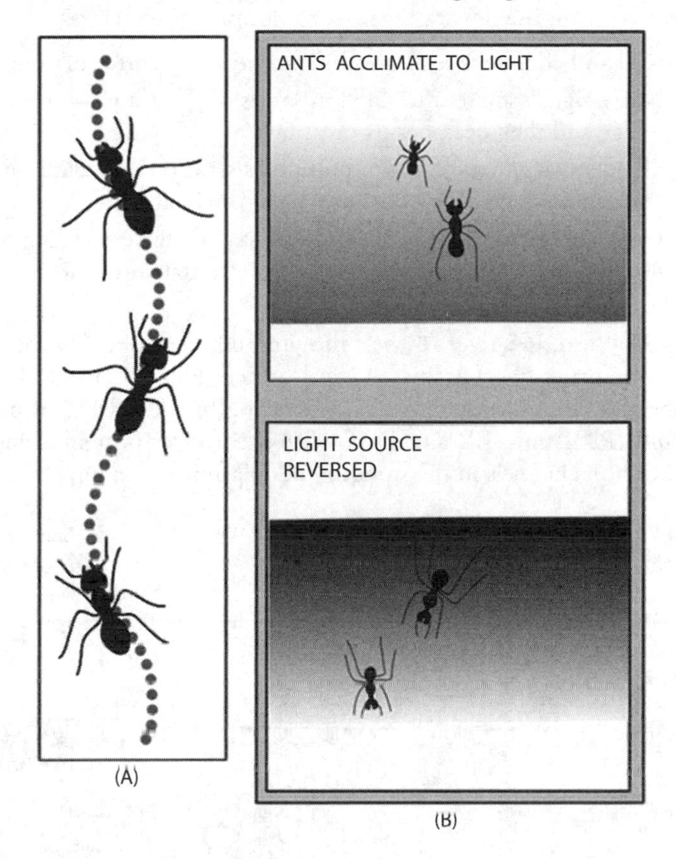

361. Which of the following is represented by the dots in diagram A?

(A) hormones
(B) pheromones
(C) neurotransmitters
(D) antibodies

362. Which of the following statements *incorrectly* describes navigation in ants as illustrated?

(A) Ants use strategies that rely on chemical and visual cues.
(B) Trail-following ants rely on information about the concentration of particular chemicals to stay on the trail.
(C) Ants learn about their surroundings by visual and olfactory (chemical) cues and can memorize the specific response required for future navigation.
(D) Ants have structures that allow them to detect the direction from which a source of light is emitted.

363. As you read this sentence, how do the photoreceptors in your eyes and the neurons in your brain work together to decipher the images?

(A) Rod and cone cells send photons to the visual cortex of your brain.
(B) Nerve signals are sent to different areas of the brain, which detect the image and then decipher its meaning.
(C) The depolarization of action potentials sent to the medulla and cerebellum depends on the content of the image.
(D) Cone cells send patterns of action potentials corresponding to the intensity of light on the page to the prefrontal cortex of the brain.

Questions 364 and 365 refer to the following diagram of a G-protein linked receptor that activates phospholipase C and protein kinase C when the receptor ligand is bound. Phospholipase C catalyzes the breakdown of a membrane phospholipid (PIP_2) into IP_3 and DAG. DAG activates protein kinase C. IP_3 opens calcium ion channels in the smooth endoplasmic reticulum.

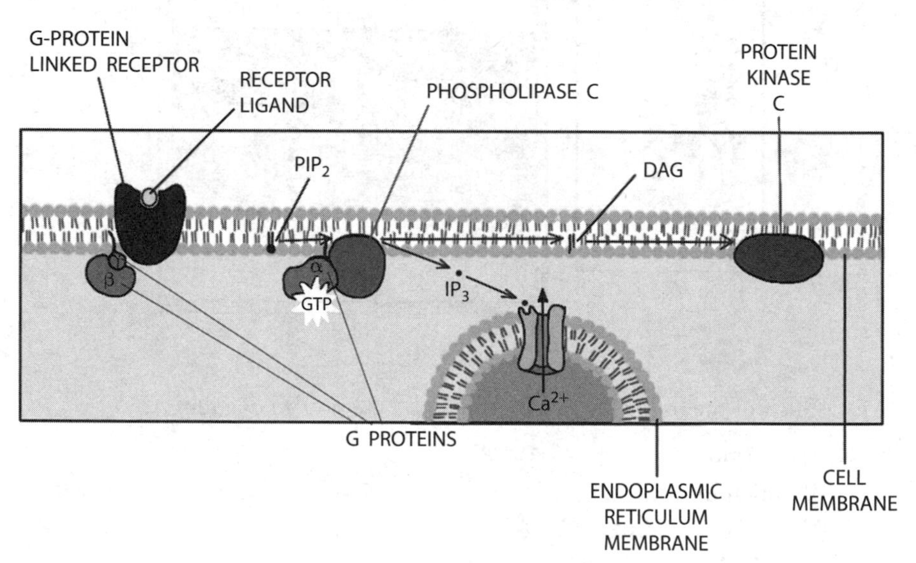

364. Intracellular calcium levels are usually very low (10^{-4} mM). Which of the following hypotheses is consistent with the mechanism in the diagram on the previous page and explains this observation?

(A) Ca^{2+} acts as a 2nd messenger in the cell.
(B) G-protein signaling requires high concentrations of intracellular Ca^{2+}.
(C) Ca^{2+} is needed as a cofactor for DAG activation of protein kinase C.
(D) Cells take up Ca^{2+} from the blood in order to maintain homeostasis.

365. Which of the following describes the likely mechanism by which DAG activates protein kinase C (PKC)?

(A) DAG phosphorylates PKC.
(B) DAG de-phosphorylates PKC.
(C) Phospholipase C covalently binds DAG to PKC.
(D) DAG induces a shape change of PKC upon binding.

CHAPTER 4

Living Systems

Big Idea 4: Biological systems interact, and these systems and their interactions possess complex properties.

Life did not take over the globe by combat but by networking.
Dorian Sagan and Lynn Margulis

Questions 366–367 refer to the following diagram.

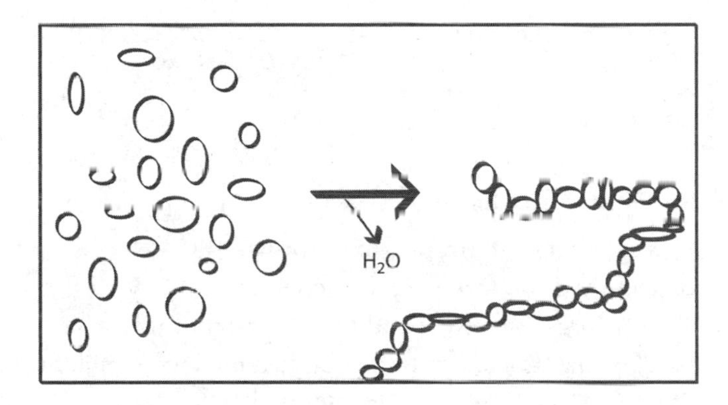

MULTIPLE-CHOICE QUESTIONS

366. Which of the following true statements most accurately describes the process represented in the diagram?

(A) A relatively small number of kinds of molecular building blocks can create a very large variety of polymers.
(B) Connections between molecules can be created by dehydration reactions in which water is a product.
(C) Complex carbohydrates, proteins, and nucleic acids can be formed from condensation reactions between their respective monomers.
(D) The three-dimensional shape of a polymer is directly determined by the specific sequence of monomers that compose it.

367. Which of the examples would be *worst* illustrated by the diagram?

(A) translation
(B) starch synthesis
(C) glycogen synthesis
(D) photosynthesis

Questions 368–371 refer to the following diagram.

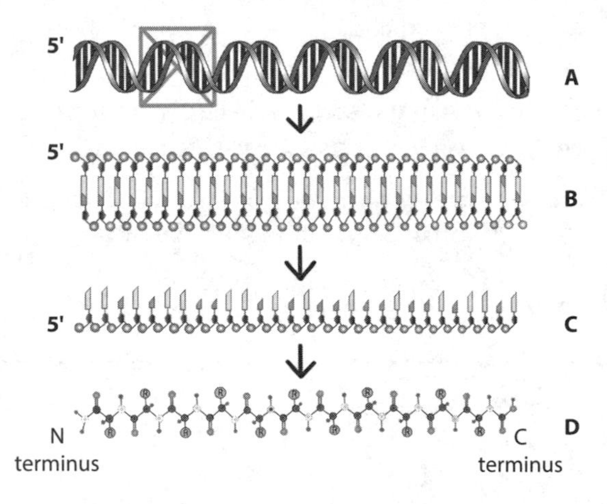

368. Collectively, the preceding processes function most *directly* to:

(A) pass on genetic information to offspring.
(B) organize and command cellular activities.
(C) allow an organism to respond to its environment.
(D) build cellular machinery from a limited number of building blocks using genetic information to organize them.

369. Collectively, the preceding processes *ultimately* function to achieve which of the following?

(A) They assist the organism whose cells perform these processes to develop and survive long enough to produce offspring capable of their own reproduction.
(B) They produce the structures within cells that perform specialized tasks.
(C) They provide the ability to change the composition of molecules within cells to produce a greater variety of structural and functional possibilities.
(D) They increase genetic variation in offspring.

370. If a change occurred in the X-marked region of molecule A, what would be expected in molecule D?

(A) no change, because the several steps in between could correct it by negative feedback
(B) likely a change in a similar position to the right of the N-term
(C) likely a change in a position the same distance from the end but at the C-term
(D) likely a change in sequence, but the location of the change (near the N-term or C-term) depends upon which strand the information on molecule A was contained

LONG FREE-RESPONSE QUESTIONS

371. **Compare** and **contrast** the structures and functions of the molecules represented in the diagram, on the previous page.

372. **Explain** how a small subset of different molecules can generate almost infinite possibilities in building an organism. Include a **description** of the importance of orientation and directionality in the synthesis of complex molecules from simple subunits.

Questions 373 and 374 refer to a hypothetical experiment in which approximately 120 cells of the same type were removed from a hypothetical multicellular organism with only one pair of chromosomes.

The nucleus was removed from 30 of the cells (the DNA donors; these cells die after their DNA is removed). Another 30 cells had their chromosomes removed but immediately replaced with a pair of chromosomes engineered from the donor DNA (assume they all lived through the procedure). Another 30 cells had their own DNA removed and replaced as a mock procedure. No procedure was performed on the remaining 30 cells.

Genetic engineering of the chromosome: One strand of each DNA molecule bound to a hypothetical complement that was unable to be transcribed by RNA polymerase. (In other words, only one strand of each DNA molecule would be able to provide the genetic information to make proteins. See the following figure.) Assume the function of gene promoters and all regulatory information would be unaffected. The newly constructed DNA would be immediately placed back into the hypothetical cells whose DNA was removed.

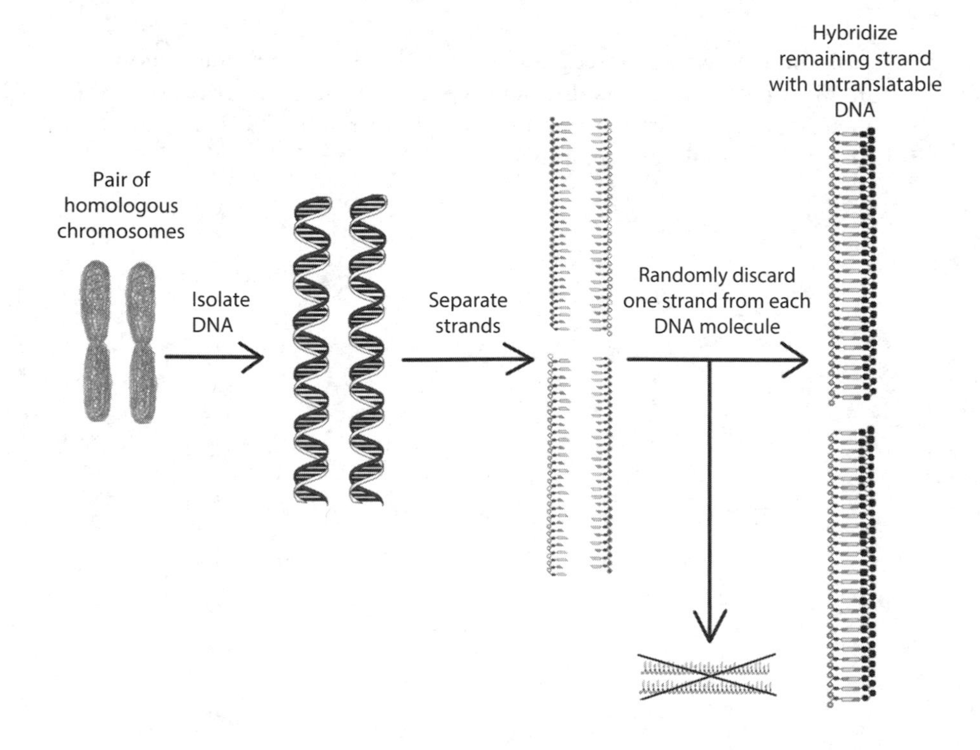

Group	Number of cells	Number of living cells 24 hours after procedure
Control cells	30	28
Mock procedure	30	26
DNA-donor cells	30	0
DNA-replaced cells	30	27

MULTIPLE-CHOICE QUESTIONS

373. What is the relationship between the nucleotide sequences among the 30 cells whose DNA was replaced and the original cells?

(A) same as each other and original
(B) same as each other but different from original
(C) different from each other and different from original
(D) approximately half would be the same as each other, half would be different from each other and the first half, and all cells would be different from the original

374. Assuming that regulatory regions were still functioning in the same way, what would be the expected result on protein synthesis?

(A) All types of proteins could be made in approximately half the DNA replaced cells compared with control cells, and approximately half of the protein types could be made in the other half of DNA-replaced cells.
(B) All types of proteins could be made in approximately half the DNA-replaced cells compared with control cells, and no proteins could be made in the other half of DNA-replaced cells.
(C) Approximately half of the protein types could be made in half of DNA-replaced cells, and no proteins could be made in the other half of the DNA-replaced cells.
(D) Approximately half of the protein types could be made in all of the DNA-replaced cells.

375. Scientists have found two amino acids in the proteins of some living cells, selenocysteine and pyrrolysine, that are not included in the 20 amino acids coded for in the genome in proteins from a variety of organisms. In the organisms in which these amino acids are found, modifications of the protein synthesis reactions produce these amino acids from their precursors, cysteine and lysine. Cysteine and lysine *are* two of the 20 amino acids coded for in the genome and used to make proteins. Which of the following would likely result from the incorporation of these nonstandard amino acids into proteins?

(A) The proteins that contain these amino acids may have novel structures or chemical features.

(B) The genetic code will mutate to incorporate the new amino acids into new proteins based on gene sequence.

(C) The proteins containing these amino acids are nonfunctional.

(D) Protein synthesis will evolve to be able to use these amino acids in all proteins.

Question 376 refers to the following table.

Nonpolar amino acids	Polar, uncharged amino acids	Acidic amino acids	Basic amino acids
Alanine	Asparagine	Aspartic acid	Arginine
Isoleucine	Cysteine	Glutamic acid	Histidine
Leucine	Glutamine		Lysine
Methionine	Glycine		
Phenylalanine	Threonine		
Proline	Tyrosine		
Tryptophan	Serine		
Valine			

376. The table lists the 20 amino acids that make up proteins organized by the chemical reactivity of their side chains. Which of the following statements correctly identifies the purpose of categorizing amino acids by the chemical nature of their side chains?

(A) The properties of a protein are a function of the shape of the protein that is determined by the composition of nonpolar amino acids.

(B) The chemistry of an amino acid determines which other amino acids it can form a peptide bond with, affecting the order of the amino acids in the proteins.

(C) The side chains of the amino acids determine the structure of a protein by determining the sequence of amino acids in the polypeptide.

(D) The function of a protein is mostly predictable by the side chain chemistry of the amino acids.

Question 377 refers to the following diagram.

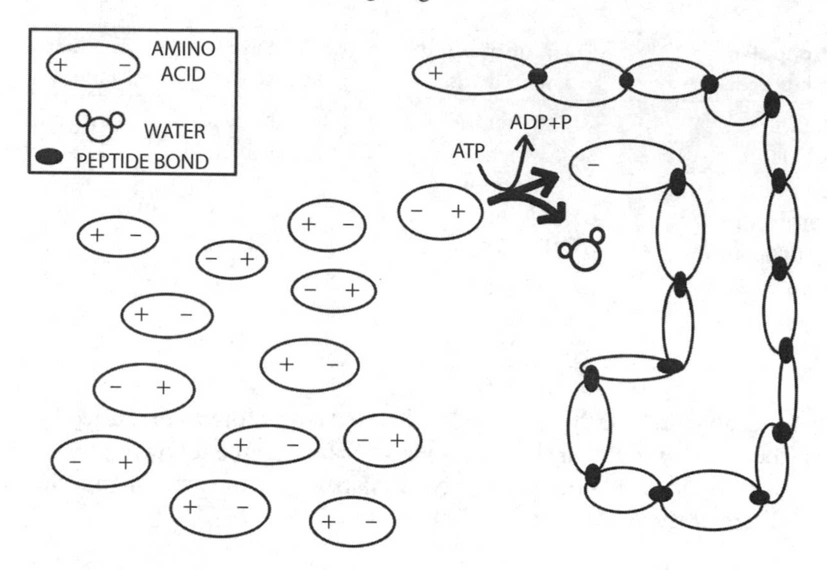

377. Which of the following is a true statement regarding the process represented in the diagram above?

(A) The amino group and carboxyl group react in a "head-to-tail" fashion, forming an unbranched chain.
(B) Protein hydrolysis specifically breaks peptide bonds to liberate amino acid monomers that may be recycled into a new polypeptide chain.
(C) The synthesis of proteins is exergonic.
(D) The synthesis of proteins requires water.

Question 378 refers to the following diagram of carbohydrate functions.

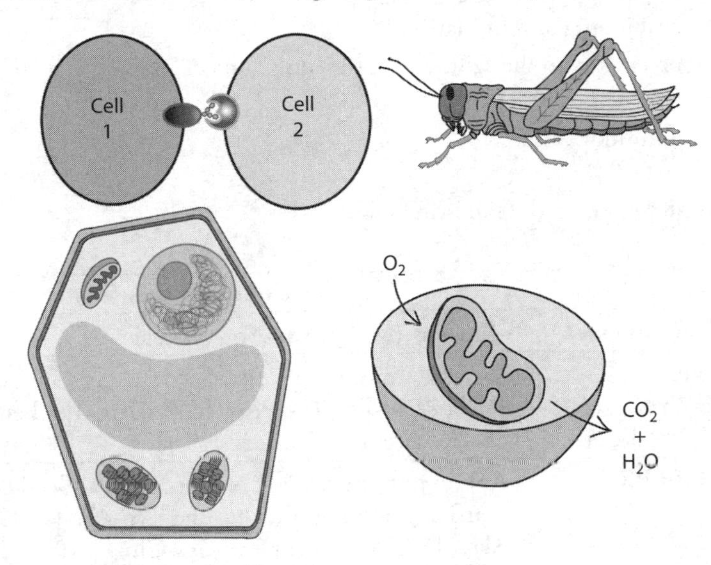

378. Which of the following functions of carbohydrates is *not* represented in the diagram?

(A) fuel (energy source)
(B) structural function at both the cellular and organismic levels
(C) recognition between cell types
(D) reversible storage of complex carbohydrates

379. Mammals are unable to synthesize eicosanoic acid, a 20-carbon fatty acid, unless they have a diet containing linoleic acid, an 18-carbon fatty acid. Which of the following most likely explains this observation?

(A) Linoleic acid is an allosteric activator of the enzyme that synthesizes eicosanoic acid.
(B) The lipid solubility of linoleic acid allows it to function as a steroid hormone that binds to transcription factors to turn on the expression of the genes that code for the enzymes of eicosanoic acid synthesis.
(C) Linoleic acid is a precursor of eicosanoic acid synthesis.
(D) Linoleic acid is a competitive inhibitor of the β-oxidation.

380. Which of the following is *not* a biological function of lipids?

(A) cushioning and insulation
(B) precursors to the synthesis of signaling molecules
(C) energy
(D) joint lubrication

Question 381 refers to the following table.

Function	Example of signal sequence
Import into nucleus	-Pro-Pro-Lys-Lys-Lys-Arg-Lys-Val-
Import into mitochondria	Amino terminus-Met-Leu-Ser-Leu-Arg-Gln-Ser-Ile-Arg-Phe-Phe-Lys-Pro-Ala-Thr-Arg-Thr-Leu-Cys-Ser-Ser-Arg-Tyr-Leu-Leu-
Import into ER	Amino terminus-Met-Met-Ser-Phe-Val-Ser-Leu-Leu-Leu-Val-Ggly-Ile-Leu-Phe-Trp-Ala-Thr-Glu-Ala-Glu-Gln-Leu-Thr-Lys-Cys-Glu-Val-Phe-Gln-

381. Most signals' sequences are cleaved from proteins once they are translocated to their final destination. Nuclear localization signals are *not* cleaved off after import into the nucleus. Which of the following is a reasonable explanation of this observation?

(A) Nuclear proteins need to be imported after each cell division.
(B) The proteins translocated into organelles other than the nucleus are not inherited by daughter cells.
(C) The signal sequence on the nuclear proteins increases their size and prevents them from diffusing through nuclear pores.
(D) Signal sequence cleavage triggers the proteolysis of nuclear proteins to prevent them from being active in other locations in the cell.

382. If the ER (endoplasmic reticulum) is completely removed from a cell, the cell can function normally for a short time but is unable to synthesize new ER during this time. Which of the following is the most likely explanation for this phenomenon?

(A) The genes for ER synthesis are present in the ER.
(B) The ER is not an obligatory organelle in most cells.
(C) The genes for ER synthesis are irreversibly turned off once cell development is complete.
(D) The information required to construct the ER does not reside exclusively in the DNA.

383. The equation $\frac{\lambda E_P}{1+\lambda h}$ quantifies the rate at which an animal obtains a food reward by foraging, where λ = the frequency with which the animal encounters prey, E_P = the amount of energy contained in the prey, h = the amount of time it takes for the animal to capture, kill, and eat the prey. Which of the following is an accurate description of this equation?

(A) The equation is a hypothesis or model.
(B) The slope of the line of the graph of the equation (when plotted against time) represents the energy spent searching for the prey.
(C) The equation cannot be accurate, because it does not include the total time spent searching for prey and the energy expended during the search.
(D) The equation is only true if the scientist believes it is true.

384. The human genome is estimated to have approximately 24,000 genes. Which of the following is a reasonable approximation of the number of types of proteins it can make?

(A) exactly 24,000
(B) at least 24,000
(C) at most 24,000
(D) The number of types of proteins an organism can make is unrelated to the number of genes.

Questions 385–387 refer to an experiment in which a specific inhibitor to enzyme 3 in a biochemical pathway was used. It is shown below:

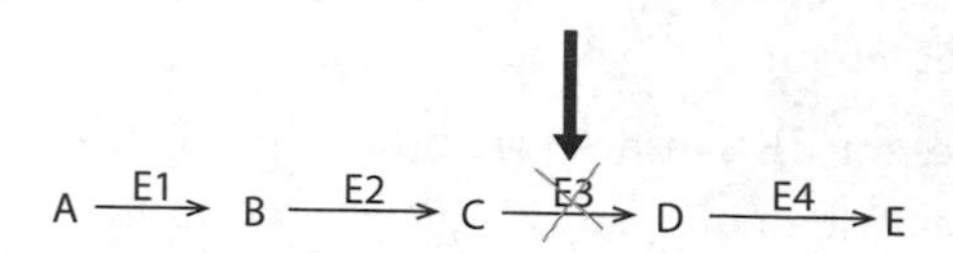

385. Which of the following profiles of metabolite concentration would be expected from the inhibition of enzyme 3?

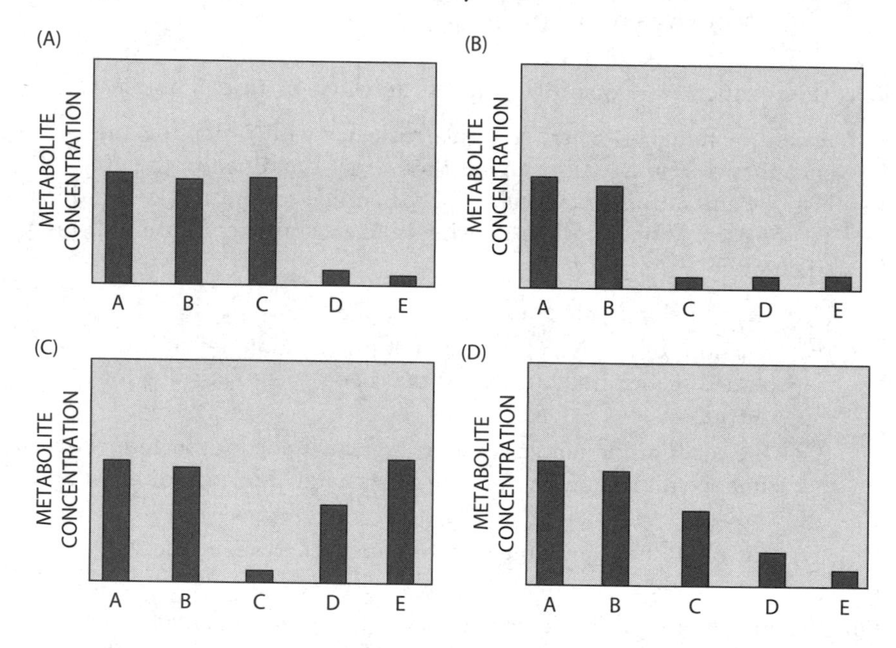

386. Which of the following is an alternative method to elucidate one or more of the steps of the metabolic pathway that does *not* require an enzyme inhibitor?

(A) Mutate (or knock-out) the gene that codes for enzyme 3 in the pathway, and measure the concentrations of the metabolites in the pathway.
(B) Increase the concentration of metabolite C, and measure the expression of the gene that codes for enzyme 3.
(C) Increase the concentration of enzyme 3 then measure the concentrations of enzymes 1, 2, and 4.
(D) Increase the concentration of metabolite E, and measure the concentration of metabolite C.

387. In an experiment, a radiolabeled analog (a molecule that mimics the substrate) for each of the enzymes is transported across the cell membrane. The analog irreversibly binds to the active site of its specific enzyme. The cell membranes are gently ruptured and the cell contents are fractionated. Which of the following statements is a logical expectation based on what is known about biochemical pathways?

(A) Some radioactivity is found in each fraction of the cell homogenate.
(B) Almost all of the radioactivity is contained in one fraction of the cell homogenate.
(C) No radioactivity is found, because the enzymes metabolized the analog.
(D) No radioactivity is found, because the enzymes were inhibited by the radioactive analog.

Question 388 refers to the following observations.

1. Genes that code for rate-limiting enzymes of gluconeogenesis (the enzymatic pathway that synthesizes glucose from nonglucose precursors) and glycogen breakdown are expressed more during sleep/fasting cycle in human liver.
2. The rates of glycolysis and cholesterol synthesis are increased during the day.
3. Disruptions in circadian rhythm result in elevated fat accumulation and other metabolic abnormalities.

388. Which of the following statements most likely accounts for these observations?

(A) Humans who hunted and gathered at night had elevated levels of fat and nutrient-poor food in their diet.
(B) Genes encoding the rate-limiting steps of the pathways are under circadian regulation.
(C) People who live on a nocturnal cycle eat more than those who live on a diurnal cycle.
(D) Being on an opposite rhythm can cause mutations in the genes that code for the enzymes of metabolism.

Questions 389–393 refer to the following answer choices of free energy (*y-axis*) versus time (*x-axis*) for four different chemical reactions.

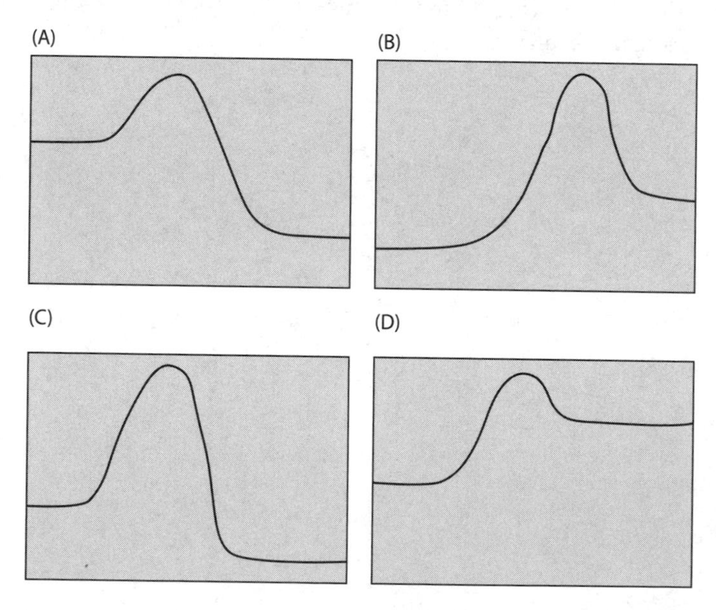

Which chemical reaction—(A) to (D)—shown on the previous page, does the following?

389. Requires the most energy to achieve the transition state.

390. Has the greatest change in free energy.

391. Has the greatest energy difference between reactants and products.

392. Requires the greatest strain in its conformational change and likely occurs the slowest.

393. Would have the highest rate in the presence of a catalyst that increases reaction rate by 100-fold.

Question 394 refers to the following data, showing the percent composition of particular lipids in rat liver cell membranes. (Adapted from Andreoli, T. E., 1987.)

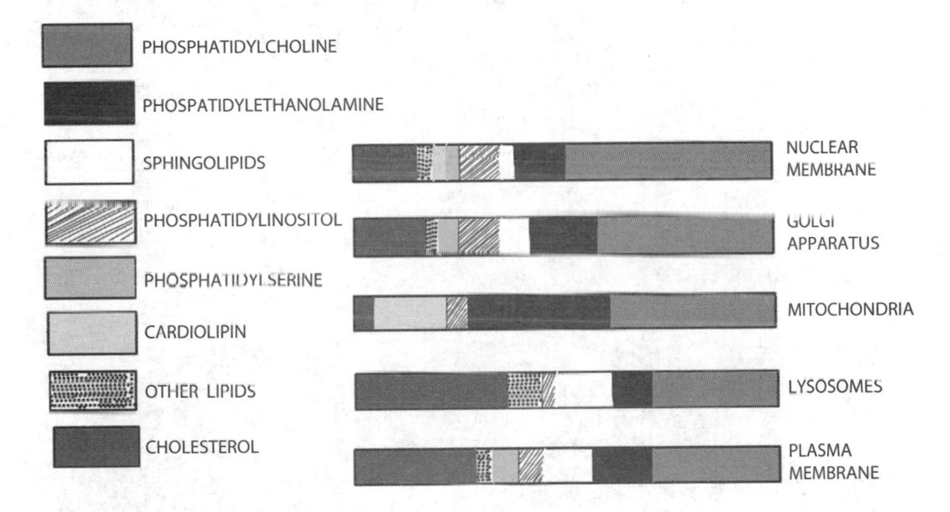

394. Which of the following is an accurate statement regarding the lipid composition of cellular membranes?

(A) Sphingolipids inhibit cellular respiration.
(B) Phosphatidylcholine is more abundant in the nuclear membrane because the nuclear membrane is an envelope of two bilayers.
(C) Cardiolipin is present only in the mitochondrial membrane.
(D) The lipid composition of the components of the endomembrane system shown are more similar to each other than to the mitochondrial membrane.

Questions 395 and 396 refer to the following data, which shows the distribution of phospholipids in human erythrocyte (red blood cell) membranes. The distributions are represented by percent by weight of each type of phospholipid in the inner and outer leaflets of the cell membrane in the bar graph below. The pie charts below show the percent weight of each type of phospholipid in the inner versus outer leaflet. (Adapted from Zachowshi, A., 1993, and Andreoli, T. E., 1987.)

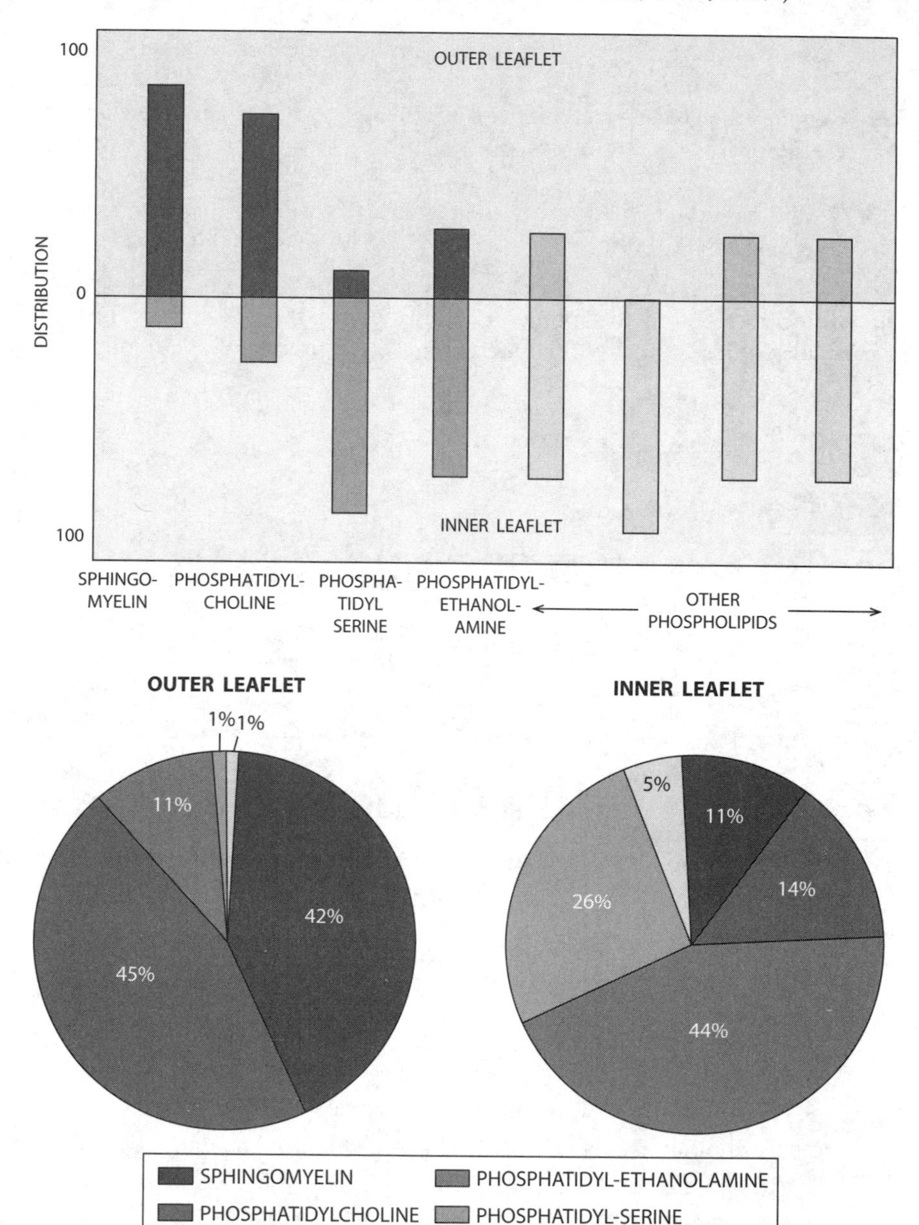

395. Biological membranes are asymmetric. They have different lipid and protein compositions in each of the two leaflets. Which of the following membrane functions does *not* rely on membrane asymmetry?

(A) cell signaling that results in the formation of a second messenger
(B) cell-cell recognition
(C) diffusion of carbon dioxide and oxygen
(D) cytoskeletal attachments

396. Which of the following true statements is evidence that membrane sidedness is important in cellular functioning?

(A) Phosphatidylinositol is the substrate of phospholipase C, an enzyme that produces the second messenger DAG.
(B) The accumulation of phosphatidylserine in the membrane triggers apoptosis.
(C) Phosphatidylethanolamine is contained in much greater percentages in the inner leaflet than the outer leaflet.
(D) Phosphatidylcholine is the most abundant phospholipid in the outer leaflet, but phosphatidylethanolamine is the most abundant phospholipid inner leaflet.

Questions 397 and 398 refer to the following information.

About 98% of the volume of red blood cells is occupied by the protein hemoglobin. Hemoglobin is a protein that carries oxygen from the lungs to the tissues. It is composed of four subunits—two identical α subunits and two identical β subunits.

Sickle-cell anemia is a heritable disease that results from a point mutation in the hemoglobin-β (HBB) gene. The mutant gene contains a thymine base instead of an adenine base and causes a missense mutation (an amino acid substitution). The mutant polypeptide contains a valine, a nonpolar amino acid, on the exterior surface of the polypeptide in the position occupied by glutamic acid in the wild-type protein.

397. Which of the following is the most likely result of the mutation?

(A) abnormal DNA structure as a result of noncomplementary base pairing
(B) altered protein properties resulting from abnormal interactions between hemoglobin molecules in red blood cells
(C) altered protein structure resulting from abnormal hydrophobic interactions between amino acid backbones
(D) altered red blood cell function resulting from an acid-base reaction between the mutant hemoglobin and the red blood cell membrane

398. Which of the following true statements most accurately and comprehensively describes how a single base-pair substitution in a polypeptide can have massive effects on a phenotype?

(A) Hemoglobin is an abundant protein in the body.
(B) Hemoglobin is the most abundant protein in red blood cells.
(C) The alteration in the protein causes oxygen transport to be compromised.
(D) People that are heterozygous for sickle-cell anemia are resistant to malaria.

SHORT FREE-RESPONSE QUESTIONS

Questions 399–405 refer to the following experiment.

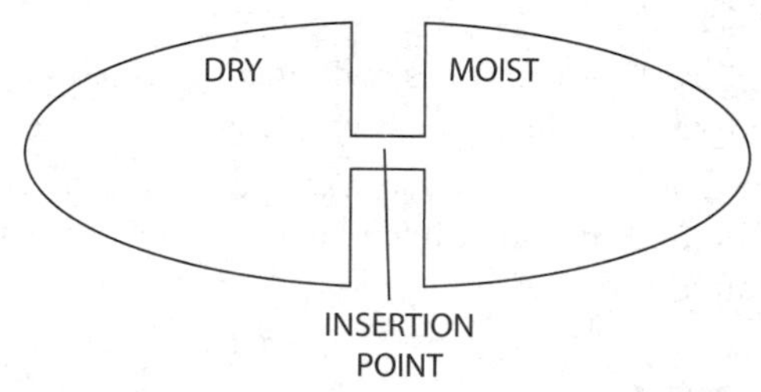

In an investigation of pill bug behavior, a covered choice chamber is used to test the distribution of pill bugs in dry versus moist environments. To test the pill bugs' preference for moist areas, 80 pill bugs were introduced into the insertion point between the two chambers. The position of the bugs was observed and recorded every 10 minutes.

399. Predict the distribution of pill bugs after 10 minutes. **Justify** your prediction.

400. Propose ONE specific **improvement** to the experimental design and **explain** how the proposed modification will affect the experiment.

401. Suppose the experiment is carried out with light and dark compartments instead of dry and moist compartments. **State** a hypothesis about the distribution of pill bugs after 30 minutes. **Specify** the null hypothesis of the experiment.

402. Assume 85 pill bugs were found in the dark chamber and 15 pill bugs were found in the light chamber after 30 minutes. **Perform** a chi-square test on data that would support your hypothesis. **State** whether your hypothesis is supported by the chi-square test and **justify** your statement.

403. **Propose** a model that describes how environmental cues affect the behavior of the bugs in the choice chamber.

404. **Describe** how animal behaviors are subject to natural selection.

405. Many animals use odors to obtain information about their environment and to communicate. **Propose a mechanism** for the detection of a chemical in an animal's environment and **explain** how it may trigger a particular behavior.

Question 406 is based on a classic experiment in animal behavior performed by Ivan Pavlov.

Before conditioning:

Food causes dogs to salivate.
Ringing of bell produces no response in the dogs.

Conditioning process:

The bell is rung and then food is immediately presented to the dogs.
Dog salivates.

After conditioning:

Dog salivates when bell is rung in the absence of food.

MULTIPLE-CHOICE QUESTIONS

406. Which of the following statements most accurately summarizes the results of the experiment?

(A) Dogs always salivate when presented with food.
(B) The dog learns to salivate when it thinks about food, even if no food is presented.
(C) An animal can be conditioned to respond to a neutral stimulus.
(D) The dog learned that when food is presented, a bell is going to ring.

Questions 407 and 408 refer to the following data from an experiment designed to determine the effects of ultraviolet (UV) light on hatchling success in three species of frogs.

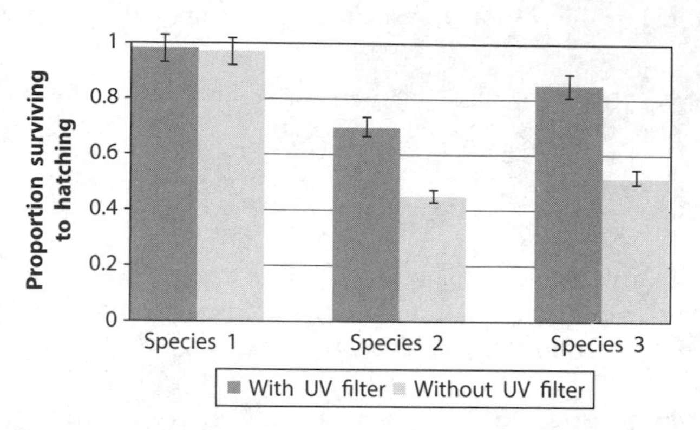

407. Which of the following is a correct interpretation of the data?

(A) UV radiation decreases hatchling success in all three species.
(B) The eggs of species 1 show little to no sensitivity to UV radiation.
(C) Species 2 and 3 are more closely related to each other than to species 1.
(D) UV radiation increases the proportion of hatchlings in species 2 and 3.

408. Which of the following is the most logical inference that can be made from the data?

(A) Species 1 lays the most eggs.
(B) Species 1 is the least likely to go extinct.
(C) Species 1 is the most fit of the three species shown.
(D) Species 1 has the highest proportion of eggs that survive to become hatchlings.

Questions 409–411 refer to an experiment in which two proteins were engineered to contain the N-terminus of the other protein. The original proteins are represented in the following diagram as either all black or all gray. The recombinant proteins were engineered with the 30 amino acids from their N-termini switched. Their cellular locations were determined and the results are shown as follows.

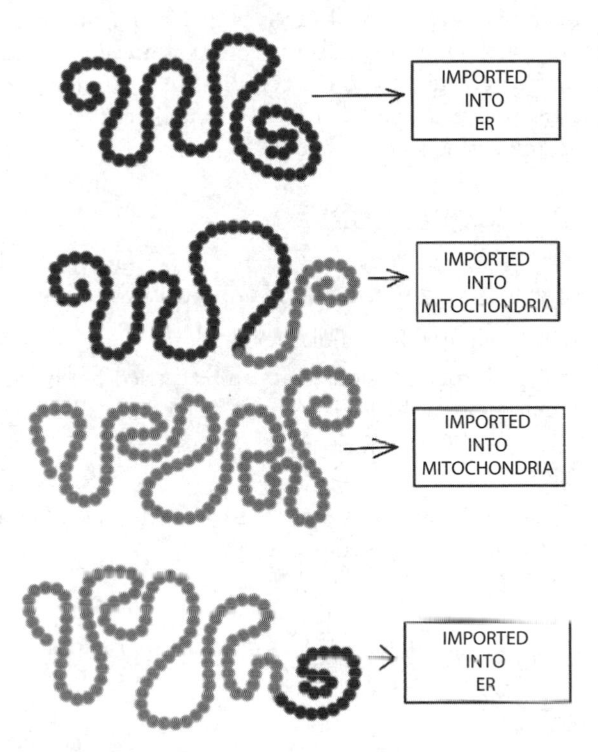

409. Which of the following is a logical interpretation of the results?

(A) The amino acid sequence of the polypeptide does not determine the location of the protein in the cell.

(B) All polypeptides of related function are translocated to the same location in the cell.

(C) A small change in amino acid sequence can alter the function of a protein in the cell.

(D) A signal sequence is necessary to target proteins to specific locations.

SHORT FREE-RESPONSE QUESTION

410. Explain a process by which the recombinant proteins can be engineered.

Question 411 refers to the following information.

Suppose that immediately after mitosis, the ER is removed from a cell. The cell does not make new ER. The cell survives for several minutes, taking in O_2 and releasing CO_2. Although the cell survived the removal of the ER, many vital functions cannot be performed and the cell dies. In a different experiment, the majority of the ER was removed from a cell immediately after mitosis. The cell survived and within days had a complete ER.

MULTIPLE-CHOICE QUESTIONS

411. Which of the following statements is supported by the experiment?

(A) The ER is required for cellular respiration.

(B) The ER performs functions that can be carried out by other organelles if the cell survives the initial trauma of having its ER removed.

(C) At least some of the information to construct the ER is contained in the ER.

(D) The ER contains proteins that are toxic to the cell when they leak out of the ER and are exposed to the cytosol.

Question 412 refers to an experiment done in the 1970s in which DNA from early mouse embryos (which do not make antibodies) were compared with DNA from mouse B-cell tumors that make one specific species of antibody.

The specific variable region and constant variable coding sequences were present on the same DNA restriction fragment in the tumor cells but on two different restriction fragments in the embryo. The experimental procedure is shown in the following diagram.

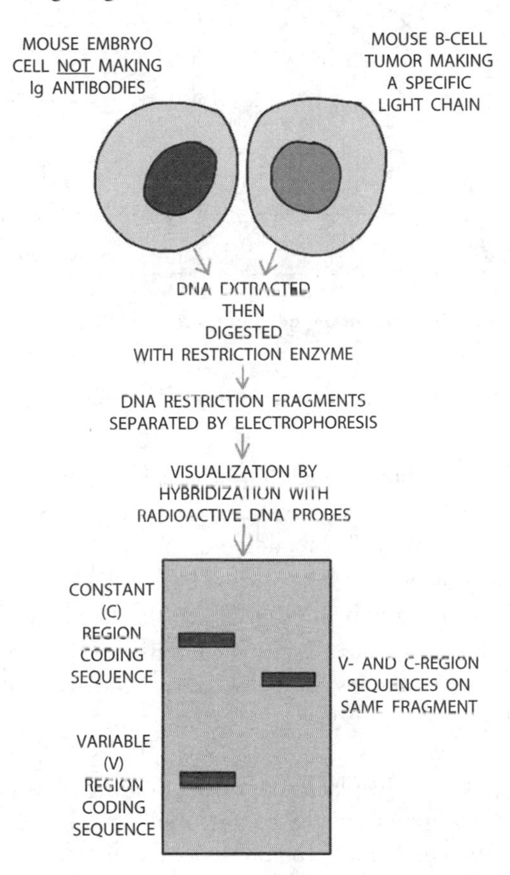

412. Which of the following is an accurate interpretation of the diagram?

(A) Tumor cells undergo meiosis to reduce the amount of DNA they need to replicate.

(B) DNA sequences encoding an antibody molecule are rearranged during B-cell development.

(C) Mouse embryo cells are diploid and heterozygous.

(D) During development, antibody genes are doubled.

Questions 413–417 refer to the following binding curves of myoglobin, hemoglobin, and fetal hemoglobin. Myoglobin is found exclusively in muscle cells. Hemoglobin is the oxygen-binding protein in red blood cells.

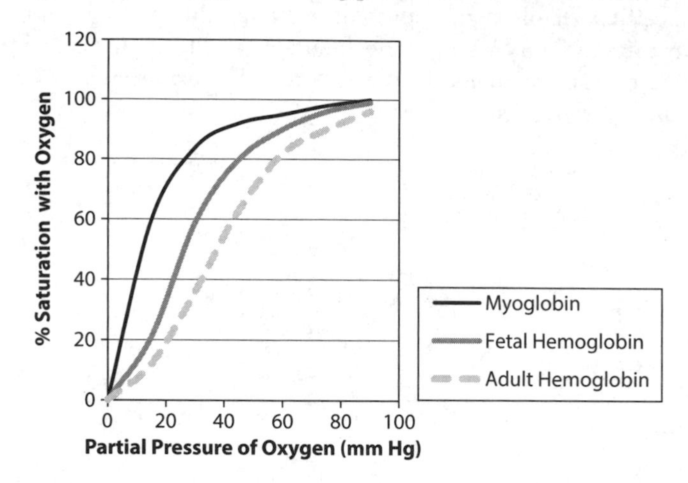

413. According to the binding curves, which of the following statements is correct?

(A) At 60% saturation, myoglobin has the least partial pressure.
(B) At 20 mm Hg partial pressure, myoglobin has approximately twice as much oxygen bound as fetal hemoglobin.
(C) Adult hemoglobin has the highest affinity for oxygen of the three proteins in the graph.
(D) Fetal hemoglobin has a lower affinity for oxygen than myoglobin, but the binding dynamics compensate for the lower percent saturation at a given partial pressure.

414. Changes in pH alter the placement of the binding curves on the graph. A slight decrease in pH causes the hemoglobin curve to shift to the right (but its shape remains the same). The physiological consequences of this are:

(A) increased affinity of hemoglobin for oxygen, allowing it to pick up more oxygen at the lungs.
(B) decreased affinity of hemoglobin for oxygen, allowing it to drop off more oxygen at the lungs.
(C) increased affinity of hemoglobin for oxygen, allowing it to pick up more oxygen at the tissues.
(D) decreased affinity of hemoglobin for oxygen, allowing it to drop off more oxygen at the tissues.

415. Which of the following correctly explains why vertebrates have two oxygen-binding proteins of different oxygen-carrying capacity?

(A) Myoglobin allows oxygen to be taken up at the lungs, while hemoglobin allows oxygen to get dropped off at tissues.
(B) Myoglobin prevents muscles from depleting the oxygen carried in the blood by hemoglobin.
(C) Myoglobin and hemoglobin allow muscles to obtain oxygen in a wide range of oxygen partial pressures.
(D) Hemoglobin transports oxygen extracellularly, while myoglobin transports oxygen intracellularly.

416. Which of the following statements correctly explains why a developing fetus requires a different kind of hemoglobin than an adult?

(A) Adult hemoglobin is inefficient at low partial pressures of oxygen.
(B) Fetal tissues need less oxygen, so their hemoglobin unloads more slowly.
(C) Fetal hemoglobin is a store of oxygen for the fetus. The tissues normally get oxygen directly from maternal hemoglobin.
(D) Fetal hemoglobin picks up oxygen from maternal hemoglobin.

417. Which of the following statements most accurately expresses the physiological importance of oxygen-binding proteins?

(A) Oxygen is not very soluble in water.
(B) Oxygen cannot cross cell membranes without a protein.
(C) Oxygen that is not carried on hemoglobin is converted to bicarbonate, which decreases blood pH.
(D) Oxygen must be dropped off at the electron transport chain by a protein.

418. Arctic foxes generally have a white coat in the winter when there is snow on the ground and a darker coat in the summer when there is no snow on the ground. Which of the following is the most likely mechanism responsible for the change in coat color between the seasons?

(A) Visual input from the arctic fox's environment (i.e., the sight of melting snow) alters the enzyme activity in the pigment cells.
(B) The diet of the arctic fox in summer includes a molecule that accumulates in the fur, making it darker.
(C) In the winter, white foxes are less prone to predation, whereas in the summer, darker foxes are better camouflaged.
(D) Cold winter temperatures significantly decrease the rate of an enzyme-mediated reaction required to make the pigment in the skin.

LONG FREE-RESPONSE QUESTION

419. Organisms must exchange matter with the environment in order to grow, maintain homeostasis and reproduce. **Compare** and **contrast** animals and plants with regard to their nutritional requirements and their mechanisms of nutrient acquisition, metabolic processes, growth, and reproduction.

SHORT FREE-RESPONSE QUESTIONS

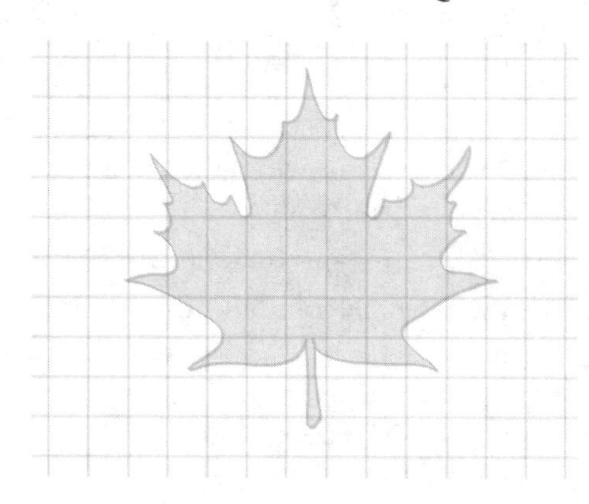

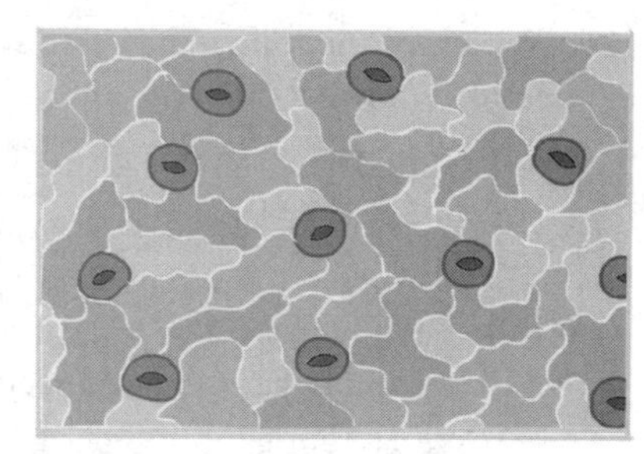

420. **Estimate** the area of the leaf. Provide a **brief description** of your method. Assume each square is 1 cm^2.

421. **Calculate** the stomatal density of the section of leaf shown on the right (the section provided is 2 mm × 1.5 mm) and use it to **estimate** the number of stomata on the leaf on the left. Assume stomatal density is constant.

422. **Briefly describe** the structure of stomata and **explain** their function.

Question 423 refers to the following data.

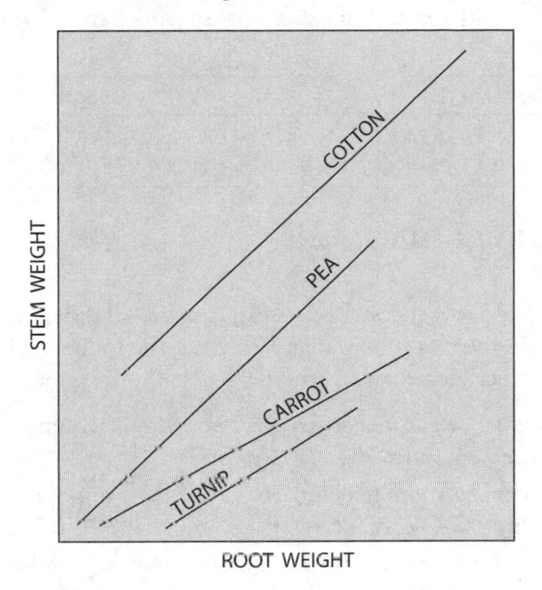

MULTIPLE-CHOICE QUESTION

423. Which of the following statements *incorrectly* describes the relationship between stem weight and root weight in the graph?

(A) The stem weight increases with root weight in all four species.

(B) Increased carrot root weight corresponds with a lesser increase in stem weight compared with pea.

(C) If a cotton and pea plant had the same wieght, they would have the same weight of root and stem.

(D) Cotton has the greatest weight of stem for a given root weight.

SHORT FREE-RESPONSE QUESTION

424. Plants have indeterminate growth; they continue to grow for their entire life. Animals have determinate growth, meaning they grow until they reach a particular size and then remain at that size for the rest of their life. **State one advantage** and **one disadvantage** of both determinate and indeterminate growth and development.

Question 425 refers to the following double-stranded DNA sequences. A-T base pairs are joined by 2 hydrogen bonds. G-C base pairs have 3 hydrogen bonds.

Segment 1	Segment 2
5′ – TATATTAATAATATATATTAT – 3′	5′ – GCCGCGAGGCTGCATGCGAG – 3′
3′ – ATATAATTATTATATATAATA – 5′	3′ – CGGCGCTCCGACGTACGCTC – 5′

MULTIPLE-CHOICE QUESTIONS

425. The nucleotide sequences of two short, double-stranded nucleic acids are shown in the preceding table. Which of the following is *not* a correct comparison between the two strands?

(A) Segment 1 would denature (the two strands would separate) at a lower temperature.

(B) Segment 2 is more likely to be part of the promoter of a gene because of its high G-C content.

(C) The two segments are from DNA because they both contain thymine, a base seen in DNA but not RNA.

(D) The two fragments are equally soluble in water.

Question 426 refers to the following diagram.

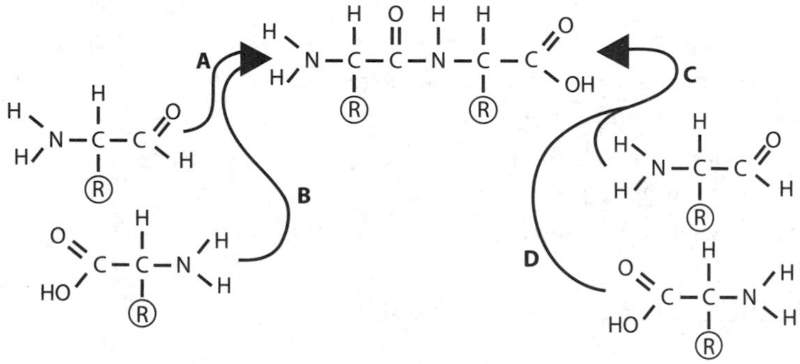

426. Which illustrates the correct positioning for the incoming amino acid to form a tripeptide from a dipeptide?

(A) A

(B) B

(C) C

(D) D

Questions 427 and 428 refer to the following graph of shell thickness in two populations of blue mussels (*Mytilus edulis*). Northern populations of mussels live among the *Carcinus* crab, while southern populations live among both *Carcinus* and *Hemigrapsus* crabs.

Young mussels that had never experienced interactions with crabs were taken from northern and southern populations and grown under one of three different controlled conditions. The mussels were exposed either to no crabs or to *Carcinus* or *Hemigrapsus* crabs in cages nearby. The crabs were kept in cages for the duration of the experiment and could not eat the mussels.

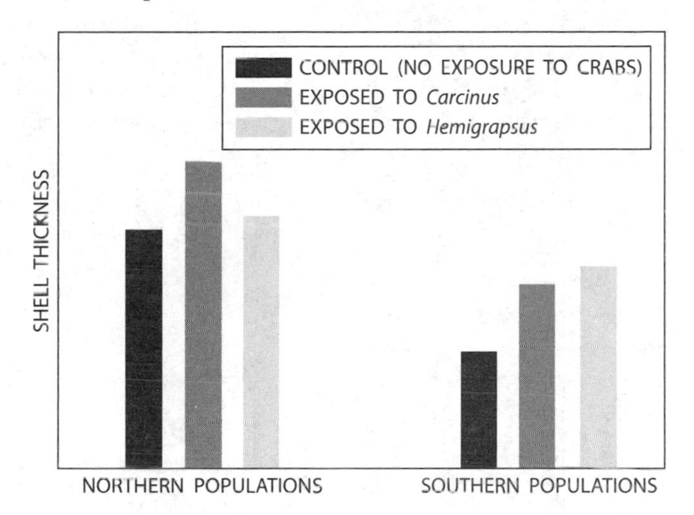

427. Which of the following statements accurately compares the response to crab exposure by the northern and southern mussel populations?

(A) The mussels exposed to *Carcinus* crabs have thicker shells than the mussels exposed to *Hemigrapsus* crabs in the northern population, but the mussels exposed to *Hemigrapsus* crabs have a thicker shell than the mussels exposed to *Carcinus* crabs in the southern population.

(B) The northern population experienced greater increases in shell thickness overall relative to the southern population.

(C) The southern population had greater increases in shell thickness because they had thinner shells at the start of the experiment.

(D) Increases in shell thickness were greatest in response to the crabs that the mussels' original populations had exposure to.

428. Which of the following statements is a biologically correct inference from the data?

(A) Interactions between individuals can result in evolution.
(B) Traits learned or acquired by parents can be transmitted to offspring.
(C) Mussels secrete chemicals that increase the thickness of shells of nearby mussels.
(D) Mussels can change their shell thickness in response to fear.

Questions 429–431 refer to the following data that shows the effects of interspecific interactions on the population dynamics of two species of *Paramecium* cultured separately or together (G. F. Gause, 1934). In the cultures where the two species were grown together, the curve indicates the growth rate of the first species listed.

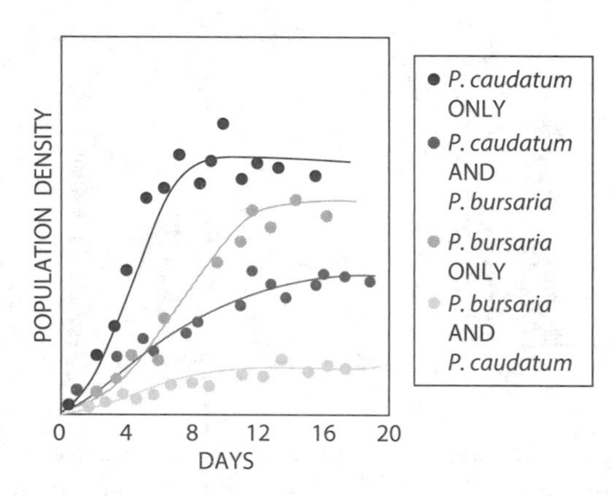

429. Which of the following is the most accurate summary of the data?

(A) *P. caudatum* is a parasite of *P. bursaria.*
(B) *P. bursaria* is a parasite of *P. caudatum.*
(C) *Paramecium* species can mutualistically occupy the same culture.
(D) The presence of a competitor reduced the growth rate of both populations.

430. In which species, if any, is the population density more adversely affected by the presence of the other?

(A) *P. caudatum*
(B) *P. bursaria*
(C) The population density of both species is equally affected.
(D) The rate of growth, not the population density, is most affected in *P. bursaria*, but the population density is most affected in *P. caudatum.*

431. Which of the following statements is a biologically correct inference from the data?

(A) The presence of a competitor will likely result in extinction of one of the species.
(B) The average population density of each species is affected by the presence or absence of other species.
(C) The per capita growth rate of all species in any community is decreased by the presence of other species.
(D) Interactions between competing species always has a negative effect on the populations involved.

432. Which of the following statements correctly explains changes in population density over time?

(A) Variations in population density depend on the magnitude of the environmental fluctuations and the immigration and emigration.
(B) Variations in population size depend solely on the diversity of the gene pool.
(C) Minor environmental changes affect population size, whereas major environmental changes affect the location of the population.
(D) Population density stays at carrying capacity as long as the source of energy remains constant.

433. Which of the following most accurately describes an effect of fluctuations in population size?

(A) Populations whose size fluctuates over several orders of magnitude are more likely to go extinct.
(B) Populations whose members are short-lived are more prone to wild fluctuations in population size within short periods than longer-lived populations.
(C) Smaller organisms have a greater capacity for homeostasis and maintain their population size within a fairly narrow range.
(D) The birth and death rates within populations of shorter-lived organisms tend to even out the effects and their population size remains fairly stable.

434. Every year, humans deforest millions of hectares of tropical rain forest. Which of the following is a likely consequence of this deforestation?

(A) Increased carbon dioxide will reduce ozone concentrations in the atmosphere, resulting in higher levels of ultraviolet radiation on the earth's surface.
(B) Lower oxygen concentrations in the atmosphere will decrease the biodiversity of terrestrial animals.
(C) Animal and plant populations will be eliminated as their habitats are destroyed, leading to a decrease in biodiversity.
(D) Increased carbon dioxide levels in the atmosphere will reduce plant and algae growth in other biomes.

Questions 435 and 436 refer to the following table showing typical concentrations of elements in soils and annual uptake by plants (from Bohn, et al., 1979).

Element	Soil content (weight %)	Annual plant uptake (kg hectare^{-1} year^{-1})	$\frac{\text{Soil content}}{\text{annual plant uptake}} \left(\frac{g}{m^2 \text{ years}}\right)$
Silicon	33.00	20.00	21,000
Aluminum	7.00	0.50	180,000
Iron	4.00	1.00	52,000
Calcium	1.00	50.00	260
Potassium	1.00	30.00	430
Sodium	0.70	2.00	4,600
Magnesium	0.60	4.00	2,000
Titanium	0.50	0.08	62,000
Nitrogen	0.10	30.00	40
Phosphorus	0.08	7.00	150

435. The most abundant element by weight in the soil is:

(A) silicon
(B) calcium
(C) aluminum
(D) titanium

436. The applicability of this data set to other soils is affected by:

(A) temperature only
(B) temperature and soil pH only
(C) soil pH and the presence of other ions only
(D) temperature, soil pH, and the presence of other ions

437. What is the meaning of the last column, $\frac{\text{soil content}}{\text{annual plant uptake}}$ as measured in $\frac{g}{m^2 \text{ years}}$?

(A) the number of years it takes for a plant to uptake its weight of the mineral
(B) the number of years it takes 1 m^2 of plant root surface to absorb 1 gram of the mineral
(C) the number of years to deplete the mineral from the soil if it is not replenished
(D) the number of years to replace the mineral from one m^2 of plant biomass

438. Which elements are most likely to be depleted from the soil if not replaced?

(A) calcium, potassium, and nitrogen
(B) silicon and aluminum
(C) aluminum and titanium
(D) titanium and iron

439. Calcium and magnesium are highly mobile ions whose absorption in plants grown in the soil from which the data are taken is limited primarily by:

(A) the absorptive capacity of the roots.
(B) the content in the soil.
(C) the weight-uptake ratio of the particular plant.
(D) the time it takes to replace after uptake.

440. In which way can a plant compensate for low mineral content in certain soils?

(A) create new biochemical pathways to synthesize molecules that do not require those minerals
(B) increase the absorptive surface area of the root system
(C) increase the stomatal density on leaves
(D) increase transpiration rate

SHORT FREE-RESPONSE QUESTION

441. Plants must strike a balance between the biomass of their roots and shoot systems. **Explain** the compromise the plant makes in balancing the two.

MULTIPLE-CHOICE QUESTIONS

442. Which of the following statements accurately explains why phosphorus often limits plant production?

(A) Phosphorus makes up the majority of minerals in the soil and blocks the uptake of other ions.
(B) Phosphorus is most abundant in the form of phosphate ions, which form insoluble complexes with other ions in acidic and alkaline (basic) soil.
(C) Phosphorus is toxic to plants in all but a narrow range of concentrations.
(D) The presence of phosphorus decreases plant growth in direct proportion to its concentration in the soil.

443. Which of the following statements correctly compares plant and animal cells?

(A) Antibiotics are toxic to plants because their ribosomes are more similar to bacteria than eukaryotic ribosomes.
(B) Plant cells, unlike animal cells, are unaffected by hypertonic solutions because of their cellulose cell wall.
(C) Plant and animal cell membranes are dynamic, constantly changing and exchanging with the endomembrane system via endo- and exocytosis.
(D) Animal cells evolved from plant cells by a mutation that neutralized the genes for the proteins of cell wall synthesis and then exocytosed the chloroplast.

444. Which of the following is most likely *not* a characteristic of plants that are adapted to nutrient-poor soils?

(A) high rate of transpiration
(B) large ratio of root surface area to shoot mass
(C) conservative growth rate
(D) symbiotic association with fungus

445. Which of the following would mostly likely be the expected response to a nutrient flush (a sudden abundance of nutrients) by plants adapted to nutrient-poor soil?

(A) increased nutrient uptake and rapid growth
(B) increased uptake and storage of the nutrient
(C) increased uptake and secretion of the nutrient
(D) no change in uptake due to decreased metabolic requirements

446. K^+ concentrations in soil are usually far less than in plant cells. However, many K^+ transporters are not completely selective for K^+ and often transport the more concentrated Na^+ ions. Which of the following is a potential consequence for plants?

(A) They may accumulate K^+ ions.
(B) They may accumulate Na^+ ions.
(C) They may not hydrolyze enough ATP to take up K^+.
(D) They may expend too much energy on Na^+ transport.

Questions 447 and 448 refer to the following table, which shows the percent composition of dissolved minerals in rivers, sea water, and the blood plasma and cells of frogs. (Data from Reid, 1961, and Gordon, 1968.)

Mineral	Delaware River	Rio Grande	Seawater	Frog Plasma	Frog Cells
Sodium	6.7	14.8	30.4	35.4	1.3
Potassium	1.5	0.9	1.1	1.3	77.7
Calcium	17.5	13.7	1.2	1.2	3.1
Magnesium	4.8	3.0	3.7	0.4	5.3
Chlorine	4.2	21.7	55.2	39.0	0.8
Sulfate	17.5	30.1	7.7	–	–
Carbonate	33.0	11.6	0.4	22.7	11.7

447. All of the numbers in the table are percentages. Why doesn't the sum of any column equal 100?

(A) Ion measurements are unreliable.
(B) Not all ions are included in the analysis.
(C) Only a small sample of each (sample) was taken.
(D) The concentration of the ions varies between samples, so comparisons are inaccurate.

448. The solubility (mass dissolved/volume of solution) of ions generally increases with temperature. Compared to a sample of solution taken from a body of water at 30°C, a sample of solution taken from the same body of water at 15°C is expected to:

(A) have the same mass of solute because it is from the same body of water.
(B) have a decreased mass of solute but the same concentration of ions.
(C) have an increased mass of solute but the same concentration of ions.
(D) have fewer ions dissolved.

Questions 449 and 450 refer to the following table of masses for *Brassica rapa* fast plants, a model plant for laboratory study.

Age	Wet mass of 10 plants (grams)	Dry mass of 10 plants (grams)
7 days	20.2	4.2
14 days	42.1	9.7

GRID INS

449. What is the percent biomass of the 14-day-old plant? Report your answer in whole number.

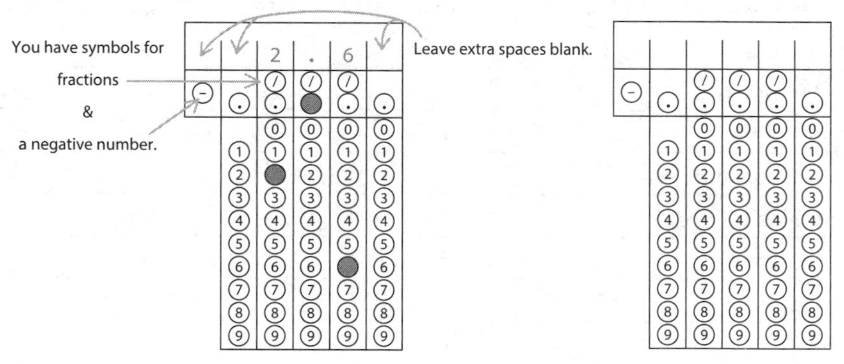

450. What is the average wet mass of one 7-day-old wet plant? Report your answer to the tenths place.

SHORT FREE-RESPONSE QUESTION

451. Keystone species have a dominating influence on the species composition of community. **Choose one** of the following species and **state** how the behavior listed may contribute to sustaining a particular species composition in their community.

- Bees: pollinator
- Hummingbirds: pollinator
- Beavers: build dams
- Sea stars: eat mussels
- Elephants: eat small trees

Questions 452 and 453 refer to the following data (adapted from McMaster, Jow, and Kummerow, 1982) in which three types of plants were grown in their native habitats. Some of the plants were then supplemented with nitrogen, phosphorus, or both nitrogen and phosphorus.

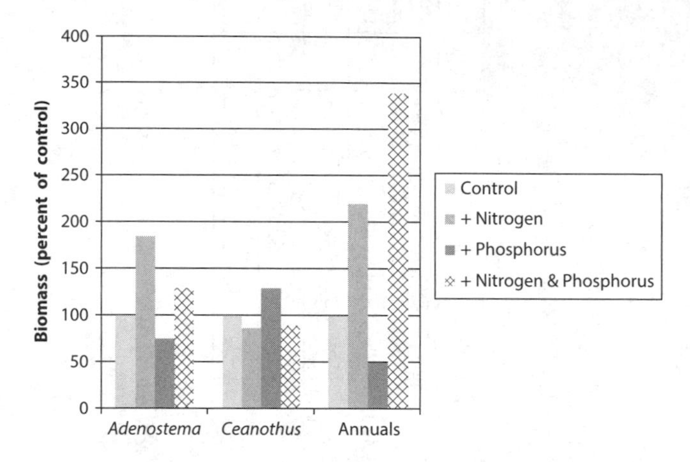

MULTIPLE-CHOICE QUESTIONS

452. Which of the following statements is most accurate regarding the growth of *Adenostema* and annuals?

(A) Phosphorus inhibits their growth.
(B) Nitrogen is the only limiting nutrient.
(C) They grow best in low phosphorus soil if nitrogen is not a limiting nutrient.
(D) They can only take advantage of increased phosphorus levels in the presence of enough nitrogen.

453. Which of the following is a reasonable inference regarding *Ceanothus*?

(A) It is poisoned by excess nitrogen.
(B) It harbors nitrogen-fixing bacteria in its root system.
(C) It has greater phosphorus requirements than *Adenostema* and annuals.
(D) Its cells contain less protein and nucleic acids than *Adenostema* and annuals.

454. Which of the following explains why the biomass of the control groups are identical?

(A) All measurements were taken as a percent of the total biomass of the plant.
(B) *Adenostema*, *Ceanothus*, and the annuals produce the same amount of biomass per year.
(C) The growth in biomass in the control groups is not limited by nitrogen or phosphorus.
(D) The biomass of the nitrogen, phosphorus, and nitrogen and phosphorus groups are relative to the control.

Questions 455–457 refer to the following graphs of reaction rates for a particular enzyme.

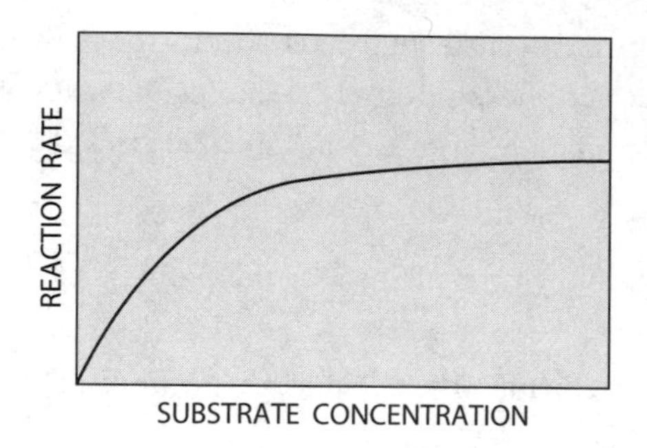

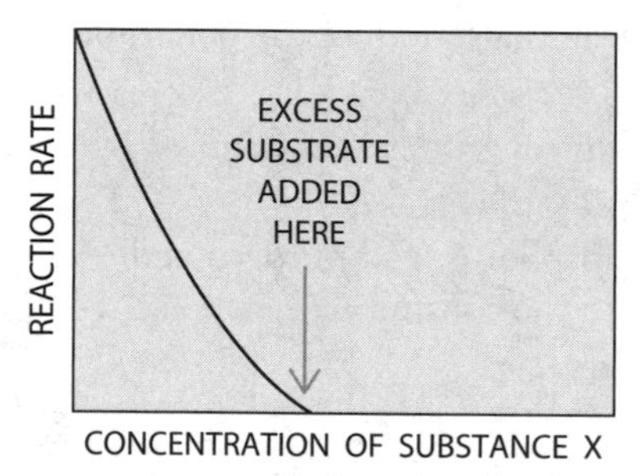

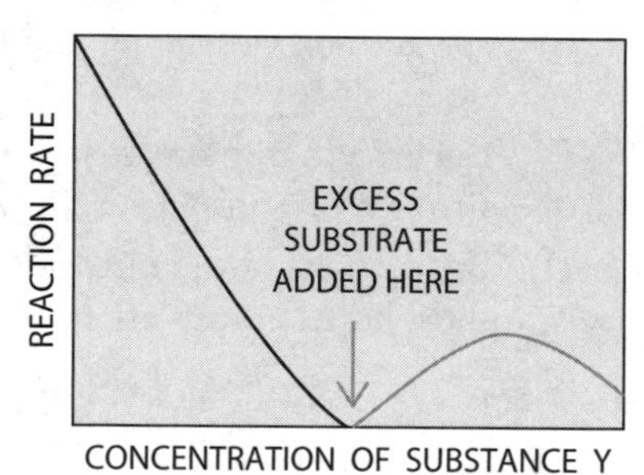

455. Which of the following statements correctly describes the enzyme?

(A) The enzyme has multiple substrates.
(B) The enzyme has at least two allosteric inhibitors.
(C) The activity of the enzyme is affected by molecules other than the substrate.
(D) The enzyme is different from most enzymes because its reaction rate doesn't continually increase with substrate concentration.

456. Which of the following is true of substance X?

(A) It competitively but reversibly binds to the active site of the enzyme.
(B) It competitively but irreversibly binds to the active site of the enzyme.
(C) It forms an insoluble complex with the substrate at low concentrations.
(D) The addition of excess substrate inactivates it.

457. Which of the following correctly explains why the reaction rate increased and decreased when excess substrate was added to the reaction containing substance Y?

(A) Substance Y reversibly binds to the active site of the enzyme.
(B) Substance Y irreversibly binds to the active site of the enzyme.
(C) Excess substrate inactivates substance Y.
(D) Substance Y allosterically inhibits the enzyme.

Question 458 refers to the following data showing the population density of phytoplankton from Lake Erie in 1962. Phytoplankton are single-celled algae that live for only a few days. (Data adapted from Davis, 1964.)

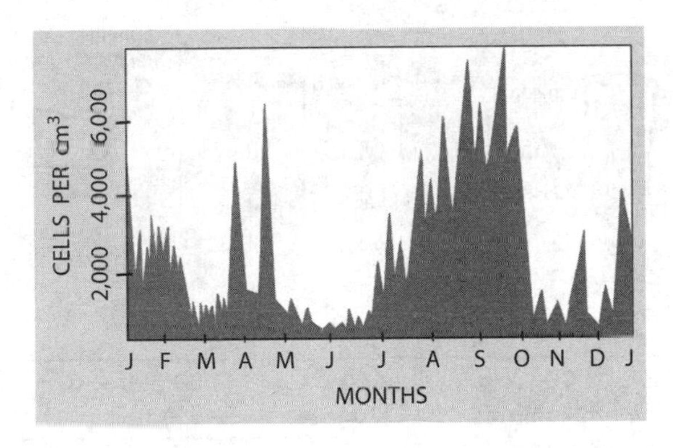

458. Which of the following is most likely the reason for the large fluctuations in population density?

(A) The population is stable but the environment is unstable.
(B) Phytoplankton are short-lived so their population turns over rapidly.
(C) The population size changes with amount of available sunlight.
(D) The density of water changes with temperature.

Questions 459–462 refer to the procedure of ultracentrifugation in which the plasma membranes of cells are disrupted leaving the organelles intact. The resulting homogenate is centrifuged, the supernatant is removed and subject to greater centrifugation. This process is repeated several times, as the following diagram shows.

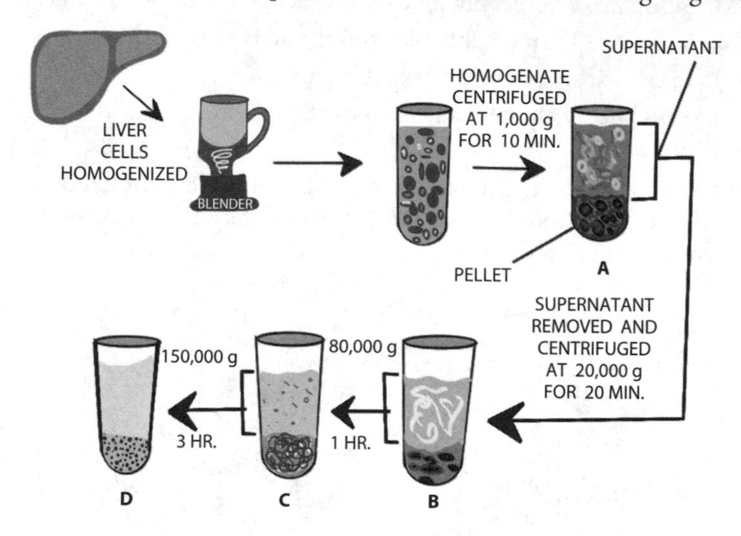

Centrifugation protocol	**1,000 × g 10 minutes**	**20,000 × g 20 min**	**80,000 × g 1 hour**	**150,000 × g 3 hours**
Fraction	**1 Pellet A Supernatant A**	**2 Pellet B Supernatant B**	**3 Pellet C Supernatant C**	**4 Pellet D Supernatant D**

459. Which of the following statements most accurately describes the purpose of the procedure?

(A) to fractionate cells into equal volumes for chemical analysis
(B) to separate cell parts based on their size and density
(C) to isolate different parts of the cell for electron microscopy
(D) to separate parts of the cell based on their function

460. Which fractions would likely contain DNA and RNA polymerase?

(A) Pellet A
(B) Pellet B
(C) Pellet C
(D) Pellet D

461. Which supernatant(s) would contain the enzymes of glycolysis?

(A) A only
(B) A and B, only
(C) A, B, and C, only
(D) A, B, C, and D

462. In which pellet would mitochondria and free ribosomes be found?

(A) Mitochondria and free ribosomes would both be found in pellet B.
(B) Mitochondria would be found in pellet B, and free ribosomes would be found in pellet D.
(C) Mitochondria would be found in pellet D, and free ribosomes would be found in pellet B.
(D) Mitochondria and free ribosomes would both be found in pellet D.

Questions 463–465 refer to the following information.

Antibodies are the one of the most abundant proteins in the blood (they make up about 20% of the plasma proteins). One of the five classes of antibodies is shown as follows.

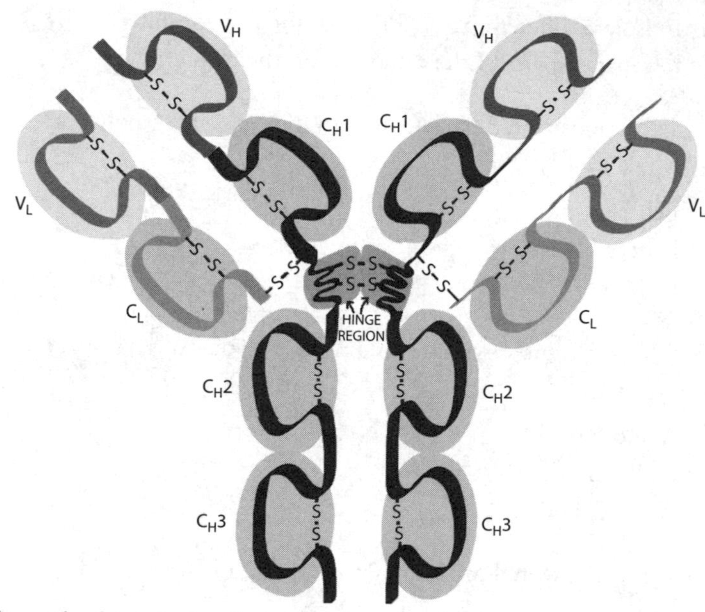

Each antibody contains four polypeptide chains: *two identical heavy chains* and *two identical light chains.* Each heavy chain has a 110 amino acid variable region (V_H) and a 330–440 amino acid constant region (C_H1, 2, and 3). Each light chain has a 110 amino acid variable (V_L) region and a 110 amino acid constant region (C_L). There are several disulfide bonds holding the structure together (indicated by S—S).

These antibodies are made by B-lymphocyte cells. During development of a B cell, the randomly chosen V gene is moved to lie next to a J gene segment (3 in the diagram). The extra J segment (J is joining segment) and the intron are transcribed and then removed by splicing, and the V3 and J3 are joined together.

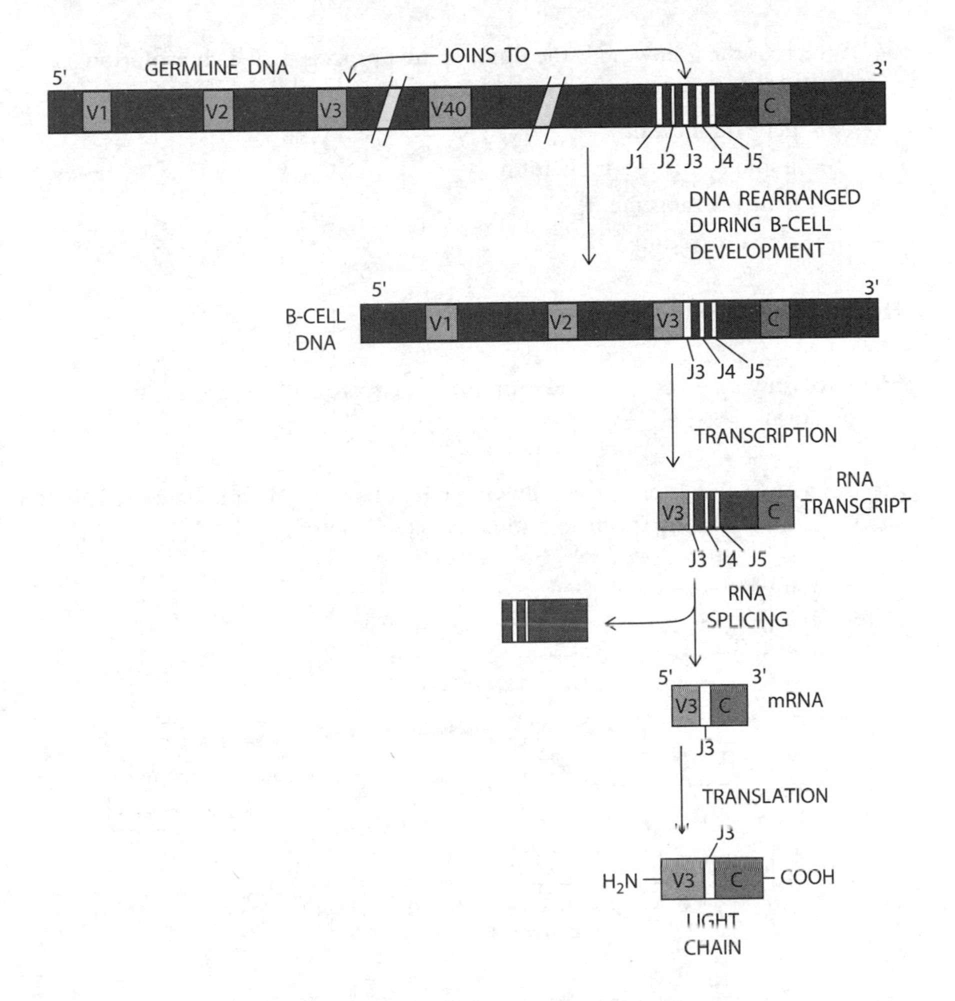

463. A human can make more than 10^{12} different antibody molecules. Which of the following most accurately explains how a human can make more kinds of antibodies than it has genes?

(A) Cells use gene duplication, mutation, and natural selection to make several kinds of proteins from a single gene.

(B) Each antibody gene is expressed multiple times during development.

(C) Antibody gene segments are assembled in different combinations.

(D) Cells use homologous recombination and gene rearrangement to make several kinds of proteins from a single gene.

464. Which of the following is the fundamental process by which evolution produces such a great variety of genes for heavy and light chains?

(A) gene duplication
(B) homologous recombination
(C) gene rearrangement
(D) clonal selection

SHORT FREE-RESPONSE QUESTION

465. Explain why diversity of the immune system is an advantage to an organism.

Questions 466–472 refer to the following data of an enzyme mediated reaction. Data was obtained using spectrophotometry (colorimetry).

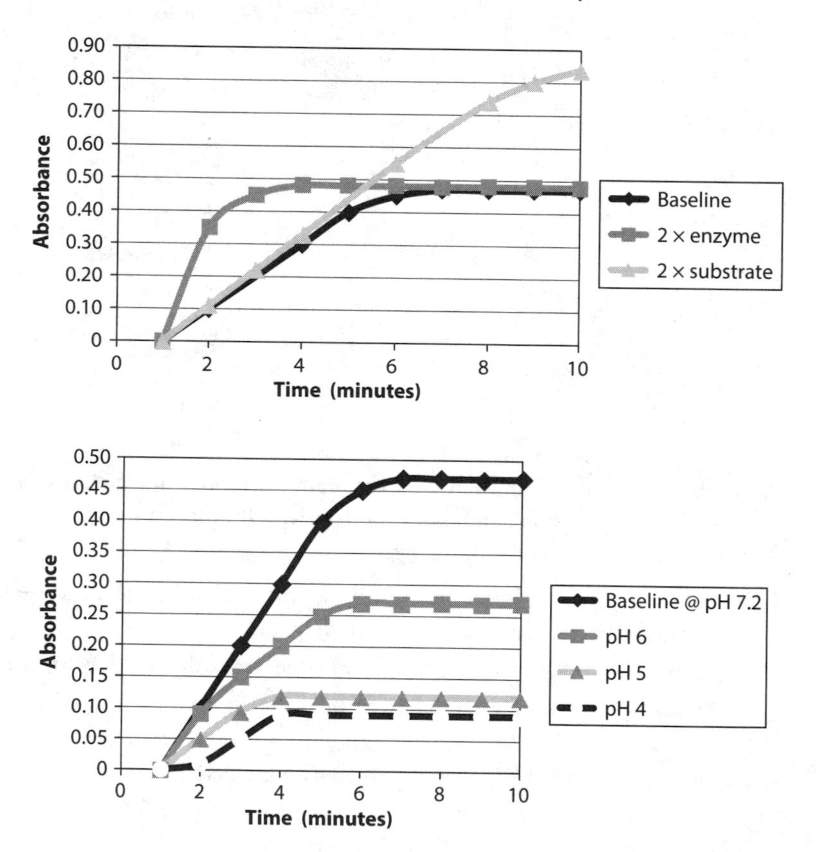

MULTIPLE-CHOICE QUESTIONS

466. Which of the following most accurately describes the purpose of the baseline curve?

(A) The baseline establishes a standard for a reaction.
(B) The baseline represents the ideal conditions for the reaction.
(C) A baseline curve is necessary to calculate absolute reaction rates.
(D) A baseline curve establishes the calibration state of the spectrophotometer.

467. At which of the following pH values was the baseline likely established for the first set of reactions?

(A) 4
(B) 5
(C) 6
(D) 7.2

468. Is there a limiting factor for this enzyme reaction?

(A) Yes, it is the substrate concentration.
(B) Yes, it is the enzyme concentration.
(C) There is no limiting factor in the reaction.
(D) There is a limiting factor, but this data set cannot be used to identify it.

GRID IN

469. Calculate the relative reaction rate from 0 and 4 minutes for the reaction that takes place at pH 5 compared to pH 6. Report your answer to the tenths place.

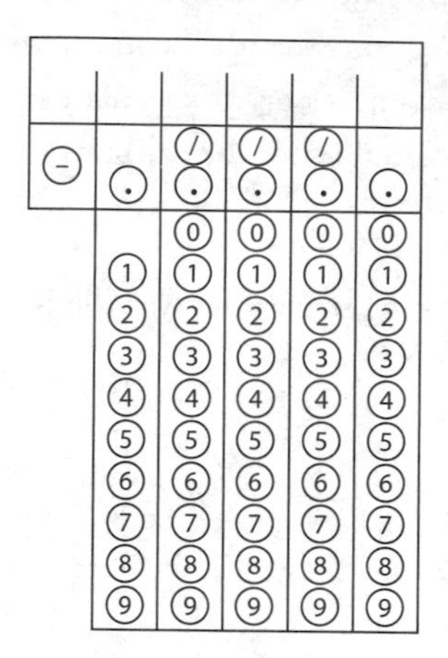

MULTIPLE-CHOICE QUESTIONS

470. Which of the following is a reasonable conclusion comparing the 2× substrate curve to the baseline?

(A) Individual enzyme molecules convert substrate to product over and over again.

(B) Increasing the substrate concentration increases the reaction rate for up to 10 minutes.

(C) Increasing the substrate concentration only increases the reaction rate at low substrate concentrations until the enzyme is saturated.

(D) This enzyme becomes saturated at approximately 5 minutes at a rate of $0.08 \frac{\text{abs}}{\text{min}}$.

471. Which of the following statements is the most accurate regarding the rate of this enzyme reaction?

(A) It is sensitive to pH but not substrate concentration.
(B) It is sensitive to pH, but the data do not provide enough information about its sensitivity to substrate concentration.
(C) It is not sensitive to pH or substrate concentration.
(D) The rate is sensitive to acidic conditions only.

LONG FREE-RESPONSE QUESTION

472. The Lock and Key model of enzyme action was proposed by Emil Fischer in 1894. Decades later (in 1958), the model was modified as Induced Fit. The refinement took into account observations that indicated that enzymes are flexible proteins, that their active site shape changes upon interaction with the substrate, and that the substrate may need to change shape to fit into the active site.

- **Explain** the purpose of models in science.
- **Describe** two observations that support the Lock and Key model.
- **State** one (benefit) that Induced Fit may provide to cells.

SHORT FREE-RESPONSE QUESTION

473. Suppose an enzyme was found in fungus that performed the same reaction as in human cells. **State** how you could determine whether the enzyme evolved once in one of the organisms and was modified later by the other, or whether the two enzymes evolved independently from each other.

MULTIPLE-CHOICE QUESTIONS

474. Many chemolithoautotrophic bacteria live at the hydrothermal vents at the bottom of the ocean. They are aerobic autotrophs that derive their energy from the oxidation of compounds such as sulfur, sulfides, hydrogen, and iron. Which of the following is likely true of these organisms?

(A) They perform cellular respiration exactly the same as in mitochondria.
(B) They can synthesize cysteine, a sulfur containing amino acid, via photosynthesis.
(C) They fix sulfur and iron into organic compounds by a type of photosynthesis.
(D) They have an electron transport system that uses oxygen as the final electron acceptor.

475. Which of the following is *not* a likely mechanism by which proteins produced by thermophilic bacteria maintain their stability at high temperatures?

(A) greater number of hydrogen bonds
(B) additional and stronger electrostatic interactions
(C) a greater number of nonpolar amino acids
(D) increased number of salt bridges (ionic interactions between amino acid side chains)

CHAPTER 5

Lab-Based Questions

Questions 476–481 refer to the following information.

The rate of photosynthesis can be measured indirectly using the **floating disk method**. In this procedure, small disks of leaves are cut using a hole-puncher (~8 mm in diameter). The air is removed from the disks with a vacuum and the disks are then immersed in a solution of bicarbonate ions, a carbon source for photosynthesis. At first, the disks sink in the bicarbonate solution, but when they are provided sufficient light, the oxygen produced during photosynthesis increases the buoyancy of the disks, causing them to float.

476. Construct a graph of the number of disks floating versus time using the information in the following table.

Time	Number of disks floating
0	0
2	0
4	1
6	2
8	4
10	6
12	6
14	7
16	9
20	10

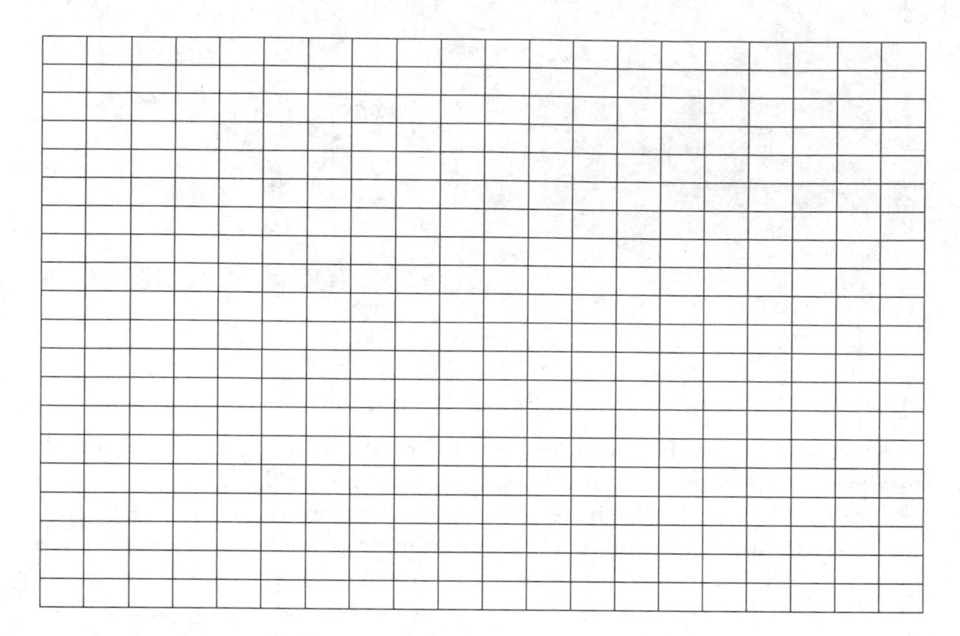

477. Calculate the ET_{50}, the time it takes for half the discs to be floating.

478. Construct a graph of the disk ET_{50} versus light intensity using the information in the following table. Light intensity is measured as FT-C. A higher FT-C value indicates a greater light intensity.

ET_{50}	Light Intensity (FT-C)
22	100
18	300
13	500
8	700
3	1,000

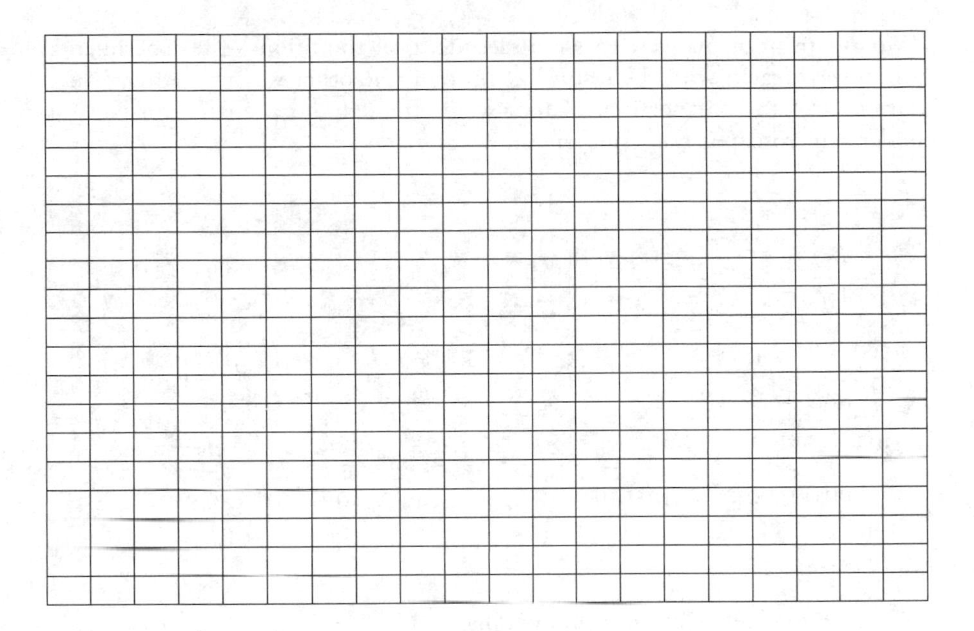

479. Calculate the slope of the line in the graph.

- **Explain** what the slope of the line represents.
- **Include a statement** about the *sign* of the slope.

480. When the disks are first punched, the spongy mesophyll layer of the leaf contains gases like CO_2 and O_2. These gases are removed from leaf disks prior to their immersion in the bicarbonate solution.

- **Explain** how the time of flotation of the disk is related to the photosynthetic rate.
- **Explain** what you would expect if the vacuum was not applied prior to the experiment.

481. State *two* environmental variables, other than light intensity, that affect the net rate of photosynthesis. **Explain** *how* they would affect the rate (increase or decrease) and **why**.

Questions 482–484 refer to the following information and experiment.

Substances present in the environment can affect mitosis. It was observed that a particular population of soybean plants (*Glycine max*) suffered decreased growth and productivity during a period of heavy rainfall. Their roots, in particular, were poorly developed. It was discovered that the secretion of a lectin-like protein by a fungal pathogen induced mitosis in some apical meristems.

Two groups of plants were grown under identical conditions. The experimental group was treated with the lectin-like protein; the other was not. After several days of growth, thin sections of root were sampled and prepared for viewing under a microscope. The results are shown as follows.

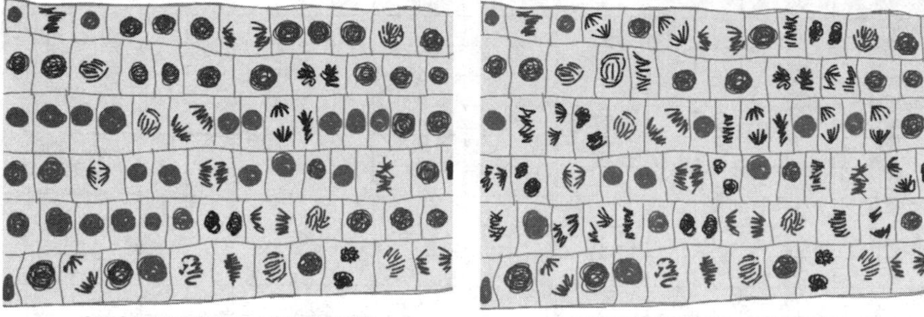

482. Estimate

- the percentage of cells in interphase
- the percentage of cells undergoing mitosis

in each of the two groups.

483. State whether the results of the experiment support the action of the lectin-like protein as a mitogen (a substance that encourages cell division).

- **Calculate** the chi-square (χ^2) value for the experiment.
- **State** the null hypothesis.
- **State** whether the null hypothesis can be rejected.
- **Justify** your response.

	Degrees of freedom							
P	**1**	**2**	**3**	**4**	**5**	**6**	**7**	**8**
0.05	3.84	5.99	7.82	9.49	11.07	12.59	14.07	15.51
0.01	6.64	9.32	11.34	13.28	15.09	16.81	18.48	20.09

484. State a hypothesis that explains how a protein that acts as a mitogen could result in poorly developed roots in *Glycine max.* **Describe** an experiment that would test your hypothesis.

Questions 485–487 refer to the following data. A *DNA ladder* is a standard used to estimate the size of DNA fragments by gel electrophoresis when run concurrently with the fragments to be measured.

485. Construct a graph of fragment size versus distance traveled using the data in the following table.

Fragment size (base pairs)	Distance traveled (mm)
10,000	6
8,000	10
6,000	13
5,000	18
4,000	22
3,000	28
2,500	37
2,000	42
1,500	54
1,000	69
700	82
500	92
300	104

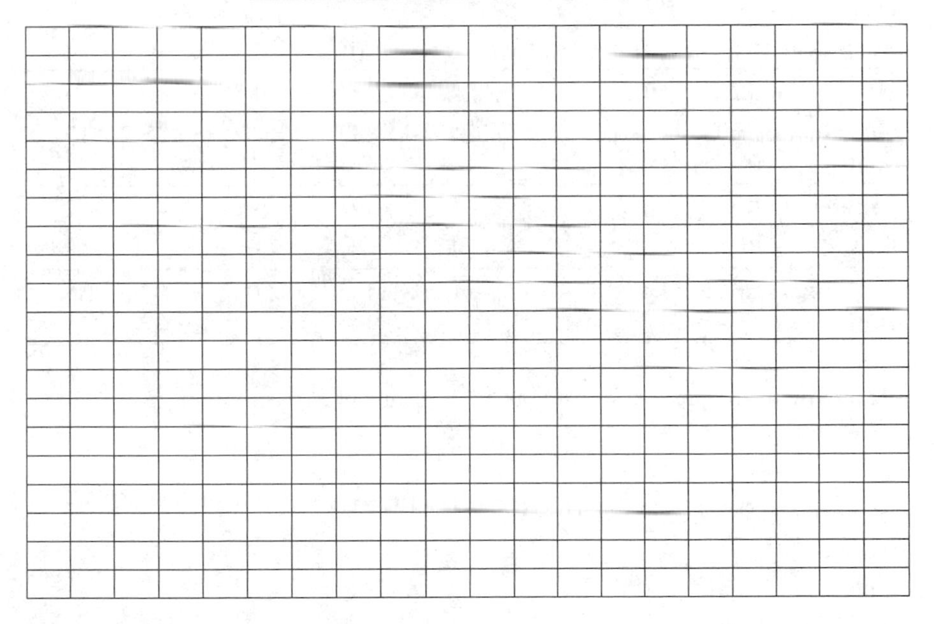

486. Estimate the fragment size of the molecule that traveled 60 mm.

487. Describe the relationship between fragment size and distance traveled.

Questions 488–494 refer to the following data. Three respirometers were prepared for three different temperatures. Each temperature had one respirometer containing pebbles, one containing dry peas, and one containing germinating peas. The respirometers were prepared with KOH to collect and solidify the CO_2 gas produced by respiration (so the gas is not responsible for any pressure changes during the experiment). The movement of water indicates a decreased pressure in the air space of the respirometer where the pebbles or peas are contained.

The following data indicate the **total number of millimeters the water** in the respirometer moved as a result of the pressure changes.

	0 min (mm)	5 min (mm)	10 min (mm)	15 min (mm)	20 min (mm)	25 min (mm)	30 min (mm)
Pebbles at 22°C	0	1	1	2	2	3	3
Dry peas at 22°C	0	1	2	3	3	4	5
Germinating peas at 22°C	0	3	6	9	14	19	23
Pebbles at 10°C	0	1	2	3	3	3	4
Dry peas at 10°C	0	2	3	3	4	4	5
Germinating peas at 10°C	0	2	4	7	11	14	17
Pebbles at 30°C	0	1	1	2	3	3	2
Dry peas at 30°C	0	2	4	5	7	8	8
Germinating peas at 30°C	0	6	11	17	22	28	36

488. Explain the purpose of using pebbles in one of each group of respirometers.

489. Adjust the data in the following tables for the germinating peas at each temperature.

	0 min (mm)	5 min (mm)	10 min (mm)	15 min (mm)	20 min (mm)	25 min (mm)	30 min (mm)
Pebbles at 30°C	0	1	1	2	3	3	2
Germinating peas at 30°C	0	6	11	17	22	28	36
Germinating peas at 30°C Adjusted							

	0 min (mm)	5 min (mm)	10 min (mm)	15 min (mm)	20 min (mm)	25 min (mm)	30 min (mm)
Pebbles at 22°C	0	1	1	2	2	3	3
Germinating peas at 22°C	0	3	6	9	14	19	23
Germinating peas at 22°C Adjusted							

	0 min (mm)	5 min (mm)	10 min (mm)	15 min (mm)	20 min (mm)	25 min (mm)	30 min (mm)
Pebbles at 10°C	0	1	2	3	3	3	4
Germinating peas at 10°C	0	2	4	7	11	14	17
Germinating peas at 10°C Adjusted							

490. Use the **adjusted data** in the tables to **calculate** the rate of respiration in the germinating peas at each temperature.

491. **Calculate** the rate of respiration in the dry peas. (Adjust the data appropriately to be comparable with the germinating peas.) **Compare** the respiration rates of dry and germinating peas at each of the three temperatures.

492. **Explain** the difference in respiration rates between the dry peas and the germinating peas.

493. **Graph** the total distance of water movement versus time for the germinating peas at each of the three temperatures.

- **Draw** a best-fit line for each of the curves.
- **Calculate** the slope of the best-fit line.

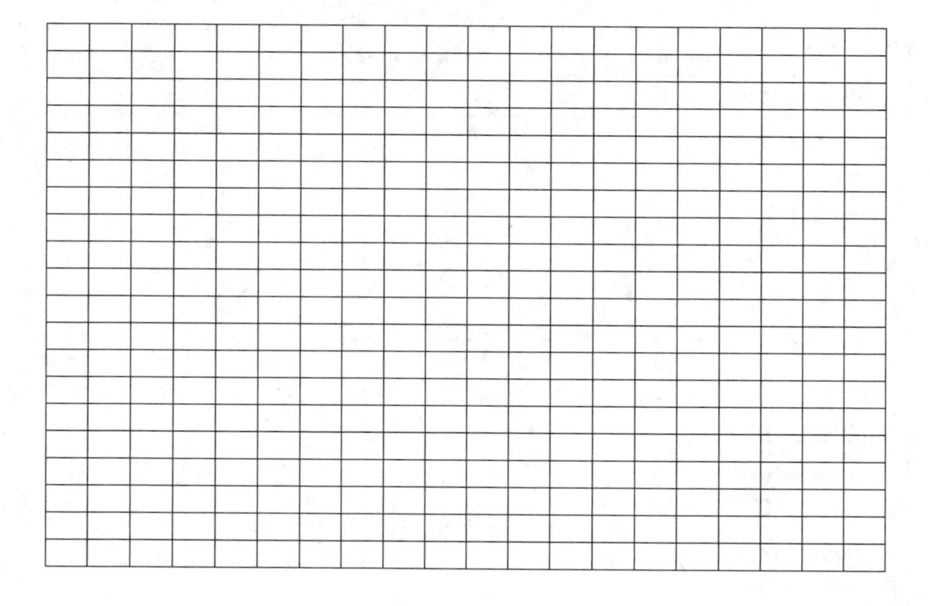

494. **Explain** how the best-fit line describes the rate of respiration.

- **Compare** the rates obtained graphically to those calculated from the table.
- **State** one reason the methods may result in different calculated rates.

Questions 495–498 refer to the following graph of bacterial growth over time. Bacteria were grown in media in a laboratory. They were maintained at 37°C for 18 hours.

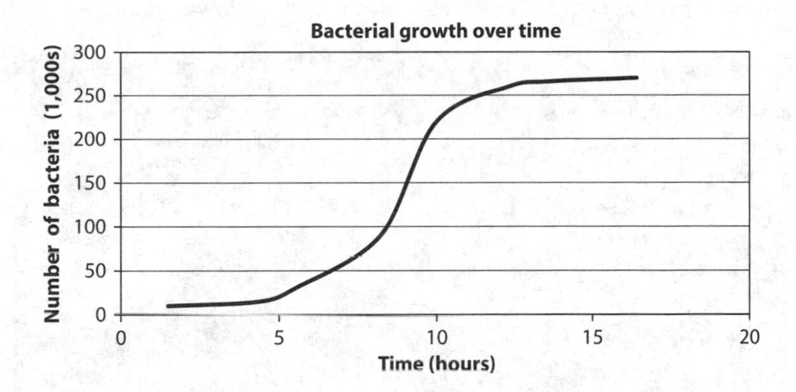

495. Explain the factors that determine the shape of the growth curve.

496. Estimate and **compare** the growth rate between:

- 6 and 8 hours
- 8 and 10 hours
- 10 and 12 hours

497. Bacteria are an important model organism in molecular biology research. **State two reasons** bacteria are useful in the study of eukaryotic cellular processes.

498. Explain *two* major limitations of the use of bacteria in studying eukaryotic gene expression.

Questions 499–500 refer to the following information and data.

The haploid fungus *Sordaria fimicola* exchanges genetic material when mycelia from two fungi meet and fuse. The resulting zygote undergoes meiosis to form spores containing asci. Each ascus contains eight haploid spores. Spore color, tan or black, is determined by one gene.

In this experiment, crosses between wild-type black and wild-type tan *Sordoria* produce asci with four black and four tan spores. If no crossing over occurred during meiosis, the asci will contain four black and four tan spores in a 4:4 pattern (parental types). If crossing over *did* occur during meiosis, the asci will contain four black and four tan spores in a 2:2:2:2 or 2:4:2 pattern (recombinants). The recombination data is used to calculate the distance of the gene from the centromere. The results of the crosses are given in the following table.

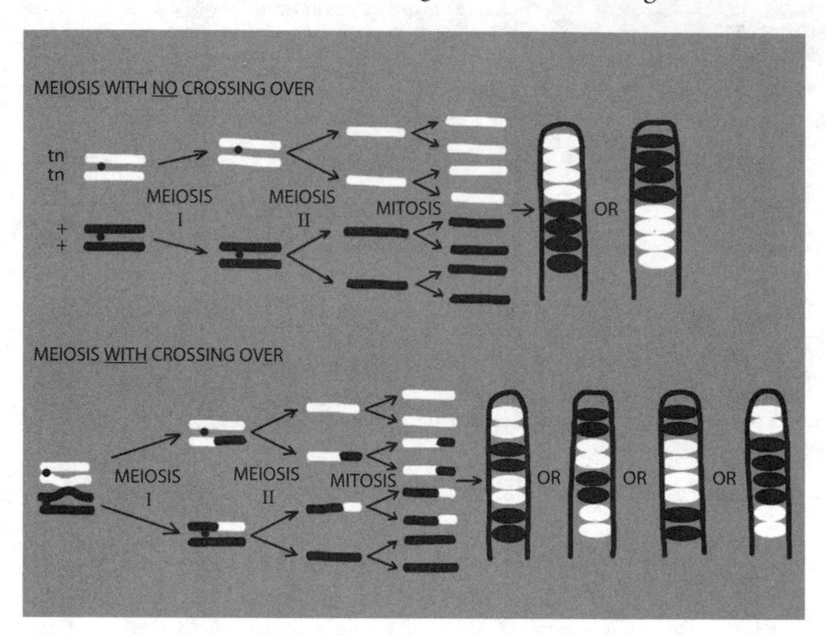

Number of 4:4 (parental types)	Number of 2:2:2:2 or 2:4:2 (recombinants)	Total	% Asci showing crossover / 2	Gene to centromere distance
240	260	500		

499. Calculate the recombination frequency and **state** the distance between the gene and the centromere in map units (one map unit = 1% recombination frequency).

500. Explain the relationship between distance between the gene and the centromere on the chromosome and the recombination frequency.

Congratulations! You've finished! ☺

ANSWERS

Chapter 1: Evolution Diversity Unity

1. (D) The passing down of DNA from one generation to the next is the mechanism by which all life is linked. Changes in DNA sequences (mutations) produce new genes and alleles, which can be inherited by offspring through reproduction. The mechanism by which genes are used inside cells is so similar across diverse forms of life that when genes from one organism are introduced into the genome of a different organism, they may be expressed: for example, genes for spider silk expressed by goats. Although all life-forms have the capacity for growth and reproduction, this fact alone supports but is not evidence that all organisms are linked by lines of common ancestry. That different species have specific traits and adaptations is a result of evolution through natural selection, but again, this fact alone is not direct evidence of common ancestry. The true statements of choices A, B, and particularly C actually require the model of natural selection to make sense.

2. (C) Natural selection works by reproduction, heritable variation, and selection. Humans lack tails *not* because tails weren't needed, but because either having one was a disadvantage and/or not having one was an advantage for our ancestors. **The lack of selective pressure for or against a particular structure can produce a vestigial structure** (our wisdom teeth and appendix are typically classified as vestigial structures). A vestigial structure is present in a rudimentary form. Typically, it has been *reduced* in size, and perhaps complexity, from a more functional structure in the ancestor. Over the course of many generations the structure has become smaller and less functional but has not been selected *against* enough to "completely disappear."

3. (B) One way to construct a phylogeny from a character table is to first **find the organism that has the least number of traits listed.** In this case it is the lamprey. That is the "first" organism on the phylogeny and represents the most ancient or ancestral species. The tuna has a vertebral column like the lamprey, but it also has hinged jaws, so the tuna is "next," and then the frog, turtle, and dog. The hinged jaw is a derived trait (or character). It is not present in the ancestor, though the ancestor may possess a character from which that trait evolved. **The organism that possesses the greatest number of derived traits listed is likely the most recent to evolve.** In this case, it is the dog.

In phylogenies, the organism that has the longest line attached to it is proposed to be one that evolved first. Of course, **all species present today have been evolving for the same 3.8 billion years,** but not necessarily in their current form. In other words, the oldest fossil of a lamprey is 360 million years old, but dogs are thought to have diverged from wolves anywhere from 15,000 to 140,000 years ago. Interestingly, lampreys have changed very little over 360 million years, while dogs have undergone a massive diversification due to selective breeding.

4. (D) A graph like this is **suggestive of a cause-and-effect relationship** but is **not proof.** There are *two things compared over time.* Importantly, the two *y*-variables, for the most part, increase and decrease concurrently (with a reasonable time lag between changes in drug use and changes in prevalence of resistant bacteria as the bacteria need time to adapt.). For example, the increased drug volume from 1986–1990 correlates with an increase in the prevalence of resistant bacteria. However, the reduction of drug volume in 1985 and 1991 is accompanied by a decrease in the prevalence of resistant bacteria.

Although the relationship is *not perfectly correlated* (e.g., in the year 1984, a decreased drug volume corresponded to very little change in resistance), the general trend shows that as drug volume increased, the prevalence of resistant strains increased with some lag time, and when drug volume decreased, the prevalence of resistant strains decreased. **This graph alone allows us to say nothing of a causal relationship between the two factors but knowledge of evolutionary theory and antibiotic resistance does allow us to select the correct answer.**

5. (C) Directional selection is the type of selection, whether from the environment or by humans (who are technically part of the environment), in which **one extreme of the phenotypic continuum is selected for (or against).**

6. (B) Generally, **the lower the percent divergence of DNA sequences between two species, the more recently they diverged from their common ancestor.** (Or, the greater the number of differences between two sequences, the further back in time they diverged from their common ancestor.) *Homo,* the genus of humans, has a percent divergence of 0.38%, which represents the percent difference between two human sequences (as noted beneath the table, humans are the only extant species of the *Homo* genus). **This difference does not reflect the difference in the entire genome, just the percent sequence difference between each individual's ψη-globin pseudogene.** The difference in the percent divergence between *Pan* and *Pongo* is 3.42%, and the difference in the percent divergence between *Pongo* and *Gorilla* is 3.39%. That's only 0.03% difference, less than the difference between the two humans.

The concept of a molecular clock assumes a steady rate in the change in DNA sequences over time. It provides at least a relative basis for estimating the time of divergence between lineages. The assumption of steady rate changes is only true under very specific circumstances. For example, **pseudogenes undergo a faster rate of change in DNA sequence because the mutations have no effect on genotype.** Some DNA sequences, considered conserved, accumulate very few mutations because they are almost always deleterious. See answers 67 and 94 for more information about molecular clocks.

Different species and groups of organisms undergo mutations at different rates because of differences in body size and generation time. It takes a lot more time for a hundred generations of elephants to pass than for a hundred generations of bacteria, so even if the same number of mutations occur in a hundred generations, the molecular clock of the elephant is much slower than that of the bacteria. The adjustment for generation time (which is correlated to body size) is one of several factors that are required for calibration of the molecular clock.

See the following box for a strategy for answering questions asked in the *negative* and using the process of elimination.

A General Strategy for Answering *Negative* Questions (questions that ask which of the choices is *not* true)

Negative questions can be tricky, even if what they are asking you about is not, because **your brain doesn't *think* in the negative.** As a result, you often end up choosing an answer that is true even though you were looking for one that was false.

Good habits to get into when practicing these questions are as follows:

(1) **CIRCLE the words *not true* or *except*** in the question to remind yourself that you are looking for a false statement.
(2) **Assess each answer choice as either *true* or *false*** (or ? if you're not sure), marking each choice T or F as you go.
(3) When you've read all the choices, **see which one doesn't fit** (the *false* one) and that's your answer!

The Process of Elimination

Sometimes *you're just not going to know the answer,* but you can probably eliminate one, two, or even three answers. **There's no ¼ point penalty for guessing on the AP Biology exam** (you just don't get the point for getting it correct), so eliminating *any* answer choices is a great start.

If you eliminate three choices but you're not sure if the remaining answer is true, then you're probably choosing the correct answer—as long as you're confident the other three are incorrect. On multiple-choice exams, knowing what's *not true* is often just as helpful as knowing what *is true*.

7. (D) Although all the choices are ways to compare, contrast, and attempt to relate organisms by common ancestry, **DNA sequence comparisons are the "gold standard."** Convergent evolution can produce similarities in anatomy and behavior (choices B and C). **Convergent evolution only makes the organisms being compared *appear* to be related** (e.g., a shark and dolphin look quite similar, but the shark is a cartilaginous fish and the dolphin is a mammal). Importantly, organisms do not become more related as time passes; they can only maintain their degree of relatedness or through future generations become more distantly related through divergence.

8. (B) Curve I has two approximately equal phenotypes; curve II shows three phenotypes, one for each genotype. A_1 is dominant to A_2 in curve I, but curve II shows additive inheritance (the heterozygote's phenotype is right in between that of the two homozygotes, like codominance or incomplete dominance). The curves indicate that the homozygotes produce approximately the same phenotype for the same amount of gene product. **Curve I shows that for the same amount of gene product, a different phenotype is produced.**

9. (A) The experiment involved *deleting a single gene,* calculating the number of traits affected by the deletion of this gene, and measuring the relative fitness of the yeast as a result.

The graph shows that *at least one gene affects the expression of over 120 traits*! However, the relative fitness of that strain is not the lowest. *The lowest relative fitness of any of the strains appears to be approximately 0.45 and as few as perhaps 5 traits affected.*

Relative fitness is the fitness of a genotype relative to a reference genotype that is assigned the value of 1. Some of the deletions produced strains of yeast whose relative fitness was greater than the original strain, and one of those deletions affected almost 100 genes!

See the box after answer 6 for a strategy for answering questions asked in the *negative* and using the process of elimination.

10. The common ancestor of all the terrestrial vertebrates, including snakes, is a lobe-finned fish. This type of fish has fleshy (muscular) paired fins joined to the body by a single bone that can rotate in the joint socket. *The ability to rotate their fleshy fins, instead of just wave them back and forth, allowed them to make the transition onto land.* This joint eventually gave rise to your shoulder socket by divergent evolution. The lobe-finned fishes have paired appendages: one pair is anterior (toward the head) and the other, posterior (toward the tail). Our arms and legs derive from these paired appendages.

Tiktaalik, a genus of extinct sarcopterygian (a group that include the coelacanths, lungfish, and tetrapods), is the most recently discovered ancestor that "walked" out of water about 360 million years ago (the late Devonian). *Tiktaalik* is considered the link between fish and amphibians.

There are a small but significant number of terrestrial vertebrates that lack legs. These include snakes, several species of legless lizards, and caecilians, but there are several more. Limblessness was selected for in these organisms. Let's not forget the cetaceans, the whales and dolphins, which descended from terrestrial vertebrates. **These species are still descendants of the four-limbed ancestor, but the reduction of limb size (even to no apparent limbs at all) would be accompanied by vestigial structures.** For example, snakes have tiny hind leg bones buried in muscles toward their tail end.

The AP Biology exam does not expect you to know this level of detail. The main point to make here is that all the terrestrial vertebrates evolved from a common ancestor with four limbs (or fins). The limbs of all the terrestrial vertebrates are homologous with each other and are derived from this ancestor.

11. (B) The replica plating technique established that mutations occur in bacteria "randomly" and without prior exposure to the antibiotic (or other selective agent). The bacteria that survived penicillin were resistant to it before they were ever plated with it (in other words, before they were exposed to it).

Although mutation rates have been shown to increase in some species in response to increased environmental stress, most biologists do not accept the concept of "directed" mutations—mutations induced in a gene to produce a specific allele that would help an organism (or, more likely, its offspring) to be better equipped to handle its environment. *If we've learned anything from biology, however, it's that the staggering diversity holds many unseen things.* All biologists have been wrong about something. So stated carefully: Although directed mutation may be *possible*, it has not been observed repeatedly or predictably and so it is not considered to be true. Whether future biologists believe this remains to be seen.

The Meaning of the Word *Random*

In science, *random* does not have the same meaning as in common language (just like the terms *theory* and *spontaneous*). ***Random* refers to a situation in which there may be several outcomes, none of which are entirely predictable.** As scientists, we test our hypotheses and theories by comparing *what we predict* with *what actually happens*. We describe the confidence we have in our predictions with statistical probabilities. For example, the probability that a coin toss will produce a tail is ½. We can never know which side will face up until the coin actually lands, but we can be 50% sure it will be tails (or heads).

Can we be 100% sure it will be one of them? No—**practically all phenomenon have stochastic (unpredictable) factors and non-random or deterministic (predictable) factors.** There is a very small (but non-zero) probability that all of the conditions would be right to cause the coin to fall exactly on its side or that while the coin is in the air someone will come by and grab it (maybe the probability of the latter is not so small).

12. Evolution by natural selection and genetic drift is a *fact*. The word *theory* in science refers to a *theoretical framework* or a *model* in which to understand natural phenomenon. It is *not* just an idea. See the box before answer 13.

Line of evidence	Examples	How it supports the theory of evolution
Fossils/ biogeography	• Almost all the mammals in Australinea (Australia and nearby islands such as Papua New Guinea) are marsupials, whereas the rest of the world's mammals are mostly placental, with the exception of the opossum. • The fossil record shows that life existed in many forms and did not always appear on Earth as it does now. • Most species that existed on Earth are now extinct. The remains of these organisms appear in rock layers according to age. • The fossil record matches many predicted sequences of events such as the evolution of wingless insects before winged insects.	• The distributions of many taxa don't make sense unless they are viewed as having arisen from common ancestors. The distribution of marsupials and placental mammals suggest that Australinea's mammals likely evolved from a single introduction of an opossum-like founder (probably via Antarctica from South America). • Critical extinct intermediates of life-forms such as *Tiktaalik* and feathered dinosaurs are evidence of both mosaic (independent evolution of characters within a lineage or clade) and gradual (phenotypic characters have evolved major changes through many slightly different intermediates over many generations) evolution.
Morphological homologies, vestigial structures	• Homologous structures result from divergence of related species due to difference selective pressures (human arm bones and bat wing bones, for example). • Vestigial structures provide evidence of ancestral features that have lost their functions but are retained through evolution.	• Similarities in structure, regardless of the differences in function, make sense if we consider that the characteristics of organisms have been modified from their ancestors. • Modifications of pre-existing designs explain why many features of organisms are suboptimal for their task.

Line of evidence	Examples	How it supports the theory of evolution
Biochemical and genetic similarities	• The genetic code is universal, yet the codon sequences are seemingly arbitrary. • ATP is the common energy source in all cells. All cells use the same 20 amino acids (and the L-isomers) to make proteins and the same nucleotides in their DNA and RNA. • All cells use double-stranded DNA to encode their genetic information and RNA to express the information contained in DNA. • Chemiosmosis is a process for energy transformation in all organisms. • Glycolysis is a metabolic pathway common to all cells. Many other metabolic pathways are common to many or most cells.	• Most of these observations only make sense as a consequence of common ancestry.

The Meaning of the Word *Theory*

A **theory** in science, unlike in common language, is a big idea, **a theoretical framework,** that encompasses many other ideas and hypotheses. A theory has **explanatory power**. It can be **used to make predictions** about scientific phenomenon and is necessary to design experiments.

You use models all the time to make predictions about the world; however, unless you have been collecting data, subjecting it to peer review and statistical analysis, your theories about the world are not scientific. Even non-scientific theories can give you insight to how valuable a scientific theory is. When a detective analyzes a crime scene, she or he will construct a theory that will help determine the motive of the individual, where other clues may be found, and so on. Your theory or model of people may help you figure out what a nice birthday gift for them would be.

In science, **a theory holds a lot more weight than a fact,** which is one tiny piece of information. **A fact, without context, is not very useful. A theory is what puts all the facts together.**

All evidence collected from multiple lines of inquiry supports evolutionary theory. That evolution happens is a *fact*. Evolutionary theory explains how evolution works and is essential to putting all the unity and diversity of life into context. It is a useful lens through which to view every subdiscipline of biology.

There is no greater unifying framework in biology than evolution.

13. Mammals and birds arose from different reptilian lineages, so endothermy likely evolved separately (independently) in the two classes, and its evolution in birds and mammals is probably **analogous**. The common ancestor between the birds and the mammals was too long ago for endothermy to be an ancestral trait in birds. However, the mode of body temperature regulation in *Saurischia* is not known.

Character versus Trait

The terms *character* and *trait* are sometimes used interchangeably and sometimes used to distinguish between a gene and an allele. Although not everyone uses the terms exactly the same way, in this book and generally:

- a **character** is a **feature** or **trait.**
- a **character state** is **one of the variant conditions of a character**.

The **character is coded for by a gene** located at a specific location (loci) or position on a chromosome. **The allele codes for the particular character state** and has a specific nucleotide sequence.

14. Evolution can be defined as a change in allele frequencies over time.

- **Genetic drift can change allele frequencies**, though the changes are not driven by "selective pressures" and are therefore, **not adaptive**. Genetic drift is *not the main driving force of evolution,* but it has a significant effect nonetheless. (See answers 22, 23, 36, and 51 for more information on genetic drift.)
- **Natural selection** changes allele frequencies because the organisms that possess certain alleles (or allele combination) have a reproductive advantage over those that don't. The reproductive advantage results in a greater number of members in the following generation that also possess the alleles (or allele combinations).

Mechanism of evolution	How they are similar	How they are different
Natural selection	Both change allele frequencies in a population.	Adaptive
Genetic drift		Random

15. (C) In the widowbirds, selection for greater tail length in males is positive because the data show that males with the longer tails have more nesting sites than males with normal length and short tails. This can also be considered negative selection for short tails in males, but that is not an answer choice. Choices A, B, and D are incorrect because there is no data regarding the reproductive success of the female birds.

This is an example of **sexual dimorphism.** This term commonly refers to phenotypic differences in secondary sex characteristics between males and females of the same species (as opposed to differences in their reproductive organs). **Sexual dimorphism is usually the result of sexual selection,** where one sex, often but not always the female, selects the opposite sex (often, but not always males) with particular characteristics. A common type of dimorphism is ornamentation, like the peacock's ornate plumage compared to the peahen's less dramatic size and coloration.

16. **(C)** The data show only the average number of nests per male, which is an indicator of the number of female mates. Choices A, B, and D are incorrect because technically, reproductive success is measured by offspring sired, not the number of nesting sites (though they are likely correlated). Because there is *no data about number of eggs or offspring,* we can infer that it is likely that increased tail length improves reproductive success in males (and the corollary that males with shorter tails leave behind less offspring). But the actual data are quite limited in showing only that males with longer tails have, on average, more nesting sites than males with shorter tails.

17. **(C)** **A control group should differ from an experimental group in only one variable,** which is why there are often multiple control groups. In this experiment, the "unmanipulated normal length" and the "cut and replaced normal length" are both controls. However, the "cut and replaced normal length" group is really a control for the "normal tail length" group, because it can be used to distinguish whether the fake tails affect female selection. Because the number of nesting sites between the "unmanipulated normal length" and the "cut and replaced normal length" groups are the same, it can be inferred that the fake tails do not affect female choice. If the number of nesting sites between these two groups were different, the way the rest of the data are interpreted would also be different.

18. **(A)** **One of the biggest problems facing terrestrial organisms is the possibility of desiccation (drying out).** Plants have the added challenge of being sessile (they don't move). Plant evolution began on land, and only after they were established in a terrestrial environment did some species begin to inhabit aquatic environments. (Remember that seaweed, kelp, and algae are not plants!)

Choice B is incorrect because mosses are plants and they live on land without vascular tissue (though their size is limited, as is their range—they must live near water to reproduce). Choice C is incorrect because **pollen and seeds** are not needed to survive on land, but they do **allow plants to reproduce in the absence of water** (swimming sperm make mosses and ferns dependent on water for reproduction), an adaptation needed to disperse into a greater diversity of territories. Choice D is incorrect because flowers are reproductive organs and fruits are seed dispersal structures. Both structures allow angiosperms to reproduce in the absence of water, but they *do not* provide energy to the plant that grew them. Fruits and flowers require energy to grow! **The stored food in the seed provides energy to the plant embryo.** The fruit helps disperse the seed, often by giving animals something to eat. **A fruit, by definition, is a structure that aids in seed dispersal.** Not all fruits are for eating! Many fruits are structures that allow seeds to get away from their parents by wind (the fruit is winged, like samaras), by water (the fruit is buoyant), or by sticking to animal fur (the fruit has hooks or burrs).

19. **(D)** Because **plants are sessile** (they can't move), choices A, B, and C are all problems the plant faces.

20. **(A)** Humans, whales, and bats are all mammals (and therefore vertebrates). The similarities in the bone structure of the forearm indicate common ancestry. As divergence occurred, the resulting populations adapted to their particular environments. For the bat, flight was an advantage and for the whale, swimming. **The differences in the structures account for their different functions in different environments.** (Though birds' and bats' wings are also **homologous**; they have slightly different structures and the same basic function.)

Stabilizing selection does not result in divergence, which is necessary for speciation and the production of homologous structures. Ecological succession is a relatively unrelated concept.

Homologous versus Analogous Structures

Type of structures	Structural comparison	Functional comparison	Example	Type of evolution
Homologous	Similar structure, "variations on a theme."	They may have the same or very similar function (bat wing, bird wing) or a very different function (human arm and whale flipper).	The bones of a human arm and a whale flipper are similar in arrangement and number.	Divergent
Analogous	Different structures.	Same function.	Bird wings have bones; insect wings do not.	Convergent

21. (B) Flight evolved in invertebrate insects (a class of arthropods) 406 million years ago and evolved completely independently in the vertebrates 160 million years ago. The common ancestor between the birds and the insects did not have wings. (Flight also evolved independently in the mammals!)

The structure of bird wings and insect wings are completely different. They share a similar function, flying, because the ability to fly was an advantage to both organisms. **Because the common ancestor of the birds and the insects did not have wings (or a wing-like structures), wings must have evolved independently in the two groups** (as indicated by their completely different morphology as well as the evolutionary history). This is an example of **convergent evolution,** which occurs when two different species adapt to their environments in similar ways (e.g., they evolve wings), but they *do not* become more genetically similar and do not have a recent common ancestor that possessed the trait in question.

- **Convergent evolution produces analogous structures.** How are you supposed to know that the common ancestor of birds and insects didn't have wings? Birds are chordates (and vertebrates), and insects are arthropods (invertebrates). You only have to trace the bird lineage back to its fish ancestors to know the common ancestor (which was hundreds of millions of years before the fish) didn't have wings. As far as the other answers are concerned, **homology is the result of divergent evolution,** so choices B and C are incorrect. (See the table after answer 20.)
- **Divergent evolution produces homologous structures.**

The AP Biology exam does not expect you to memorize the tree of life but given a subset of information, you should be able to infer common descent.

Of course we're all related to the universal common ancestor, a prokaryote that inhabited Earth 3.8 billion years ago, but multiple divergences allowed different populations to take different evolutionary paths to the present day.

22. (D) Genetic drift is the null hypothesis (the default explanation). Only if there is evidence that the differences within or between populations is due to another force acting on allele frequencies can another hypothesis, like natural selection, be considered. In other words, **unless evidence exists to the contrary, the allele frequency changes are assumed to occur by chance.** (See answers 14, 23, 36, and 51 for more information on genetic drift.)

23. (C) If no stabilizing force (such as selection) returns the allele frequency to 0.5, the frequency will wander or drift, eventually reaching 0 (lost) or 1 (fixed). If the allele is lost, it can return to the population only through mutation or gene flow (new alleles introduced into the population). Alternatively, once an allele has been fixed, the population is considered monomorphic for that allele, meaning there is only one version of the allele present and all members of the population are (effectively) homozygous. The frequency is 100%. **The rate of decline of heterozygosity is often used as a measure of the rate at which genetic drift is proceeding.**

An important assumption is that all alleles have an identical effect on fitness—they are not subject to natural selection. **At any time, an allele's probability of getting fixed or lost is equal to its frequency.** Just as the probability of tossing tails is ½ with every toss and all previous tosses can be ignored, with **each generation the allele frequency can be considered a "new starting point" from which to calculate or predict the probability of the allele being lost or fixed in the future.**

An allele is more likely to become fixed in a small population. If an allele has just arisen by mutation and is present in just one heterozygote individual, its frequency as a function of time is $p_t = {}^{1}/_{2N}$. You don't have to know that formula, but it is clearly going to become fixed in fewer generations at a small N as compared with a large one. **Generally, allele substitution proceeds more quickly in small populations.** (See answers 14, 22, and 36, for more information on genetic drift, and see the Random Walk under answer 51.)

24.

	–	1	5		

–15, with an acceptable range of –10 to –20.

The frequency of left-mouthed individuals in 1988 is greater than 0.4 but less than 0.5. The frequency of left-mouthed individuals in 1985 is 0.6, so the difference is between

0.4 – 0.6 = –0.2 (–20%)

and

0.5 – 0.6 = –0.1 (–10%).

This effect is known as **frequency-dependent selection, a phenomenon in which the fitness of a phenotype depends on how common it is in a population**. In this example, a left-mouthed fish always attacks on the right side of the fish it's feeding on. The prey fish become more adept to defending against attack from the most common phenotype of scale-eater in the lake. If most of the attacks come from the right, they guard attacks from the right. The result is that selection works *against* the phenotype that is most common (or favors the phenotype that is least common). This oscillation in selective pressure between two phenotypes based on their frequencies results in a balancing selection which keeps each frequency close to 50%.

What's a species?

The **biological species concept (BSC)** was introduced in 1942 by Ernst Mayr. He stated that "Species are groups of actual or potentially interbreeding populations, which are reproductively isolated from other such groups."

Most biologists do not require 100% reproductive isolation for qualification as a species; there is often "leakage" between populations, though the extent of which is usually relatively small. **The biological species concept emphasizes the process by which species arise** and is the most commonly utilized definition of a species.

The phylogenetic species concept (PSC) emphasizes new species as the outcome of evolution. They are the products of divergence.

Populations or species are reproductively isolated if the biological differences between populations greatly reduces gene exchange between them, whether or not they are geographically separated.

25. (C) The three essential elements of natural selection are reproduction, heritable variation, and selection. Without variation, there are no options to select from (eliminating choices A and B). Choice D is incorrect. Variations alone *cannot* result in a new species: It is necessary (sine qua non), but not sufficient, for speciation.

26. (B) There is a clear, direct (linear) relationship between mean spring temperature and first mean flowering for the six species studied. The higher the mean spring temperature, the earlier the first mean flowering. Choice A is incorrect because the graph compares the first flowering with the temperature, not the time it takes to flower. Choice C is incorrect because the length of the winter is not determined by the mean spring temperature.

The term *winter*, like all seasons, has two meanings: the *astronomical meaning* (the location of the sun as viewed from Earth, winter starts on December 21st) and the *meteorological meaning* (the winter is the three coldest months of the year, winter starts on December 1st). A short winter, which could be seen as an early spring (or a long autumn), just means that warmer temperatures started early. However, that does not indicate the mean spring temperature is higher.

Choice D is incorrect because there is no data showing that higher temperatures are preferred or are better. The data simply show that **for the six species studied, higher spring temperatures resulted in earlier flowering.**

27. (B) Biological variation creates enormous diversity, and **the particular environment a genotype finds itself in will determine which variations succeed.** Biology is considered a "messy" science because it is the study of an emergent phenomenon (life) that does not lend itself to strict categorization. This is due to the astounding variation created by sexual reproduction.

Most trends in biology will have at least a few outliers, and they are interesting to study along with the trend-fitters to understand the relationships involved. Remember that **geneticists, at least initially, found many genes by identifying and studying the rare mutants, not the wild types**.

28. (C) Choice A is incorrect because there is no way to infer a positive or negative impact from the data provided. Choice B attempts to confuse pollination with flowering. Although the statement may be correct, it is not relevant to the flowering process. There is also no information to indicate that the six species studied were pollinated by insects. Choice D states the opposite conclusion of what the data show. If the timing of flowering was completely regulated by genetics, the mean spring temperature should have no effect. **The formation of flowers involves a change from vegetative to reproductive growth.** This change requires a shift in the energy resources of the plant, a transition that, according to the data, is triggered by environmental cues (specifically, the temperature).

29. Flowering is an energetically expensive process for a plant. **Reproductive effort is the proportion of energy (or biomass) that an organism allocates to reproduction** (instead of growth or maintenance). Ill-timed flowering maximizes the reproductive effort while minimizing the number of offspring. For example, if flowering occurs at the wrong time of year, other members of the same species may not have flowered yet, resulting in a loss of reproductive opportunity for the plant. If flowering occurs in inclement weather, it may reduce or eliminate the number of offspring that result. If the plant is an annual plant, it may have forfeited reproduction entirely.

Flowering plants typically depend on animals to assist in the pollination process. If a plant flowers at an unusual time of the year, the conditions for pollination may not be met. For example, the animal pollinators may not be present or active at that time of year.

30. (C) The x-axis shows the possible allelic combinations in the gametes. There is no indication that allele 1 or 2 for either gene is dominant or recessive. It doesn't matter—the question is whether or not the alleles assort independently. **There are four possible combinations of gametes, each of which appear with equal frequency,** so it can be assumed that the genes for A and B are on different chromosomes, or that they are located far enough apart on the same chromosome to be separated by crossing over during meiosis about 50% of the time (they appear to assort independently). (See the following figure on independent assortment and answer 31 to contrast with gene linkage.)

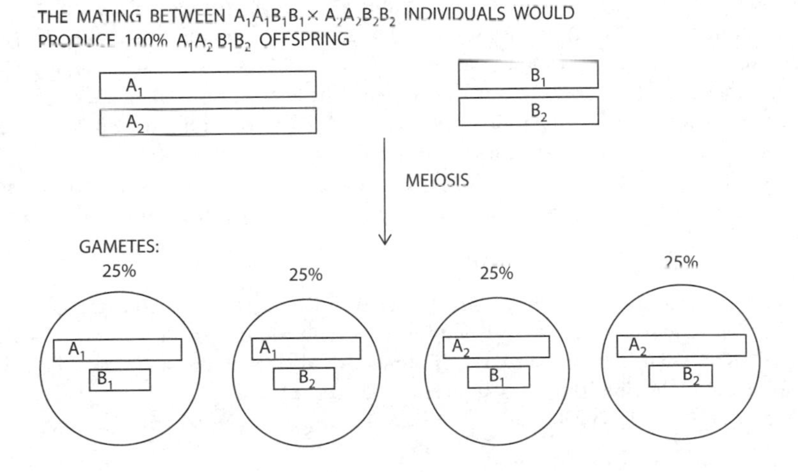

31. (D) Gene linkage *can* refer to any genes that are on the same chromosome, but it **usually refers to genes on the same chromosome that are located close enough to be inherited together more frequently than they are inherited separately** (appear to assort independently). **If they do not segregate independently during gamete formation, they are functionally linked** (i.e., not just on the same chromosome, but close enough to be inherited together more than would be expected if they were on different chromosomes).

Notice that the A_1/B_1 and the A_1/B_2 combinations appear with much greater frequency than the A_1/B_2 or A_2/B_1 combinations. This suggests that the parents (F1) had allele 1 for both genes on one chromosome and allele 2 for both genes on the homologue (shown as follows).

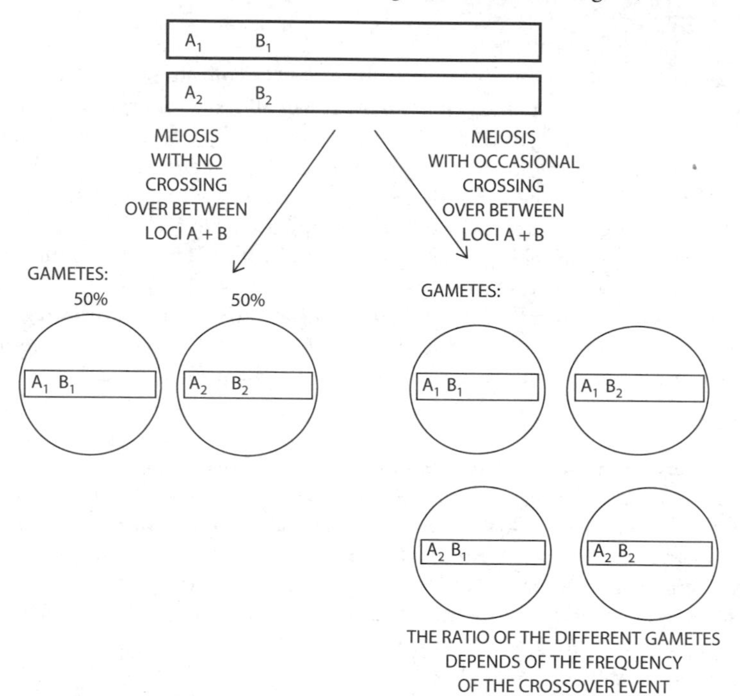

Unless crossing over separates the alleles during meiosis, the two alleles for the traits that are located on the same chromosome will be inherited together, since the gamete gets all the alleles on each chromosome it inherits.

Note that the F1 parents did not have to have the allele arrangement indicated in the diagram. Meiosis in the P generation ($A_1A_1B_1B_1 \times A_2A_2B_2B_2$) would have produced some recombinant chromosomes (not shown, for simplicity).

A significant degree of variation generation occurs during meiosis, when an individual's alleles are recombined to produce new combinations of alleles on chromosomes.

32. Natural selection is unlikely to create a perfect species (and the definition of *perfect* is not universally agreed upon). If ever "a perfect species" could exist, it would be "perfectly adapted" to its environment. But **environments are always changing**, slowly or quickly. And so, natural selection is unlikely to create a perfect species.

The two greatest limitations are (1) lack of planning and (2) structural limitations.

(1) **Lack of planning: A population *must* seed its next generation to be present in the future.** Whatever phenotype is "good enough" to survive long enough to reproduce gets some of its genes represented one more time, in the next generation. **Natural selection acts only on the gene pool that is present**. There appears to be **no short-term planning** (the next generation) **or long-term planning** (several generations down the line). *Selection favors what works right now* to continue this lineage for one more generation. To continue to exist, this process must continue to be repeated. However, nothing is guaranteed. Finally, random factors can have a great impact on what survives, regardless of how well adapted they may be. For example, the K/T mass extinction was caused (at least much in part) by a meteor strike and massive volcanism.

(2) **Structural limitations: Features of organisms almost always evolve from the pre-existing features of their ancestors.** Although mutation and genetic rearrangement have clearly produced "new" alleles and structures throughout time, these processes have generally done so by the **modification of pre-existing structures.** For example, nails, hair, talons, skin, and scales (of bony fish) are all made of the same protein, keratin. No known organisms have structures made of titanium: There's simply no genetic program or biological mechanisms to build it. **Organisms are also limited by their ability to accumulate and manipulate certain materials.** Without transporters to allow the specific uptake of titanium from the environment or the cells that could build structures with it, there is no way to exploit an environment in which it is available. Think of all the nitrogen in the atmosphere that for billions of years only bacteria could "fix" into forms usable by other life-forms. (Fritz Haber figured out the process in a laboratory in 1909.)

33. (D) The parameters that constrain the program are the five conditions required for Hardy-Weinberg equilibrium. **An equilibrium situation is established when nothing appears to be changing.** In a chemical equilibrium, *the rate of the forward reaction is equal to the rate of the reverse reaction, and so the concentration of products and reactants in the reaction vessel remains unchanged.* **The size of a population is at equilibrium if the number of individuals remains constant.** This occurs when the **birth rate and death rate are equal** and the **rate of migration in and out of the population are equal.** Even though the specific individuals that make up the population change, the number of individuals within the population remains the same.

In Hardy-Weinberg equilibrium, the allele frequencies in a population remain the same from generation to generation. In order for this to occur, five conditions must be met:

(1) **large population (no genetic drift)**
(2) **random mating (no sexual selection)**
(3) **no mutations (no new alleles)**
(4) **no gene flow (no immigration or emigration that can introduce new alleles)**
(5) **no natural selection (the trait or traits in question are selectively neutral)**

If these conditions are in place, allele frequencies will not change over time and therefore evolution will not occur.

You can also think of it in the reverse way: **genetic drift, sexual selection, mutations, gene flow, and natural selection can all produce changes in allele frequencies and therefore, evolution.**

34. (D) Our long-term understanding is this: **Acquired characteristics cannot be passed on to offspring and therefore play no role in evolution by natural selection.** Each breeding produced mice with tails, so the offspring never inherited characteristics acquired by the parents (so choice B is not the answer).

The mouse tail experiment performed by August Weismann in 1890 was designed to refute the ideas of Lamarck (specifically, inheritance of acquired characteristics). Weismann made an important contribution to evolutionary biology by stating that **hereditary information in the body cannot pass into gametes or into the next generation.** This is called the **Weismann barrier**.

35. (B) Fitness of an organism is not measured by its reproductive potential, but by **differential reproductive success. Only offspring that have been produced matter.** In the case of choice D, no offspring have been produced yet. Choice B, the old bear, has 2 offspring, but each offspring has 2 offspring, so there are 6 bears in later generations carrying his genetic material. The bear in choice A has 5 cubs, so she's the next most fit. The young male with 3 cubs gets the bronze.

How do you measure fitness?

Fitness is defined with respect to a genotype or a phenotype. The greater the contribution made to the gene pool of the next generation, the greater the fitness of the genotype or phenotype.

36. (A) The changes in allele frequency due to genetic drift are higher in smaller populations (see the graph above question 23). In the northern elephant seal, the effective population size is smaller than the actual population size because most of the males do not sire offspring. **The "effective" size is determined by the number of individuals who actually reproduce** because the next generation is composed only of the genetic information put there by reproducing individuals. Also noteworthy is that the population of northern elephant seals does *not* engage in random mating, and so it does *not* meet the conditions of Hardy-Weinberg equilibrium. (See answers 14, 22, 23, and 51 for more information on genetic drift.)

37. (D) Choices A, B, and C are all mechanisms by which new genes and alleles can be created by the modification of pre-existing genetic information. Choice D is incorrect (therefore, the correct answer to the question) because DNA sequences are not assembled "from scratch" (*de novo*) in cells; they can be made only from a template or a modification of an existing sequence.

38. (D) A population is in Hardy-Weinberg equilibrium when its allele (and, therefore, phenotype) **frequencies do not change from one generation to the next.** This occurs when the M allele frequency shows no changes. From 1975–1985, the M allele frequency is 0.3; therefore, the m allele frequency is 0.7. From 2000–2010, the M allele frequency is 0.9, and, therefore, the m allele frequency is 0.1.

39.

	8	2			

82%, with an acceptable range of 80–85%. **82%** The M (or p) allele frequency in 1980 is 0.3; therefore, the m (or q) allele frequency is 0.7. The red insects are homozygous dominant (MM or p^2) and heterozygous (Mm or 2pq). Therefore, the percentage of red insects is 0.09 (p^2) + 0.42 (2pq), or 0.51, or 51%.

The question asks for the percent of *red* insects that are heterozygous, *not* the percentage of red or the percentage of heterozygotes: 42% of the *total insect population* are heterozygous, but only 51% of the insects are red. If the population size was 100, 9 are homozygous red, 42 are heterozygous (∴ red), and 49 are black.

$$42 + 9 = \text{\# red insects}$$

$$\frac{42 \text{ red insects}}{(42 + 9) \text{ total insects}} = \text{percent of reds that are heterozygotes}$$

$$\frac{42}{51} = \sim 82\%$$

40.

	8	2			

82%, with an acceptable range of 80–85%. By 2005, the allele frequencies have changed, so the M (or p) frequency is 0.9 and the m (or q) frequency is 0.1 (as given by p + q = 1).

This question asks which insects will be homozygous, but does not indicate dominant or recessive, so we must consider both MM (or p^2) and mm (or q^2)

$$\therefore 0.9^2 + 0.1^2 = 0.81 + 0.01 = 0.82 \text{ or } 82\%.$$

41.

	9	9			

99%, with an acceptable range of 98–100%

By 2005, the frequency of M (or p) = 0.9 = 90%
The frequency of the m (or q) allele = 1 − p = 0.1 = 10%
The dominant case: MM = $\mathbf{p^2}$ = 0.9^2 = **81%**
The heterozygous case: Mm = **2pq** = 2(0.9)(0.1) = **18%**
81% + 18% = **99%**

42. (A) Each population was seeded by one individual; therefore, it was genetically uniform at the start. **Fitness increased because mutations created new alleles that conferred advantages to the bacteria that had them.** The bacteria that lacked them were "outreproduced" by those who did. Over time, the bacteria with advantageous alleles "took over." Remember that **bacteria are capable of horizontal gene transfer, which may hasten the spread of advantageous alleles.**

43. (D) The *Cit*$^+$ mutation occurred in addition to another mutation that occurred previously. The previous mutation either facilitated the origin of the *Cit*$^+$ mutation or made it advantageous. The evolution of citrate usage was contingent on a previous genetic change. **This adaptation enabled the use of a new ecological niche** ("speciation by niche-iation," though whether or not this strain is a new species is debatable). **The evolution of new characteristics may rely on rare combinations or sequences of mutational events.** In other words, **evolutionary change may be limited by the mutational process.**

44. New alleles are made from pre-existing alleles by mutation. The duplicated gene may diverge in sequence, become a nonfunctional pseudogene, undergo a deletion, or undergo gene convergence.

Gene duplication is a mutational event (strictly speaking, a mutation is any change in the DNA sequence). There is nothing inherently positive or negative about it, and there is no constraint on the size of the change. **It results in a single gene present in two or more loci** (locations on a chromosome). Over time, changes may occur in one or more of the duplicates (as long as one copy of the gene remains intact, there should be no harmful effect of the mutations in the duplicates).

If the duplicates are not transcribed and translated, they are called *pseudogenes*. They do not code for RNA or proteins and do not perform regulatory functions, so mutations are presumably selectively neutral. They may persist as pseudogenes or may eventually provide the sequence information for a new protein.

Gene duplications often arise from unequal crossing over. This produces tandem duplication on one chromosome and a deletion on the other.

Recombination (during crossing over) begins with precise alignment of the sequences on homologous chromosomes. If the homologous sequences differ by two or more base pairs, new sequences can be generated by intragenic recombination. ***Intragenic recombination* is recombination within a gene.** Just as crossing over creates new combinations of alleles on chromosomes, **new combinations of gene segments can arise through intragenic recombination.** Because a particular segment of a gene often codes for the amino acid sequence of a specific domain of a protein, new combinations of gene segments can theoretically produce new combinations of protein domains, creating proteins that have new functions, new combinations of functions, or the ability to recognize new substrates or ligands.

Unequal crossing over is one of the processes that has generated the enormous number of copies of nonfunctional sequences that constitutes much of our DNA, as well as the DNA of most eukaryotes. It is extremely important in the evolution of new functional genes and the total increase in the amount of DNA in eukaryotes as compared with prokaryotes.

Mutation rates vary among genes and between regions within genes. Mutation rates are typically measured by the effect on phenotype (which means many mutations will not be "registered") and average about $10^{-6} - 10^{-5}$ mutations per gamete per generation. If measured by the average mutation rate per base pair, one mutation is introduced in every 10^{-10} base pairs in prokaryotes and 10^{-9} base pairs in eukaryotes (during meiosis).

45.

	4	3	8		

438, with an acceptable range of 350–450.

Population on day 4: ~250 individuals
Population on day 8: ~2,000 individuals
$2{,}000 - 250 = 1{,}750$

$$\frac{1{,}750 \text{ individuals}}{4 \text{ days}} = \frac{438 \text{ individuals}}{\text{day}}$$

*One of the 7 **Science Practices** the College Board expects students to engage in is **the appropriate use of mathematics.** You are expected to use mathematics to solve problems, analyze data, make predictions, and describe natural phenomenon. You should be comfortable quantifying data, estimating answers, and describing processes symbolically.*
Be ready to justify your choice of mathematical tools—they should be appropriate for the situation. Hardy-Weinberg and chi-square are two particularly useful mathematical tools for understanding the data in Big Idea 1.

46. At the beginning of a population cycle, there are few to no density-dependent limiting factors to limit population size. Once the population size increases, the resources available for each individual become more limiting. There are many factors that influence population growth, but the most important factor that limits a particular population's size is available resources. This is the **logistic growth pattern.** The graph asymptotes are at the carrying capacity (N), the number of individuals that a particular area can support over time.

47. Your answer could have been as low as or as high as:

	6	9	2		

692, with an acceptable range between 690 and 700. There are **120 homozygous recessive phenotypes** in a population of 1,000 individuals so $\therefore q^2 = 0.120 \therefore q = 0.346$ (or 0.35). (See the following box if you don't know how to get $q^2 = 0.120$.)

You won't have a calculator for this portion of the exam, so either you'll get easy numbers to crunch or you'll have to do some estimating. Taking the square roots of decimals (and even squaring them) can be tedious, so use values greater than 1.

(1) Convert the decimal into a percent (ultimately, you want a number greater than 1).

$$0.120 = 12\%$$

(2) Find the square root of that number, or estimate it.
What is the square root of 12? Off the top of your head, you may not know.
But since $3^2 = 9$ and $4^2 = 16$ and 12 is in between, you can estimate.
Choose one number in between if needed to get closer to the actual number:
$3.5^2 = 12.25$ . . . Close enough! (The actual square root of 0.120 = 0.3464.)

(3) Convert the number back into a decimal or whatever form the question asks for.

The question asks you to estimate the *number* of recessive alleles in generation 1. If the allele frequency is 0.346, or 0.35, that means 34.6%, or 35%, of the alleles are recessive. In a population of 1,000 diploid individuals, there are 2,000 alleles for the trait. That means there are approximately 700 copies of the q allele in the population.

Quick check: In a population with 120 recessive individuals there are a minimum of 240 recessive alleles (2 for each homozygous recessive individual). So 240 is the absolute minimum number you would expect. Each heterozygote has 1 recessive allele. If q = 0.35 and p = 0.65 (1 − q = p), the frequency of heterozygotes is 2pq (or 0.455).

Number of q alleles in homozygous recessive individuals = ~240
Number of q alleles in heterozygous individuals = ~455
Total = ~695

The AP Biology exam will typically allow a range of answers for grid in calculations. Don't take too many liberties with rounding and estimating, but don't be overly worried about the exact number you grid in.

ALLELE versus GENOTYPE Frequencies

To calculate allele frequencies use p + q = 1.

The equation states that there are only two forms of the gene, p and q, for the given trait and that the sum of the two alleles accounts for 100% of the alleles for that particular trait in the population. Typically p represents the dominant allele, and q represents the recessive allele.

To calculate genotype frequencies use $p^2 + 2pq + q^2 = 1$.

This equation states that there are three genotypes and their sum constitutes all of the genotypes for that trait in the population.

p^2 = frequency of homozygous dominant genotype
q^2 = frequency of homozygous recessive genotype
2pq = frequency of the heterozygous genotype

Because homozygote recessive genotypes are often the easiest to identify by their appearance (in the case of simple dominant/recessive inheritance patterns), quantifying the frequency of q^2 is often the first step in calculating the allele frequencies:

$$q = \sqrt{q^2}$$
$$p = 1 - q$$

Once the allele frequencies are known, the genotype frequencies are easy to calculate using $p^2 + 2pq + q^2 = 1$.

48.

	2	4			

24%, with an acceptable range between 23 and 25%. **The q allele frequency in generation 1 is 0.35** (or 0.346, calculated in answer 47).

In generation 7 there are **345 recessive individuals** ∴ $q^2 = 0.345$ ∴ $q = 0.587$. The change is 0.587 − 0.346 = 24.1% or **24%**.

49.

	4	9			

~**49%**, with an acceptable range of 48–50%. Generation 5 has **340 recessive individuals** $\therefore$ $\mathbf{q^2 = 0.340}$ $\therefore$ $\mathbf{q = 0.58}$.

$$\mathbf{p = 1 - q} = 1 - 0.58 = \mathbf{0.42}$$
$$\therefore \mathbf{2pq} = 2\,(0.42)\,(0.58) = 0.4872 = 0.49 = \mathbf{49\%}$$

50. Phylogenies are hypotheses that attempt to explain the evolutionary relationship among organisms. Several types of data are used to construct a phylogeny, but sometimes it is useful to construct two or more phylogenies, each of which is based on only one type of data, in order to compare the different predictions made by the different types of data.

No one knows exactly what happened from 3.8 billion years ago to now, so everyone's reconstruction of life's history is a hypothesis. **A good hypothesis is consistent with the data available and provides testable predictions.**

Species	Sequence 1 2 3 4 5 6 7 8 9 10	Age of oldest fossil (mya = millions of years ago)
A	A A C G C T T A A G	75 mya
B	C T T A C T T C C G	
C	G T T A C T T C C G	150 mya
D	G T T A C C T C C G	
E	A A C G A T T A A T	
F	A A C G T T T A A T	
G	A A T G C T T C A G	100 mya, extinct
H	A T T G C T A C A G	400 mya, extinct
I	A T T A C T T C A G	
J	A T T G C T T C A G	200 mya, extinct
K	A A T G C T T A A G	150 mya, extinct
L	A T T A C T T C C G	120 mya, extinct
M	A A C G C T T A A T	

To handle sequence data (refer to the preceding sequence data and the following phylogeny):

(1) Evaluate sequences by position number. For example, vertically scanning position 1 reveals that A is more common than G or C.

(2) Mark nucleotide differences so they can be tallied later. Sequences G, H, I, and J differ by only one nucleotide (they each have only one mark), so one (or more) of them is the ancestral species.

Assumption 1: Species H is ancestral. The oldest known fossil for species H is 400 million years old (myo). There's no fossil data for species I. Fossil data for species J and G are <400 myo. The evidence supporting H as the ancestral sequence is weak.

The process involves some trial and error, so expect to reconfigure your phylogeny a few times.

> **It's OK to make assumptions** based on limited data because sometimes it's the only way you can "move on" with an analysis.
>
> However, **you must be aware of the assumptions you've made** as you continue your analysis. If your analysis starts to fall apart, you probably need to **reevaluate** your assumptions.

(1) **Find the sequence with the least number of nucleotide differences with the ancestral species.** H and J have one difference at position 7. All other species have a T in that position, so a reasonable hypothesis is that H is ancestral to J, and J gave rise to the other species. Remember that this is *just a story* until there is sufficient data to support the hypotheses.

(2) **Link sequences in "time" by locating sequences that differ in only one (maybe two) positions, particularly if it is a change in the same position.** The sequences that are most similar to J are I and G, which each differ from J by only one nucleotide, but the differences are present at different positions (position 4 in I and position 2 in G). This could have resulted from a few different histories, one solely due to divergence, and another that includes divergence and a back mutation.

Assumption 2: No back mutations. A *back mutation* is when a mutation occurs and then in later generations, another mutation reverses the original. Many sequence phylogenies assume no back mutations (at first).

(1) **Branch points may be identified by single substitutions in descendant species that occur at different sequence positions from the common ancestor.** The sequences between ancestor species J and descendants I and G differ by one nucleotide but at two different positions (position 4 in I and 2 in G).

(2) **Branch points may be identified by a different nucleotide substitution in the same position.** The sequence from species L differs from species C and B at the same position but at a different location (position 1 changed to a G in species C and a C in species B). The sequence from species M differs from species E and F at position 5. In species E, C was substituted with A, and in species F, C was substituted with T.

Assumption 3: A different, single nucleotide change at the same position indicates the divergence of two populations from the common ancestor. It is possible that the sequence change occurred in the transition from M to E (or F) and then another substitution at the same location from the divergence of F from E (or E from F).

The following phylogeny shows the nucleotide and position numbers for each change. **It is not the only phylogeny that makes sense from the data.** There are also **numerous assumptions made in the construction of the phylogeny that were not specifically listed**, such as the assumption that the sequences with the least number of differences *in this specific piece of sequence* are more closely related and that nucleotide substitutions always indicated divergence.

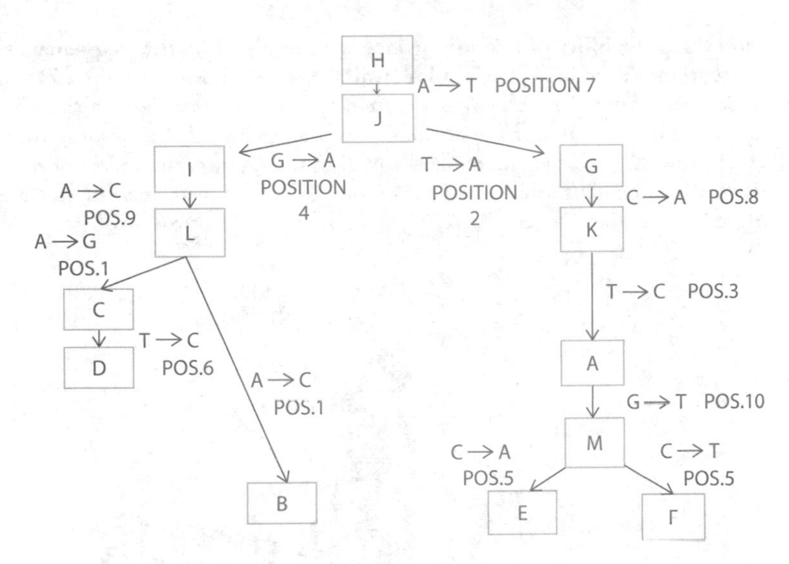

51. (A) The frequency of alleles that have little to no effect on fitness (neutral) can fluctuate randomly in a population because only a sample of a population's genes are transmitted to the next generation. This process of genetic drift reduces genetic variation and **leads to a loss or fixation of an allele** unless counteracted by another process such as gene flow or mutation, or for some reason selection begins to act on that allele.

The probability that a particular allele will be fixed in the future is dependent on the frequency of that allele at that time. If the frequency changes, the probability of it becoming fixed or lost changes. **The time it takes for the allele to become fixed or lost depends on the population size. The smaller the population, the more rapidly genetic drift operates.** The "effective" population size is often smaller than the actual population size and may hasten the rate at which genetic drift occurs.

The heterozygosity of a population (the frequency of the heterozygous genotype) is an indicator of genetic drift. As an allele frequency shifts toward fixation (frequency = 1), the alternate allele frequency shifts toward loss (frequency = 0). **The mathematical theory of genetic drift predicts a decrease in a population's heterozygosity,** which was seen in this experiment.

When allele frequencies drift toward fixation or loss, which is what genetic drift implies, **the frequency of heterozygotes decreases over time.**

Imagine a population where $p = 0.5$ and $q = 0.5$.

As the frequency of the p allele decreases, the frequency of the q allele increases. The frequency of individuals with both alleles necessarily decreases with the reduction of the allele frequency of one of the two alleles.

Genotype frequencies given by $p^2 + 2pq + q^2 = 1$.

p	**q**	**p^2**	**q^2**	**2pq**	
0.500	0.500	0.25	0.250	0.50	
0.100	0.900	0.01	0.810	0.18	
0.001	0.999	1×10^{-6}	0.998	1.996×10^{-6}	
0	1	0	1	0	**p is lost, q is fixed**

Because the probability of fixation or loss is determined by the frequency at that time, the situation is described as a "random walk" (see the following picture). *The smaller the population size, the more quickly the "fall" from the balance beam.* Which side she falls on is determined by the initial frequency of one of the alleles—if she's leaning (to your) right (her left, toward loss), she's more likely to fall to the right, meaning loss of the allele (frequency = 0), and if she's leaning left, she's more likely to fall left, meaning the allele is more likely to be fixed. (See answers 14, 22, 23, and 36 for more information on genetic drift.)

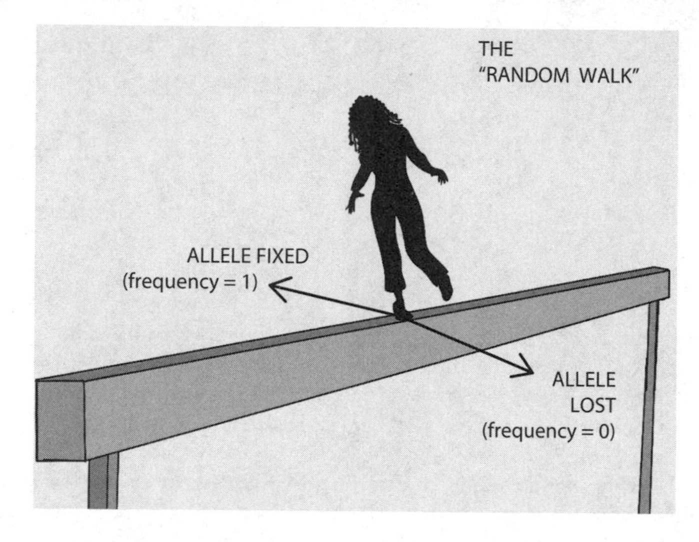

52. (A) One criteria for evolution in a population is an allele frequency change from one generation to the next (or over time). The **strength of selection** (as well as **the frequency of the allele)** determines how quickly change in its frequency change occurs. *A deleterious allele will be removed from a population at a higher rate than a neutral allele,* for example, because there is no selective pressure on the neutral allele (but that doesn't mean its frequency won't change due to drift). Notice that the rate of allele removal changes with frequency. It generally increases with frequency until frequencies of approximately 0.8, when the rate of removal decreases rather abruptly.

The rate of removal of a lethal allele will increase with increasing frequency until the frequency is "fixed" at a frequency of 1. The rate of change in allele frequency is highest in this population, the strength of selection is highest, and its rate of evolution is expected to be greatest.

53. The figure shows that the **strength of selection affects the rate of removal** of an allele with the greatest removal rate for an allele with a selective strength of 1. The **frequency** has an effect, as well, with the largest effects occurring at frequencies at about 0.8 for selective strengths less than 1.

An important difference between dominant and recessive alleles is that the dominant phenotype is exposed to selection even at very low frequencies, and so selection always acts on it. **A harmful recessive allele can "hide" from selection in heterozygotes**, so at low frequencies it can avoid complete removal from the population.

54. (A) An allele is considered dominant when the phenotypes of the heterozygote and homozygous dominant genotypes are identical. Because dominant alleles cannot be masked in the heterozygote condition as recessive alleles can, they are more prone to selection than recessive alleles.

55. (B) It is often easier to think "positively," in other words, asking "What evidence would support contamination by amino acids from Earth?" may be easier to answer than "Which of the following is evidence that amino acids did not contaminate the meteorite?".

If the amino acids *were* contaminants from Earth, they would be present almost exclusively as L-isomers, the specific isomer (specifically, enantiomer) of all the amino acids that make up proteins in living things. When amino acids are synthesized by abiotic means, they are produced in a racemic mixture (50% D, "right-handed" enantiomers and 50% L, "left-handed" enantiomers, except for glycine, which doesn't have D- or L-isomers).

See the box after answer 6 for a strategy for answering questions asked in the *negative* and using the process of elimination.

56. (A) The fact that cells contain a greater mass of RNA than DNA is *not* evidence that RNA was the first genetic material. Water is the most abundant molecule in the cell, for example, but that does not support its role as a molecule of inheritance. The information contained in genes is used to make functional RNA molecules such as tRNA and rRNA and proteins.

See the box after answer 6 for a strategy for answering questions asked in the *negative* and using the process of elimination.

57. (B) Layer D seems to contain only fish, but layer C appears to contain tetrapods with fully developed limbs. **The transition fossils are most likely to be found between these two layers.**

58. (C) It is crucial to know where the rocks of the appropriate age are exposed and accessible. Radioactive dating is not typically done on site. You want to know the location of the rocks of the age you're studying before you get there. Choice B is tempting but what is coastal now was not necessarily coastal 354 million years ago, so knowing where the rocks that were coastal back then are now, and making sure you can get to them, is the most crucial aspect of planning of the four things listed.

The details of radioactive dating methods are beyond the scope of the AP Biology exam.

59. (D) If the iridium-rich cloud layer blocked out the sun, plant matter would be expected to decrease. The iridium layer settled to Earth, but depending on the depth of the layer, the deposition rate, and some other factors, a reduction in plant life is expected over the time the iridium was in the atmosphere. When the atmosphere cleared, the plants that survived would be expected to produce an increase in plant biomass over time, but they would have to "rebuild" the biomass they lost in the initial reduction. **After an extinction event, diversity tends to increase slowly, at least at first.**

The hypothesis that the sun was blocked out and caused the mass extinction, or was significant, is not generally accepted. It appears that the impact, the temperature fluctuations that ensued, and the concurrent massive volcanism all conspired to make Earth uninhabitable for about half the kinds of organisms living on it. **The exact causes of the massive extinction has not been conclusively determined, but it is generally agreed that the meteor played a large part in it.**

The End-Cretaceous extinction (also called the **K/T extinction**) ended the lineages of the nonavian dinosaurs, the ammonoids, rudists, many families of invertebrates and planktonic protists, and most marine reptiles. Approximately 155 families of marine animals and 47% of all genera became extinct.

See the box after answer 6 for a strategy for answering questions asked in the *negative* and using the process of elimination.

Memorization of the names and dates of the major extinction events will not be assessed on the AP exam.

60. (A) The high carrier frequency of the allele in Ashkenazi Jews is probably caused by a combination of founder effect, genetic drift, and differential immigration patterns. Genetic analysis links the bottlenecks to the 70 CE (AD) diaspora and the Black Death in 1348.

It was hypothesized that Tay-Sachs carriers may have had a resistance to tuberculosis, but evidence to support the hypothesis is inconclusive. *Grandparents of Tay-Sachs carriers die from proportionally the same causes as grandparents of noncarriers. In other words, carriers are not less likely to die from tuberculosis.* **This supports the null hypothesis: that the allele confers no advantage to heterozygotes.**

61. (C) The graph shows that as a population size increases, the frequency of heterozygosity increases. Maximum heterozygosity occurs when the allele frequencies (of two alleles) is 0.500. See the following table.

p	q	2pq
0.500	0.500	**0.500**
0.250	0.750	**0.375**
0.001	0.999	**1.200×10^{-6}**
0.000	1.000	**0.000**

Choices A, B, and D all illustrate that smaller populations tend to have less variation and a lower frequency of heterozygotes. Larger populations tend to exhibit greater diversity than small populations, as well. In fact, **average heterozygosity is used by scientists as a proxy for genetic diversity.**

Choice C is the least illustrative of the data because with frequency-dependent selection, the fitness of a particular genotype varies as a function of its frequency. but whether the genotype is homozygous or heterozygous is irrelevant to the concept of heterozygosity and diversity being tested.

62. (C) Dark moths are not necessarily preferred by birds, since in the polluted woods the light moths are eaten more. The populations of each color are not known, so choice B is not necessarily true. It may be that dark moths are so common that they get eaten more just from higher frequency (though not likely). The most probable explanation of the data is that the **dark moths are well camouflaged in the polluted woods and the light moths are better camouflaged in the unpolluted woods.**

The coloring of the moths matters because the birds use the sense of vision to find the moths. It's easy to imagine this since humans are very visual animals, but it's important to remember that **vision is not necessarily important to every animal species**, even predators. However, **vision is the most important sense for birds, particularly because it is essential for safe flight (though bats use sonar)**. Birds of prey tend to have the greatest visual acuity.

Interestingly, birds have four types of color receptors (as opposed to three, as in humans). The extra "color" receptor confers vision in the UV range of the electromagnetic spectrum.

63. (C) The data are a "snapshot" of the current situation. There is **no time variable** included. The percentage of light moths eaten by birds was higher in the polluted woods versus the unpolluted woods. The total number of moths eaten is lower in the polluted woods, as well. However, *with no other information about the diet of the birds, the effect on their biomass cannot be determined.*

In the polluted woods, light moths are eaten more than dark ones, which could decrease the population over time. **The dark moths are eaten more than the light moths in the unpolluted forest, so the ratio of light to dark is expected to increase over time,** regardless of the starting population size. Because only a few trees were samples, it is not possible to tell if the polluted woods will have a lower number of dark moths, especially since the dark moths are *less* likely to be eaten by birds in the polluted woods.

64. There are many pieces of biochemical and genetic evidence supporting the hypothesis that all organisms on Earth share a common origin of life. The following is an (incomplete) list:

- All organisms use the same genetic code to assemble polypeptides from DNA nucleotide sequences.
- All organisms use the same enantiomer of amino acids (L-) and sugars (D-).
- Gene sequence data shows successive, cumulative changes in gene sequences in lineages (calibrated through time by the fossil record).
- All organisms share the same basic mechanisms of transcription, translation, and DNA replication.
- Double-stranded DNA is the genetic information of all organisms.
- Prokaryotes and eukaryotes share similar cell structure.
- Eukaryotes share very similar cell structures.
- Organisms show remarkable similarity in their metabolic pathways even though the possibilities within organic chemistry are almost unlimited.

This is strong evidence that all life on Earth descended from a common ancestor, thought to be a prokaryote. Almost all metabolic pathways and all forms of nutrition evolved prior to the appearance of eukaryotes (1 billion years ago). Glycolysis is common to every organism and almost every cell. It evolved before the appearance of oxygen in the atmosphere.

When the AP Biology exam asks you for two pieces of evidence, provide only two pieces. If you give them more, they will not consider them.
They will assess only the first two (or whatever number they asked for).

65. Cladograms, like phylogenies, are hypotheses about the evolutionary relationship between organisms. Like all hypotheses, cladograms should be supported by some amount of evidence, whether little evidence or a lot. For many organisms, evidence is lacking, making it impossible to place great certainty in any one cladogram. **A good hypothesis is consistent with a larger framework, and a good model allows for the creation/production of good hypotheses.**

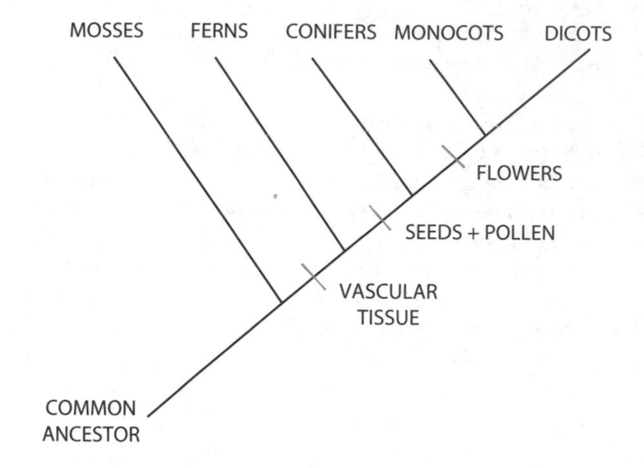

There are usually many hypotheses one cladogram (or phylogeny) will "contain." For example, one "sub-hypothesis" (if we consider the cladogram as the hypothesis) is that **the mosses, ferns, conifers, monocots, and dicots all descended from one common ancestor.** There's a lot of evidence to support that (not shown here), but it's still a hypothesis that is put forth as true in order for the rest of the cladogram to make sense.

A second hypothesis is that **none of the traits were the result of convergent evolution.** In other words, the ancestral traits were all retained and the derived traits "accumulated" in the most recent descendents.

66. (D) Of the 665 original mutations, only 28 increased fitness relative to the ancestral genotype, so the vast majority of the mutations made in this experiment were deleterious. The fitness effects of the 28 mutations that conferred an advantage are plotted in the graph. The data show that 8 of the 28 of the mutations had a fitness effect of 5%, approximately 5 of them had a fitness effect of 1%, and 1 of them increased fitness by 25%.

67. The lack of fossils is not as strong evidence that something didn't exist in a particular time or area than the (having) of fossil evidence is to support that they were at a particular time and place or coexisted with other species. The genetic data are not "bulletproof." Data to support pre-extinction diversification would be the existence of several fossils of diverse mammals dated over 65 million years ago.

Molecular clocks may need different calibration for different organisms. Large body size and **long generation times**, for example, **tend to slow the molecular clock. This is because the same period of time has fewer generations for a longer-lived organism, so not as many replications and mutations would have occurred in gametes**. See answers 6 and 94 for more information on molecular clocks.

Evidence that would support post-extinction diversification is a new piece of information that recalibrated the molecular data to place the diversification earlier than 100 million years ago. Observations that support, but are not technically evidence, would be the continued failure to find older placental mammal fossils and the construction of a phylogeny (based on additional fossil discoveries) that completely explains the data without the requirement of a pre-extinction diversification event.

Why are life, Earth, and the universe like they are now, and not any of the other almost infinite ways they could have been?

No one knows.

Science is a process of continued observation, experimentation, data collection, and analysis. Since new information and data are continually being discovered, all interpretations are subject to refinement and some (or many) may need complete revaluation.

68. (D) Transmission of the plasmid to daughter cells requires time and energy in replication of the plasmid. Bacteria that transmit the plasmid are not conferring any immediate benefit to their offspring. Over time, there would be fewer bacteria carrying the plasmid relative to those not carrying the plasmid, because those without the plasmid would take longer and require more energy to reproduce, although by not that much. The plasmid for ampicillin resistance does not confer any other benefit or have any other deleterious effect. So it is unlikely to disappear completely from the population, especially in the "several generations."

69. (A) The *y*-axis shows the number of individual mutations out of the 665 that have an effect on phenotype. Approximately 130 of the single mutations reduced fitness to 0.7 from 0.95 (the fitness of the ancestral genotype). A very small number increased fitness (to 1, 1.1 and 1.2), and a very small number drastically reduced fitness (to as low as 0.35).

70. Darwinian selection can occur in nonliving things as long as there are three essential ingredients (necessary and sufficient): **reproduction, heritable variation, and selection**. Viruses can reproduce within cells. They are obligate cellular parasites. They contain genetic information in the form of RNA and/or DNA, so when they are replicated, variation can be produced, usually by mutation. They are subject to selection because only those viruses that have the ability to find, enter, and manipulate host cells into replicating them will produce new viral particles.

The Null Hypothesis

The null hypothesis is a statement that "the effect (or difference) is due to chance," or "nothing interesting is going on."

For example:

- 70 out of 100 coin tosses end up tails . . . The null hypothesis states, "The coin is not weighted, it was just a *coin*cidence."
- A 1:1:1:1 phenotype ratio was expected; a 4:4:1:1 phenotype ratio was observed . . . The null hypothesis states, "The genes assort independently, they are not linked."

71. (B) The null hypothesis assumes that the observed difference from the expected is due to chance, not the different environments. The null hypothesis is assumed to be true unless there is evidence to indicate otherwise.

The number of degrees of freedom is the total number of "options" minus one. If there were 2 possible phenotypes, there is one degree of freedom. *You always subtract one from your range of options because there is an option that is selected, and so that is removed from the pool of options.* If there were four possible phenotypes, there would be three degrees of freedom because the organism would be one of the four, leaving three "options" behind.

In this case there is one degree of freedom, so a σ^2 value of 3.84 means that there is a 5% probability that the observed difference in height is due to chance. It also means there is a 95% probability that the difference in height is due to the environment.

72. (A) The null hypothesis is simple dominance, so only two phenotypes are expected: dominant for both A and B, or recessive for A and dominant for B (there is no cross that will make a recessive B offspring). That means there is only one degree of freedom (two possibilities minus 1, because one phenotype will be "chosen"). A p value of 0.05 is the largest that is assumed significant, so with one degree of freedom, 3.84 is the χ^2 value.

73. (D) A ring species, in which two reproductively isolated taxa are connected by a chain of interbreeding populations, provides a rare opportunity to use spatial variation to reconstruct the history of divergence. If A is the ancestral species, then A and B, and B and C are more closely related. If there is gene flow between A and B, and A and C, then there must be gene flow between B and C. A and H probably won't interbreed if "given the chance" but G and H are the least likely since they are the two most recent divergences from two lineages that separate the longest time ago.

74. (B) Answer choices A, C, and D are all true and were part of the construction of the proposed ring structure. **The differences among groups in the ring gradually increases as the distance increases.** For example, the difference in plumage between groups A and B are less dramatic than the differences between groups A and G. Choice B may be true, but it does not support the *ring structure*. It suggests that species E and D diverged from the ancestor A at about the same time. If E and D had the same number of differences between each of the species and A, it would support E and D as being more closely related.

See the box after answer 6 for a strategy for answering questions asked in the *negative* and using the process of elimination.

75. Evolution requires a change in allele frequencies from one generation to the next. The two main mechanisms by which evolution occurs are natural selection and genetic drift. In natural selection, phenotypes are differentially suited for their environment. **In artificial selection, humans control the reproduction of a particular group of individuals.** Differential breeding of organisms by traits allowed humans to change hundreds of species over time.

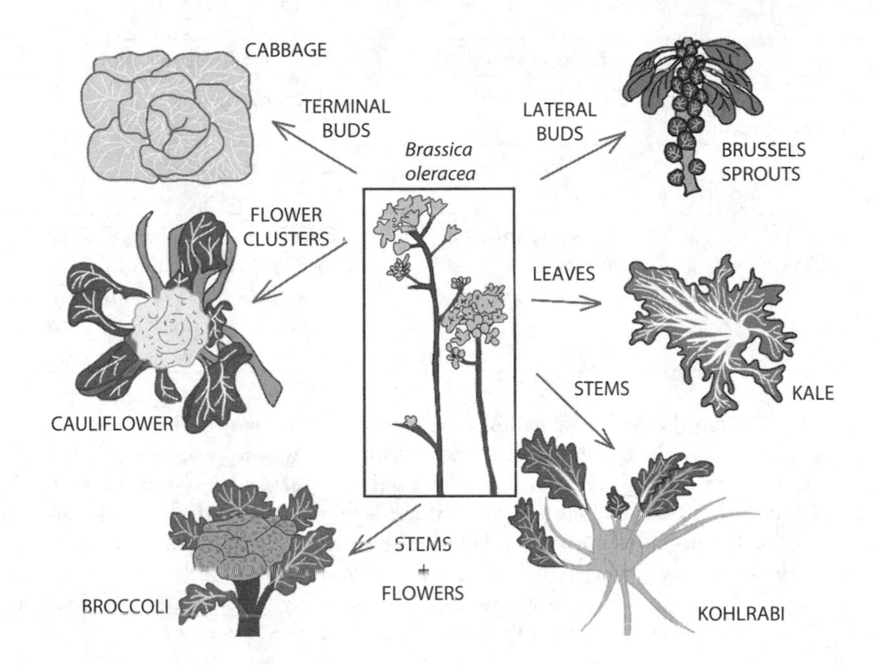

Artificial selection is human-driven natural selection. It has been used to produce practically all of our food crops and domesticated animals. Artificial selection was used to produce all the varieties of dogs. Because the word *breed* has a wide range of interpretations, it is difficult to quantify the number of dog breeds. However, dogs have been selectively bred by humans for thousands of years. Over this time, hundreds of traits have been selected for in dogs.

Natural selection is driven by the environment that selects for a particular combination of traits that succeeds in that particular environment. The preceding diagram shows the six traits selected for to produce the six agricultural plants. In the case of artificial selection, the selecting is done by the person growing the food.

In reality, there is not much of a difference between artificial and natural selection. For example, when pollinators choose the flowers or when animals choose the fruit they eat, they are selecting which plants get to breed. Perhaps the major difference is the scope of the selection and the time course over which it occurs. Because humans can essentially grow huge plots of one crop, the change in allele frequencies between generations can be quite dramatic. An argument could be made that natural selection created us and our ability to learn, cooperate, and manipulate our environment, so the specific selective pressure we place on a particular population in order to manipulate their reproduction to serve our purposes is still the product of natural selection. It's quite amazing . . . **The human species is a product of natural selection that uses the mechanism of natural selection (reproduction, heritable variation, selection) to change other species to suit its purposes.**

	Natural selection	Artificial selection
Who does the selecting?	• The environment	• Humans only
Time course	• Varies, but often long	• Can be very short
Unit of selection	• **Whole organism** is selected.	• **One or a few traits** are selected for "in the context" of the other genes that accompanied the alleles for the traits in the organism that were selected.

76. (D) Although the genes for hybrid sterility are best known in *Drosophila* and not well-characterized in most other species, they are not thought to have any effect on the nonhybrid organism that possesses them. **The genes in diverging populations likely differ enough that their recombination in the offspring creates sterility or inviability because of disharmonious interactions between the different sets of genetic information supplied by the parents.** Alleles at different loci within the same population are said to be **co-adapted.** Their harmonious combination has presumably been selected for. **The gene pool of this population, and any population, is a co-adapted gene pool.**

77. Your position on genetic engineering is not going to be judged. What is most important is the reasoning that supports your position.

Genetic diversity is essential for a population's long-term success. For example, the southern corn leaf blight fungus of 1970 was the consequence of planting genetically uniform corn. The corn contained a genetic factor that increased its susceptibility to the fungus. Because no individuals existed that lacked the genetic factor, the corn was completely wiped out. To preserve genetic diversity, several seed banks of as many strains of wild-type plants should be maintained.

Useful genes are found in wild species and can be crossed or engineered into others. At least 40 genes for resistance to various diseases have been found among several species of South American tomatoes. At least half of these have been crossed (by hybridization) into commercial crops.

A major concern is that the transgenes will spread into wild crops. This may make more vigorous weeds or may permanently eliminate wild-type plants from the area.

Engineered pesticide resistance could reduce or eliminate the use of chemical pesticides. Evolution of pests makes resistance to chemical pesticides an ongoing problem for farmers, which contributes to their greater, more widespread, and diverse use. Humans exist at the trophic level in which **biological magnification** is a concern.

A risk-versus-benefit analysis needs to be conducted that includes the short-term and long-term effects. The effects included should cover health, environmental, social, and economic. **The benefits need to outweigh risks.** Long-term studies of genetically engineered (GE) food should be conducted in animals before introducing it into the human food supply. Computer simulations can be used to predict the effects of cross-breeding GE plants with other species. It may be safer to take a conservative approach and consider using only genes from same or very closely related species until more data are collected.

	Natural selection	Artificial selection	Genetic engineering
Who does the selecting?	• The environment	• Humans only	• Humans only
Time course	• Varies, but often long	• Can be very short	• One generation
Unit of selection	• Whole organism is selected.	• **One or a few traits** are selected for "in the context" of the other genes that accompanied the alleles for the traits in the organism in which it/they were selected.	• Only **one or a few genes** are changed at a time. • The genes are introduced "out of context." • An organism that contained a particular mix of genetic information based on phenotype was not selected for.

78. (C) If there is no genetic differentiation between two (or more) populations, speciation cannot occur. Reproductive isolation allows two (or more) gene pools to diverge. If there is gene flow between populations (in other words, they are not reproductively isolated), their allele frequencies to differ enough to maintain separate, distinct populations.

In order for speciation to occur between two populations (according the biological species concept, see the box under answer 24), the two populations must not be able to interbreed, or at least not significantly (so choice B is totally incorrect). Choice A states two mechanisms that keep different species from interbreeding *after* they have already speciated. Choice D is incorrect because adaptive radiation is the result of the lack of gene flow, not the cause of it.

79. (B) In recently diverged populations or species, prezygotic, often premating isolating, mechanisms are a greater barrier to gene flow than post-zygotic mechanisms. (Song and mating behaviors, for example, are mechanisms of identification of possible intraspecific mates.) The data clearly show that **prezygotic isolation is stronger among allopatric pairs compared with sympatric pairs of taxa.** Although C may seem accurate, the data are specifically comparing prezygotic isolation, a specific type of reproductive isolation, and **the data show only correlation**, they do not show cause and effect. **There must be isolating mechanisms in place for isolation to occur**, but it is not clear that isolation evolved *to prevent* hybridization.

80. (C) Iteroparous and semelparous refer to the age schedules of reproduction. If survival contributes to fitness only for as long as reproduction continues, why not start reproducing as early in life as possible and continue to reproduce for as long as possible? As with all processes shaped by natural selection, the answer is **trade-offs.**

The cost of reproduction is generally high, and early reproduction is correlated with lowered subsequent reproduction. Risks associated with a substantial reproductive effort at an early age are an increased risk of death, decreased growth, and decreased subsequent fecundity. Suppose you decide to invest in fixing up your home before selling it. This is expected to increase the value and therefore the selling price of the house. However, if the house loses value (for whatever reason), the cost of the repairs may never be recovered. Selling the house as soon as possible greatly increases the chance that you'll walk away with a payout, but you may not get the best price for the house.

Reproduction requires that the reallocation of energy and resources used for growth, maintenance, and self-defense be used for the reproductive effort. Delaying reproduction until you are large enough, for example, may pay off greatly later. However, every day that reproduction is delayed presents another opportunity for another factor, like a predator, parasite, or random event like a volcano or mudslide, that may kill you.

Certain traits make **semelparous ("big bang" reproduction) an advantage** for a particular species.

- The rate of growth of body mass declines as an individual grows larger.
- The probability of survival increases with body mass.
- There is an exponential relationship between body mass and reproductive output.

Iteroparity is an advantage if greater fecundity can be achieved by deferring reproduction. **Repeated reproduction is more likely to evolve if adults have high survival rates from one age class to the next and if the rate of population increase is low.** A major advantage of iteroparity is that the risk of failure is spread out over many bouts of reproduction.

81. (A) Each bout of reproduction requires a substantial investment of energy and increased risk to the parent. Generally, the reproductive effort reduces, at least temporarily, parental survival. The yearly resources associated with semelparous versus iteroparous reproduction is greater because **the semelparous individuals have only one opportunity to reproduce.** Any semelparous organism that is not successful in its only attempt to reproduce will not get its lineage into the future. **It is therefore worth the price and the risk evolutionarily for a semelparous species to funnel as much energy and resources into the reproductive effort as it can afford.**

For iteroparous species, the proportion of energy devoted to reproduction increases with age; the cost of reproduction adds up with the number of reproductive efforts. However, iteroparous species can allocate their energy and resources back into growth, maintenance, and survival because they have time between successive bouts of reproduction.

82. Natural selection favors minimizing reproductive effort. Reproductive effort is the proportion of energy and other resources that an organism allocates to reproduction as opposed to its own growth and maintenance. Reproductive effort is greater per year in annual plant species because they have only one opportunity to reproduce. Annual plants reproduce once and then die.

Perennial plants will reproduce many times over the course of a lifetime, so each reproductive effort is typically lower than the one reproductive effort made by the annual species. The **cumulative cost** of reproduction in perennial plants increases with time and **can exceed that of an annual species over its lifetime**. However, **the perennial plant can continue to acquire resources over its life span, whereas annual species will typically have less time to obtain the energy to devote to reproduction**.

83. (B) Guppies from the streams where *Crenicichla* live are smaller, mature faster, and reproduce more frequently than those that live in streams where *Crenicichla* are absent. This hastened development and reproduction result in a higher reproductive effort for the guppies that live in *Crenicichla*-dominated streams but is selected for because those guppies that have a longer development and reproductive schedule are less likely to live long enough to reproduce.

In a population where the predator is present, the "cost" of living long enough to delay reproduction (and the risk that it requires) does not offset the benefit of delayed reproduction (lower reproductive effort). In *Crenicichla*-dominated streams, guppies that reproduced at a younger age clearly left behind more offspring than guppies that delayed reproduction, as demonstrated by the decreased size and increased rate of maturation in later generations.

84. (A) Reproductive effort is the proportion of energy and other resources that an organism allocates to reproduction as opposed to its own growth and maintenance. In the case of the guppies, the reproductive effort is measured by comparing the relative weights of the embryos to their mother. By itself, the embryo weight is not a true indicator of the effort the mother puts into creating the embryo. For example, a mother of low mass would put a larger percent of total resources than a higher mass mother into producing an embryo of the same mass. **Maximizing resources is a trait that will typically be selected for, so a lower reproductive effort for the same quality of offspring would be selected for.**

85. (C) Selection favors smaller size and early maturation and reproduction in the guppies preyed upon by *Crenicichla*. However, this requires a larger reproductive effort from the guppies. *Reproductive effort* is the proportion of energy and other resources that an organism allocates to reproduction as opposed to its own growth and maintenance. **Because natural selection favors a lower reproductive effort, the removal of the predation is expected to delay maturity and reproduction, which results in a larger size and lower reproductive effort,** which is what was observed in the experiment. This is an example of evolution occurring on a time scale that could be observed by humans.

86. (B) In many animals, including mammals and *Drosophila*, **more mutations enter the population from sperm as compared to eggs because many more cell divisions have transpired in the germ line before spermatogenesis as compared with oogenesis for two organisms at the same age.**

Whereas the female is born with all the potential eggs she will ever produce, the male continues to produce sperm through much or all of his life. The male germ cells divide by mitosis to produce diploid spermatogonium. It is these spermatogonia that continuously divide to produce a steady supply of spermatocytes, the cells that undergo meiosis. The average male will produce 525 billion sperm over the course of his lifetime.

***A note on the word* accurate** . . . *Accurate* means **"correct, true or widely accepted as true."**

An explanation that **directly** accounts for an observation would be one that provides a reason or cause that *directly precedes the phenomenon it attempts to explain*. (See the following box.)

The direct cause or explanation is linked in time and may require physical contact. Choice A is less of a direct cause than choice B, because choice A is the cause of choice B. Choice B is the direct cause of the greater number of mutations in sperm as compared with eggs—a greater number of divisions (therefore, genome replications, where mutations are likely to occur). The production of a large number of sperm needed to maximize the likelihood of fertilization requires a lot of cell divisions.

Why do you need to breathe oxygen?

An illustration of DIRECT versus ULTIMATE causation (and consequences)

The **direct** cause, or reason, for why you need to breathe oxygen is to supply your **electron transport chain** with the final electron acceptor (oxygen) it requires. Oxygen deprivation prevents the electron transport chain from functioning (**a direct consequence**), and for humans, more than a few minutes of oxygen deprivation is typically lethal (**the ultimate consequence**).

You have an electron transport chain because a primitive eukaryote engulfed a free-living bacteria (the mitochondria) about 2 billion years ago that used oxygen to generate a free-energy potential for electrons from organic molecules and oxygen to generate proton gradients for chemiosmosis.

Why did the mitochondria use oxygen? Oxygen became abundant at a time when it was poisonous to most organisms but also at a time when a biochemical system (the oxygen-driven electron transport chain) was or became available that could use it to extract energy from substrates in the environment.

Why was oxygen used? Nitrogen was also in the atmosphere. However, oxygen is more highly electronegative, so the movement of electrons from glucose to oxygen results in a large decrease in their free energy, which can be transformed into chemical energy in the cell. Also, the double bond between the two oxygen atoms in the O_2 molecule is easier to break than the triple bond between the nitrogen atoms in the N_2 molecule. And so on . . .

Ultimate causes are often related to evolution, natural selection, and the laws of chemistry and physics as applied to biology.

87. Before we construct an answer, let's make sure we understand what the graph is saying. There are two lines that tell a similar story: The dark grey line indicates the total extinction rate in number of families per million years and the light grey line indicates the number of extant families. The large peak in the dark grey line (the extinction rate) at 252 million years shows a rise from about a 2.5 families per million years to about 27.5 families per million years, an increase of 25 families per million years. The light grey line shows that the number of extant (not extinct) families declined from about 400 families to about 200 families, a 50% drop.

In taxonomic terms, a family contains one or more genera and each genera contains one or more species. So a family can contain a minimum of one species but can contain a large number of species. For example, the *Coccinellidae* family of beetles (commonly known as the ladybug beetles) contains approximately 6,000 species of beetles. **Only one species of the family has to survive in order for the family to remain extant**. That alone can account for the apparent discrepancy in the number of families declining by 50% but the number of species declining by 96%.

Another less important but still significant issue that could cause the apparent discrepancy is that the lost 96% of species were *marine*. By 252 million years ago, there were many terrestrial and freshwater species, as well.

88. (A) Two processes, vicariance and dispersal, can explain the distribution of many related organisms.

- ***Vicariance* is the process by which the geographical range of one or more taxa is split by the formation of a physical barrier that reproductively isolates them.** The process of vicariance *typically fragments a group,* resulting in a widespread distribution of descended species. The vicariance hypothesis is supported by data that indicate the time of origin of the new taxa of *Cichlidae* coincided with the separation of the land masses (choice A). This suggests that the diversification was the result of the geographical change. The times of origins of these clades that were estimated by calibrating DNA sequence divergence by many different fish fossils matches the sequence and times of separation of these land masses in the Cretaceous.
- ***Dispersal* is the process by which a species' range increases over time.** The *Psychotria* genus established populations on all the islands of Hawaii by means of dispersal. Choice D suggests, though is not strong evidence for, dispersal. However, you shouldn't assume that a taxon originated in the place where it is presently most diverse. See question 87 for a species whose distribution was (formed) by dispersal.

Although these two processes are usually contrasted with each other, there are many species in which they both occurred (often the break up precedes the dispersal).

Speciation does not require that the populations that dispersed or were fragmented from the original population evolved by adaptation to a new environment (natural selection). **If there is no gene flow between the populations, they may genetically drift independently of each other.**

89. (C) The lack of armadillo fossils predating the formation of a connection between North and South American land masses implies that the armadillos dispersed into North America from South America after the land bridge formed. However, it is *not definitive proof* (choice A), because it is possible that armadillos or an ancestor could have inhabited the region long before the land bridge formed but the fossil evidence was destroyed or just hasn't been discovered yet.

The absence of fossils is not typically "proof" that something didn't exist in a particular area at a particular time. If the fossil record of a particular area at a particular time is very sparse, for example, the absence of a particular type of fossil may not be useful in the attempt to reconstruct the biota of the region during that time.

The presence of fossils in a particular place and at a particular time (determined by its existence in a specific rock layer) is much stronger evidence for the existence of that organism in the proposed time and place than the absence of a fossil suggests that the particular species did not exist at the time and place. That said, a rich fossil record of a particular area and time may represent a true sample of the biota at that time. *Because fossils are rare and their capacity to survive intact through time is subject to many random events, fossil evidence can be difficult to interpret.*

90. (A) The pharyngeal arches are embryonic structures that differentiate into many parts of the head and neck in vertebrates. The exact structure a particular arch develops into depends on the species. In agnathans (jawless fish), the 1st pharyngeal arch gives rise to teeth, and their development is under the control of the Hox genes. (Vertebrate teeth originated in the posterior pharynx of jawless fish over 500 million years ago!)

The development of jaws in vertebrates does not require Hox gene expression. A reasonable hypothesis is that the absence of Hox gene expression in the first pharyngeal arch is what allowed the oral jaw to appear many, many years later. None of the other answer choices make sense given the information in the question.

This question lends itself well to the **process of elimination.** See the box after answer 6 for suggestions on when and how to use it.

91. (B) Development of first teeth was under Hox gene regulation in agnathans (in the pharynx) but this network came under different control during the evolution of the teeth in gnathostome oral jaw. Because most gnathostomes do not possess pharyngeal teeth (some sharks and bony fishes do), the cichlids are an interesting case to study. They possess secondary jaws (the pharyngeal jaws) where the first vertebrate teeth evolved and have teeth where the first vertebrate jaw evolved (oral jaw). This suggests that **the similar patterns of dentition (teeth) in the two jaws appear to be controlled by a common gene regulatory circuit.**

92. The double jaws may have primed them for their great diversification in the three lakes.

Speciation rates vary, especially when adaptive radiation occurs, because new habitats become available . . . "*speciation by niche-iation.*" Extinction rates increase under times of ecological stress.

Convergent evolution **is the evolution of similar features in two different species that develop independently from one another.** The similar features are part of different lineages and usually develop through different pathways, genes and/or developmental programs. *Although the genotype is diverging, the phenotype is converging.*

Adaptive radiation **is the evolutionary divergence of members of the same lineage into a variety of adaptive forms as a result of the inhabitation of different habitats.** Typically, adaptive radiation infers that the speciation results from different use of resources in new habitats that occurs over a relatively short period of time. The three lakes are all very large, so each lake has many "neighborhoods"; in other words, each lake has a diverse range of environments within it, so a cichlid at one end of the lake may never wander out of its neighborhood into other environments. *Cichlid species with very different food sources may never interact with each other.*

93. ***Convergent evolution*** **is the evolution of similar features in two different species that develop independently from one another.** Because the three lakes provide similar environments to their inhabitants, the populations of the different lakes are under similar selective pressures. For example, every ecosystem has decomposers, but the specific species that do the decomposing vary between the ecosystems. Each lake will have algae growing on rocks and at least one population of cichlids that will adapt to using that as a food source.

Adaptive radiation **is the evolutionary divergence of members of the same lineage into a variety of adaptive forms as a result of their inhabitation of different habitats.** In the case of the cichlids, whichever ancestral population or populations initially populated the lakes underwent adaptive radiation as cichlids moved into different areas of the lake and adapted to the local conditions.

94. (C) All phylogenies and cladograms are hypotheses that are revised when new data or techniques for analyzing data become available. Sequencing the entire genomes of all the species in a phylogenic tree is an unrealistic goal. Genomes are complex and not fully understood, so the excess sequence information would not necessarily be useful.

When biologists compare sequences between species, they choose homologous sequences and need to calibrate the "molecular clock" of the particular regions they are comparing. Some genes or pseudogenes accumulate sequence changes faster than others, and species with longer generation times usually have a slower molecular clock because they produce less generations in a given period of time. See answers 6 and 67 for more information about molecular clocks.

See the box after answer 6 for a strategy for answering questions asked in the *negative* and using the process of elimination.

95. (D) Genetic similarities and differences among human populations have been used to trace the dispersal of humans from Africa throughout the rest of the world. **The amount of variation within human populations is lower the farther they are from Africa, as would be expected from successive colonizations.**

The amount of DNA sequence variation within human populations decreases with increasing distance from Africa. Fossils of modern humans date back 195,000–170,000 years ago in Africa. Fossils of modern humans outside of Africa are not found until 80,000 years ago. **The amount of genetic variation within populations of hunter-gatherers south of the Saharan desert, particularly in southern Africa, is much greater than in non-African populations.** This is expected if **colonizing populations carried only a sample of genes from the ancestral population (similar to the founder's effect).** The genetic divergence of some of these African populations suggests they have been diverging for up to 150,000 years. *Genetic difference among non African populations are mostly quite small.*

96.

		4			

4, with an acceptable range from 3–5.

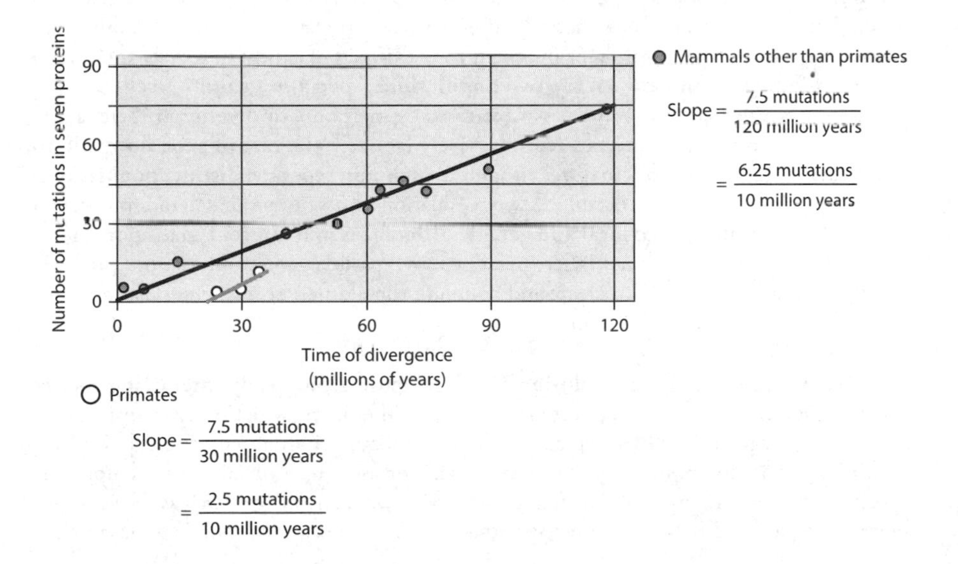

The calculations in the diagram on the previous page show how to calculate the number of mutations (in seven proteins) per 10 million years. The slope for non-primate mammals is straightforward,

$$\frac{(75 \quad 0\ \textit{mutations})}{(120-0\ \textit{million years})}=\frac{(75\ \textit{mutations})}{(120\ \textit{million years})}=\frac{(0.625\ \textit{mutations})}{(1\ \textit{million years}\ )}\times 10=\frac{(6.25\ \textit{mutations})}{(10\ \textit{million years})}$$

The slope of the trend line connecting the dots for primates will likely vary from person to person, but the values should be in the same ballpark of about 2.5 mutations per 10 million years. Because the numbers on the *x*- and *y*-axes are difficult to estimate due to small values and large dots, where the trendline crosses the 30 million year mark gives a good estimate of the number of mutations (in seven proteins) in 30 million years. That number may be as large as 7.5 or as low as 5 depending on your trendline, which would make the mutation rate per 10 million years as high as 2.5 or as low as 1.7.

The difference can then be calculated:

6.25 – 1.7 = 4.55 mutations per 10 million years, round up to 5

Or if you round first, 6 – 2 = 4.

6.25 – 2.5 = 3.75 mutations per 10 million years, round up to 4

Or if you round first, 6 – 3 = 3.

97. The gene sequence can have more differences than the polypeptide sequence but not the reverse. **A change in the DNA sequence will change the nucleotide sequence, but it may not change the amino acid sequence.** However, **all amino acid sequence changes are caused by nucleotide sequence changes**.

98. ***Reproductive isolation* is the reduction or elimination of gene exchange between populations.** It can result from several factors that typically arise from biological differences between populations.

Generally, **gene flow contributes to genetic variation within populations** and can promote adaptation of the population. If gene flow is greater than the pressure of selection, a population will not become genetically differentiated from the surrounding population, even if the particular environment in which they exist significantly differs. **If there is no genetic differentiation between the two populations, speciation cannot occur.**

Reproductive isolation allows two (or more) gene pools to diverge. If there is gene flow (i.e., **the populations are not reproductively isolated, the rate of gene flow will not allow their allele frequencies to differ enough to maintain separate, distinct populations**).

In order for speciation to occur, the two populations must not be able to interbreed, or at least not significantly. **Behavioral and genetic differences may prevent mating from ever occurring.** If mating does occur, there must be incompatible gene combinations produced during the mating. In order to create and maintain these differences, the populations must remain genetically distinct and separate.

99. ***Coevolution* is the joint evolution of two or more ecologically interacting species.** Each of the species imposes a selective pressure on the other, which evolves in response. There are a large variety of interactions these interacting individuals could have: They could be competitors, mutualists, or have a predator–prey relationship, for example. The specific nature and strength of the interaction depend on many factors such as the genotype, environmental conditions, and other species with which the two (or more) species interact.

There are many examples of coevolution in which only one species adapts to the interaction. For example, the orchid *Ophrys apifera* produces a chemical that mimics a bee pheromone that attracts male bees of the species *Tetralonia cressa*. The shape of the flower is such that, when a male bee attempts to copulate with it, pollen adheres to his body. This is considered coevolution because the orchid exploited the ability to specifically manipulate the male *Tetralonia*. As of yet, the bee has not evolved to defend against this exploitation.

In specific coevolution, two species undergo adaptation in response to each other, a sort of "evolutionary arms race." Examples of relationships that likely coevolve include predator and prey, parasites and their hosts, and herbivores and their host plants. The "arms race" can escalate indefinitely, result in a stable equilibrium, a cycle of adaptation, or the extinction of one or both species.

Any adaptation that increases the offensive capacity of the predator or the defensive capacity of the prey clearly has benefits, but the cost of maintaining the adaptation may eventually outweigh its benefits. For example, the production of anti-herbivore compounds may require up to 10% of a plant's energy. There may be other consequences, as well. For example, cucumber plants that produce *cucurbitacins* are able to resist spider mites, but the compound attracts certain leaf beetles.

Taricha granulosa, the rough-skinned newt, produces tetrotoxin (TTX), one of the most potent defenses against predation known. One newt has enough TTX to kill 25,000 mice! Most populations of rough-skinned newts have high levels of TTX in their skin. *Thamnophis sirtalis* is a species of garter snake that feed on these toxic newts but is resistant to the toxin. However, a few populations of rough-skinned newts produce practically no TTX and the *Thamnophis sirtalis* populations sympatric with the nontoxic newts have little to no resistance to TTX.

Finally, there are many examples of coevolution that are also **parallel diversifications**. For example, aphids have evolved specialized cells that house endosymbiotic bacteria that supply them with tryptophan, an essential amino acid.

100. Learning is a change in behavior due to experience. Instincts and fixed action patterns, which are not learned and therefore cannot be unlearned or changed, do not require learning to occur. This allows some animals to respond to specific triggers in their environment without learning and in a mostly fixed way. *Learning is particularly important in animals because most animals move and therefore must be able to navigate their environment and respond to changes relatively quickly.*

Learning is a necessary part of being able to fly, run, or walk. The ability to learn would be an advantage to animals that hunt other animals because they could associate their prey with a particular time and place, for example, so they are more likely to find their prey for a given amount of hunting energy. Animals learn when they get sick from eating something toxic. An animal that learns to avoid harmful food is more likely to survive because it will not expose itself repeatedly to a noxious substance. **The ability to learn is essential for most animals to survive and reproduce.** It is likely that only because animals can learn could they evolve and diversify to the extent that they have. Once the program for learning evolved, the ability to learn complex things became possible. Many birds and mammals demonstrate complex social behaviors and innovation in acquiring skills and making tools.

Chapter 2: Free Energy and Homeostasis

101. (B) Living things are highly ordered and require a constant input of free energy to do work, much of which is designated to maintain order. The loss of order in cells and organisms eventually results in death. Organisms continually take in energy to perform biological work, much of which involves **entropy** reduction. (See the "Entropy" box following this answer for a more in-depth description.)

Processes that increase entropy (spontaneous) can be coupled to those that decrease entropy (nonspontaneous, these processes "do" work). Most energy transformation produces **heat**, the most entropic form of energy, which **cannot be used for work in biological systems**.

A spontaneous (exergonic) process can do work and requires no action from the surroundings.

The terms **exergonic** and **endergonic** refer to the spontaneity of a process. **A spontaneous, or exergonic, process can do work.** No action from the environment is needed to make a spontaneous reaction occur, although in biological systems, enzymes (biological catalysts) may make the process occur faster. A **nonspontaneous, or endergonic, process requires work to be done on it**, in other words, action from the environment is needed, to make the process happen. **An erg is a unit of work** (equal to 10^{-7} joules; you don't have to know what an erg is for the AP Biology exam).

A nonspontaneous (endergonic) process will not occur unless the surroundings take action (do work) on it.

Entropy

Entropy is **a measure of energy dispersal** in a process. A process that results in the increase of entropy can theoretically do work. For example, concentration gradients are more orderly, or less entropic, than the result of their dispersion (after diffusion, the substance would be present in equal concentrations on both sides of the membrane). However, **the diffusion of the gradient can be used to do work**!

The second law of thermodynamics states that the entropy of the universe tends to increase over time. Things *spontaneously* become less orderly ∴ work (or energy) must be input into a system to maintain order.

Maintaining a body temperature of 37°C requires the production of heat through metabolism. Approximately 58% of the energy in the fuel oxidized in the mitochondria is stored in the bonds of ATP. The remaining 42% is lost as heat to the environment, which increases the entropy of the universe (because heat is the most "disorderly" form of energy)!

No matter how much entropy you reduce in your body through cellular work, a greater amount of entropy will be created in your environment as a result.

102. (A) Graph A represents an endergonic process (cannot do work), and graph B represents an exergonic process (can do work). Because the free energy of the products in graph B is less than the free energy in the reactants, the reaction "lost" free energy. This reaction could be coupled with a reaction that requires free energy (a nonspontaneous or endergonic reaction, a reaction that needs work to be done in order to occur).

The synthesis of ATP is endergonic—it requires the oxidation of an organic molecule such as glucose (an exergonic process) to fuel its synthesis. **The hydrolysis of ATP "releases" free energy,** in other words, the products of ATP hydrolysis (ADP and P_i) have less free energy than ATP.

The **activation energy**, E_A, of an endergonic reaction includes the free energy change (see the diagram after answer 104).

Choice C is correct—**the energy changes of the forward and reverse reaction are equal in magnitude with the opposite sign.** This is because the value of the free energy change of a process (ΔG) is calculated by subtracting the free energy of the reactants (initial) from the free energy of the products (final). Since the reactants of the forward reaction are the products of the reverse reaction, and the products of the forward reaction are the reactants of the reverse, the numbers will be the same but the order of subtraction will be reversed:

$$\text{Reaction A: } \Delta G_{RXN} = \Delta G_{PRODUCTS} - \Delta G_{REACTANTS} = 8 - 2 = \mathbf{+6}$$
$$\text{Reaction B: } \Delta G_{RXN} = \Delta G_{PRODUCTS} - \Delta G_{REACTANTS} = 2 - 8 = \mathbf{-6}$$

See the box after answer 6 for a strategy for answering questions asked in the *negative* and using the process of elimination.

103.

–	6				

The value of the free energy change of a process (ΔG) is calculated by subtracting the free energy of the reactants (initial) from the free energy of the products (final):

$$\mathbf{\Delta G_{RXN} = \Delta G_{PRODUCTS} - \Delta G_{REACTANTS}}$$
$$\mathbf{\Delta G_{RXN} = \Delta G_{FINAL} - \Delta G_{INITIAL}}$$

$$\text{Reaction B: } \Delta G_{RXN} = \Delta G_{PRODUCTS} - \Delta G_{REACTANTS} = 2 - 8 = \mathbf{-6}$$

If the free energy of the products is *less than* the free energy of the reactants, the free energy change is negative (−ΔG) and the process is exergonic (or spontaneous). An exergonic (spontaneous) process can do work.

If the free energy of the products is *greater than* the free energy of the reactants, the change is positive (ΔG) and the process is endergonic or nonspontaneous because free energy is needed to get the process to occur. Endergonic processes cannot do work; in fact they need work to be done on them in order to get them to occur. (That wording may sound awkward but is necessary to clearly communicate what's happening in a complex process.) To illustrate, imagine a simple process:

Your pen falls on the floor (spontaneous).
You do work on the pen when you pick it up (nonspontaneous).

104.

	1	0			

To find the activation energy (E_A) from a graph of energy as a function of time, draw a horizontal line (parallel to the *x*-axis) out from the energy of the reactants (as shown in the following figures). Then drop a vertical line from the highest point on the curve (the high-energy transition state) to the horizontal line of the reactant energy. The length of the vertical line represents the activation energy.

- The activation energy (E_A) of reaction A is 10 kJ (12 kJ at the highest point − 2 kJ, the energy of the reactants).
- The activation energy (E_A) of reaction B is 4 kJ (12 kJ at the highest point − 8 kJ, the energy of the reactants).

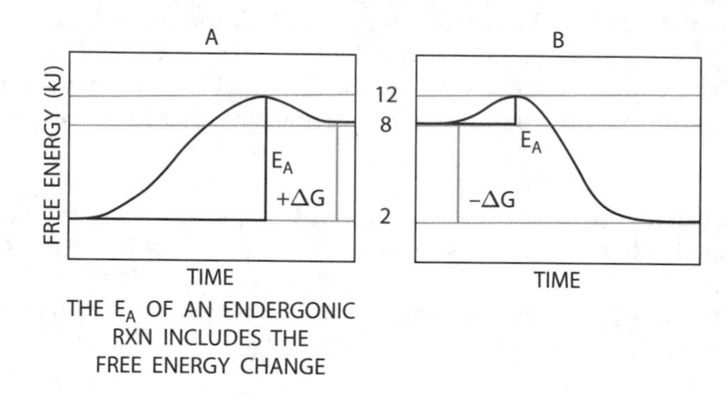

Endothermic (+ΔH) and Exothermic (−ΔH)	**vs**	**Endergonic (+ΔG) and Exergonic (ΔG)**

Endo- and exo**thermic** processes absorb or release **HEAT**.
Ender- and ex**ergo**nic processes require or can do **WORK**.

They are related through entropy and the absolute temperature by the formula:

$$\Delta G = \Delta H - T\Delta S$$

ΔG = free energy change. A **−ΔG** is ex**ergo**nic (spontaneous).
ΔH = enthalpy change. A **−ΔH** is exo**therm**ic.
ΔS = entropy change. A **+ΔS** means the process increased entropy.
T = absolute temperature (in Kelvin). Humans maintain a body temperature of ~310 K.

Do NOT confuse EXOTHERMIC with ECTOTHERMIC!
ECTOthermic describes the mechanism of body temperature regulation of an **ECTOtherm**, an animal that, through its behavior, exploits the environment to gain or lose heat.

105. (B) The viscosity increased because actin and myosin combined in some way to form a complex. The reduction in viscosity by the addition of ATP was likely due to the disassembly of the acto-myosin complex.

You may have learned that ATP hydrolysis during the muscle contraction cycle is coupled to the release of actin from myosin, which is why the absence of ATP results in a rigor-like state (stiffening). If ATP increased the affinity of actin for myosin (choice A), the viscosity of the solution would *increase*. The hydrolysis of ATP during muscle contraction is catalyzed by the ATP-ase activity of myosin, but this experiment alone cannot distinguish which protein, actin or myosin, has the ATP-ase activity, only that one of them does (choice C). Although conformational changes are crucial for the association and dissociation of actin and myosin, it is the hydrolysis of ATP that is "powered" by ATP hydrolysis (choice D).

This is another example of **coupled reactions**—the **endergonic conformational changes** required for actin and myosin to dissociate from each other is **coupled to the exergonic hydrolysis of ATP.**

106. (B) Radiolabeled atoms are useful as tracers because they allow the atom to be followed through a series of reactions. In this experiment, it is important to understand that the carbon and oxygen atoms were provided to the plant in three separate experiments and as substrates—the carbon in CO_2 in one experiment, the oxygen in H_2O in a second experiment, and then the oxygen in CO_2 in the third. The results show that both **the carbon and oxygen atoms in CO_2 end up in glucose** and **the oxygen in H_2O ends up in molecular oxygen (oxygen gas).** See the following reaction.

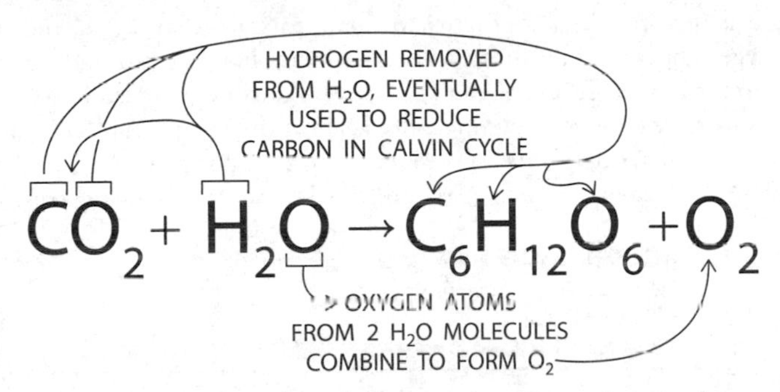

Choice A is incorrect because first, the balanced equation given in the question shows that 2 water molecules are required for each carbon dioxide molecule (on the right side of the arrow). However, 6 water molecules are also produced by photosynthesis, so the equation can also be written as

$$\mathbf{6}\ CO_2 + \mathbf{6}\ H_2O + \text{light energy} \rightarrow C_6H_{12}O_6 + \mathbf{6}\ O_2$$

Choice D is incorrect because the data show that oxygen from carbon dioxide gets incorporated into glucose.

Choice C is incorrect because hydrogen is never traced through this experiment, so the data can't tell you the fate of the hydrogen atom in water.

That's all you need to know to answer this question, but it may be illuminating to understand why using a radiolabeled water molecule would yield difficult results in this experiment.

The radiolabeled atoms are specific isotopes of the atom they are "replacing." *Isotopes* are two or more atoms of the same element (∴ they have the same number of protons) that differ in their number of neutrons.

For example, all carbon atoms have 6 protons:

- ^{12}C contains 6 neutrons (98.9% of all carbon atoms on Earth).
- ^{13}C contains 7 neutrons (~1.1% of all carbon atoms on Earth).
- ^{14}C contains 8 neutrons ($\lll$1% of all carbon atoms on Earth).

To label the water molecule, ^2H or ^3H would need to be used. Remember that in the photolysis, the electrons from the hydrogen are removed, leaving the proton behind (which may get used to create the proton gradient). Because the tiny atom gets broken into subatomic pieces, the isotope is almost impossible to recover at the end of the experiment.

Autotrophs must take in carbon from the environment (usually as CO_2) in order to build organic molecules.

Memorization of the steps of the Calvin cycle, the names of the enzymes, and the names and structures of the intermediates will not be tested on the AP Biology exam! Neither will the names of the specific electron carriers of the thylakoid electron transport chain.

107. (A) Radiolabeled atoms are useful as tracers because they allow the atom to be followed through a series of reactions. In this analysis, three different experiments were performed, one for each source of oxygen.

Choice B is incorrect because there is no data to show that any of the atoms from glucose or oxygen appear in ATP. In fact, the experiment shows that the radiolabeled carbon and oxygen atoms in glucose are recovered in carbon dioxide (why choice C is incorrect). Choice D is incorrect because the atoms of oxygen in the oxygen molecule are recovered in water with hydrogen, originally part of glucose but transported to the electron transport chain via the reduced **coenzymes** NADH and $FADH_2$. See the following equation.

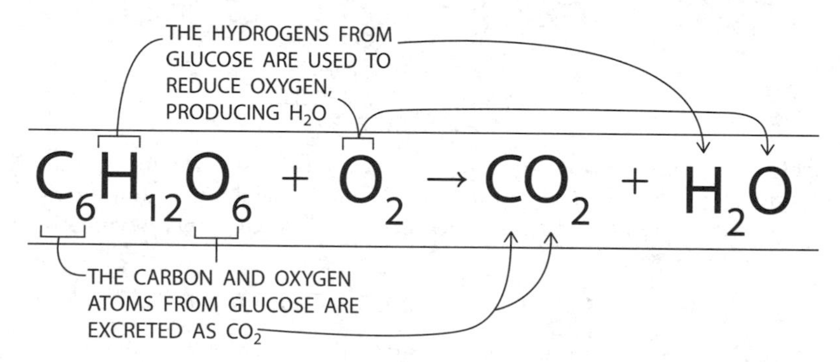

The specific steps, enzyme names, and names and structures of the intermediates of glycolysis and the Krebs cycle will not be tested on the AP Biology exam! Neither will the names of the specific electron carriers of the mitochondrial electron transport chain. You should know ATP synthase, and you should be able to recognize glucose and carbon dioxide.

108. (B) Plants build organic molecules during photosynthesis and oxidize them in cellular respiration. When in light, there will be net photosynthesis, so the CO_2 concentration will decrease and the concentration of O_2 will increase. Plants respire in the light, too, but will produce more O_2 than they use. In the dark, however, the light reactions that produce O_2 will not occur, but the plant will continue to respire. Respiration consumes O_2, reducing its concentration in the sealed container.

That plants use the energy of the sun to fix carbon into organic compounds only to "combust" them later in mitochondria for energy may seem like a complicated way to harness energy: Why not just use the ATP generated in the light reactions directly for energy? **Bacteriorhodopsin, a light-driven H^+ pump in bacteria**, does! But plants need to make all their organic molecules by photosynthesis, and some cells, like root cells, can't do photosynthesis but still require energy. Perhaps the most important reason is historical: The mitochondria and chloroplast were free-living bacteria about 1.5 billion years ago, but were engulfed, sequentially (mitochondria first), by the ancestor of the plants (and photosynthetic "protists"). **See answer 128 for a detailed explanation of endosymbiosis.**

109. (B) Plants appear green because they do not absorb green light, they reflect it. Putting a plant in green light is similar to keeping it in the dark. The light reactions can't use green light to excite electrons; there aren't even pigments that sufficiently absorb it. If the light reactions don't occur, oxygen isn't produced. The oxygen produced during photosynthesis is all from the light reaction and occurs by the **photolysis** of water catalyzed by photosystem II.

Regardless of whether or not the plant is photosynthesizing, **most cells will still respire ∴ O_2 is consumed and CO_2 produced.** In the absence of photosynthesis, cells use molecular fuel sources they already have on hand to produce ATP. Cells with chloroplasts, photochemical cells, can take care of most of their energy requirements by photophosphorylation in the light, but in the dark they respire.

110. (C) Acids typically taste sour. If you've eaten yogurt, you've consumed **lactic acid** and the **bacteria** that make it (*Lactobacillus acidophilus, Streptococcus thermophilus,* and *Lactobacillus delbrueckii* subspecies *bulgaricus*). Lactic acid, as its name suggests, is an acid, that **reduces the pH and causes proteins in the milk to denature**, making them lose their shape, stick to each other, and become clumpy. Cottage cheese is an example of denatured milk proteins (the curds) called casein. Whey is another milk protein, but does not curdle as much as casein.

Almost all (if not all) **bacteria can produce lactic acid via fermentation**, but these species are specific in that they can tolerate the low pH that results from the accumulation of lactic acid in the yogurt environment.

Yeast can also carry out fermentation (they are facultative anaerobes), but they produce ethanol and CO_2 instead of lactic acid.

That all living cells contain the enzymes of glycolysis, from bacteria to yeast to plants and animals, supports the common ancestry of all organisms and their metabolic pathways.

111. (D) The *direct* answer is that the hydrolysis of ATP is highly exergonic ∴ its synthesis is highly endergonic. See the diagrams that accompany question 125 for the structure of ATP and the reaction profile of ATP hydrolysis.

The *ultimate* answer is because all living things are descended from a common ancestor. This foundation of metabolism is evolutionarily conserved.

See the box after answer 86 for an illustration of the uses of **direct** versus **ultimate** causes and consequences.

112. (B) Limiting factors in ecosystems place many limits on the system. Limiting factors exist because **organisms must acquire specific atoms and compounds from the environment in order to function.**

Limiting factors can be identified in several ways, but the most **direct method** is to supplement the system with the suspected limiting factor. If the effect of **supplementation** was an increased rate of growth or synthesis, or an increase in population size, then the added factor was most likely the limiter. If supplementation is then discontinued, the rate of growth or synthesis (or population size) is expected to decrease again. **Limiting factors (determine) the carrying capacity of the environment.**

If phosphates were the limiting factor in this culture, the synthesis of phosphate-containing compounds would be expected to increase after the addition of phosphates. Graph B shows a significant *increase in the synthesis of nucleic acids.* It also shows a slight increase in protein synthesis, which could be the result of increased nucleic acid synthesis (several different kinds of nucleic acid are required for protein synthesis). There is also a slight increase in the rate of lipid synthesis. Phospholipids, the specific class of lipids that compose membranes, contain both nitrogen and phosphate.

Nucleic acids, polymers of nucleotides, contain a large amount of phosphate. Every nucleotide contains one phosphate. Nucleotide-like coenzymes, like NAD^+, $NADP^+$, and FAD also contain at least one if not two phosphate groups. The ribo-nucleotide ATP contains three phosphate groups (technically, it's a nucleoside triphosphate). **Phosphates are often the limiting factor in an ecosystem** (nitrogen is another common limiting factor). **Phosphates are usually absorbed by organisms whole as PO_4^{3-}, or as part of phosphate-containing compounds.** They don't typically get broken down into phosphorus and oxygen atoms.

Graph A shows significant increases in protein and nucleic acid synthesis rates, and a slight increase in the rate of lipid synthesis. Amino acids and nucleic acids contain a significant amount of nitrogen, but only the nucleic acids contain phosphates. However, phospholipids, the specific class of lipid that composes membranes, contain both nitrogen and phosphate.

Graphs C and D both show decreased rates of carbohydrate synthesis. This might result from a lack of light, CO_2, or H_2O for photosynthesis. However, they both increased the rate of lipid synthesis, which would not be expected if rates of photosynthesis were low. The decreased protein and nucleic acid synthesis rates shown in graph C could result from a deficiency of nitrogen (nitrogen is the limiting factor).

Limiting Factors in Ecosystems: The Redfield Ratio

Phytoplankton account for much of the primary productivity of aquatic ecosystems. The ratio at which they assimilate (absorb and incorporate) nitrogen and phosphorus into their cells is called the Redfield ratio.

Redfield ratio ~ 16:1 (nitrogen: phosphorus)

Redfield ratio $<$ 16:1
Nitrogen is limiting.

Redfield ratio $>$ 16:1
Phosphorus is limiting.

Redfield ratio in freshwater lakes usually $>$ 16:1 ∴ phosphorus limiting
Redfield ratio in estuary and marine biomes usually $<$ 16:1 ∴ nitrogen limiting

The College Board does not expect you to know the Redfield Ratio. It is a useful demonstration of a way to identify an ecological abiotic limiting factor. That is, for a given species, in a given environment, growth and reproduction may be limited by the amount of nutrients available.

113. (A) Limiting factors in ecosystems place many limits on the system. Limiting factors exist because **organisms must acquire specific atoms and compounds from the environment in order to function.**

If nitrogen were the limiting factor in this culture, the synthesis of nitrogen-containing compounds would be expected to increase after the addition of nitrates.

Nitrogen taken from the environment (typically in the form of nitrates, NO_3^-, and ammonium, NH_4^+ in autotrophs) is used to build many important biological molecules, including, but not limited to, nucleic acids and amino acids. **Nitrogen is often the limiting factor in aquatic ecosystems.**

See answer 112 for a detailed explanation of each graph and the previous box, "Limiting Factors in Ecosystems: The Redfield Ratio."

114. ***Biological oxygen demand*** **(BOD)** is the amount of dissolved oxygen needed by the organisms in a particular body of water to decompose the organic material present in a specified amount of time. ***Eutrophication*** (or hypertrophication) is the enrichment (or over-enrichment) of water by nutrients. It results in excessive bacterial and/or algae growth, often called *blooms*. When these organisms die, the oxidative breakdown of the detritus leads to an oxygen shortage. Depletion of surface oxygen can suffocate fish and other obligate aerobes.

Human activities that enrich bodies of water are dumping raw sewage and farming in places where the runoff enters the drainage basin.

115. The most direct method to determine a **limiting factor** is to supplement the system with the factor you suspect is the **limiter**. If **supplementation** increased the rate of growth or synthesis of a particular compound or group of compounds, or increased the population, then the added factor was most likely the limiter. Supplementation should then be discontinued. If the rate of growth or synthesis (or population size) decreases, the supplemented factor was limiting.

Experimental conditions:

The community is equally divided into three separate aquariums and allowed to recover for several days (or weeks, depending on what kind of organisms are present).

The constants at the start of the experiment (should be the same for all three tanks):

- concentrations of dissolved nutrients, oxygen, and carbon dioxide
- pH
- temperature
- tank volume
- number and type of organisms
- illumination

Tank C will be the control and receive no supplementation.
Tank N will receive nitrogen supplementation.
Tank P will receive phosphorus supplementation.

The biomass of algae will be determined one week before supplementation, every week during supplementation (for 8 weeks), and once each week for several weeks after supplementation has been suspended.

The biomass of algae will be estimated by both visual inspection (and photographic "recording") of the tank and spectroscopic analysis of the water (water with a high concentration of algae will absorb more red light). If the N or P tanks appear to have a greater concentration of algae compared with the control after supplementation, then nitrogen or phosphorus is the limiting factor.

One source of error in measurement is that the visual inspection of the growth of algae is likely to be inaccurate, imprecise, and difficult to quantify.

See answers 112 and 113 for a detailed explanation of phosphorus and nitrogen as limiting factors in aquatic systems and the **box after answer 112.**

116. (D) Process I represents **glycolysis, a 10-step process** in which the **6 carbons in glucose are rearranged into two 3-carbon molecules of pyruvate.** Glycolysis requires 10 enzymes (which you don't have to know for your AP exam), one for each step of the pathway. **These enzymes are located in the cytosol,** which means that **glycolysis happens in the cytosol.**

Choice A is incorrect because process I occurs in the cytosol. Processes II and III occur in the mitochondria.

Choice B is incorrect because none of the atoms from the glucose molecule become part of the ATP molecule during cellular respiration. In other words, **glucose is not *converted to* ATP during cellular respiration.**

Choice C is incorrect because fermentation is not shown in the diagram. Although molecule B in the diagram has 3 carbons, like lactate, the molecule is further broken down into a 2-carbon (acetate, which combines with coenzyme A to form acetyl-CoA) and a 1-carbon compound (CO_2). Lactate, on the other hand, is eliminated from cells to get metabolized by the liver and heart.

See the following diagram.

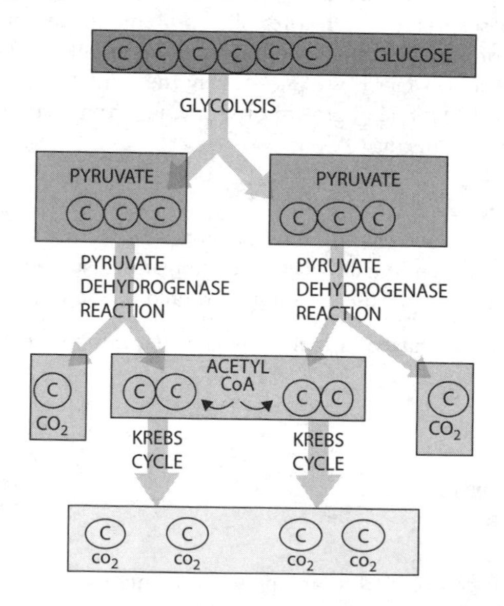

117. (B) Process II is **the pyruvate dehydrogenase reaction, the step between glycolysis and the Krebs cycle that converts pyruvate to acetyl-CoA and carbon dioxide.** Pyruvate must be transported into the mitochondria for this reaction, and the Krebs cycle, to occur (choice A).

Oxygen is *not directly* required for the Krebs cycle. However, it is considered aerobic because the reduced coenzymes $FADH_2$ and NADH must drop off their hydrogen atoms (get oxidized) at the electron transport chain before they can pick up more hydrogen atoms from the Krebs cycle. Although NADH can reduce pyruvate to lactate during fermentation to regenerate NAD^+, $FADH_2$ must get oxidized at the electron transport chain. **The Krebs cycle is considered aerobic because it requires a functioning electron transport chain to continue at "full throttle."**

Processes I, II, and III all require enzymes and coenzymes (NAD^+, FAD, or both) to accept the atoms of hydrogen removed from the glucose fragment by some of the enzymes.

See the box after answer 6 for a strategy on answering questions asked in the *negative* and using the process of elimination.

118. (A) Molecule D is **carbon dioxide, the most oxidized form of carbon** (see the chemical structures that accompany answer 123).

- Molecule A is glucose. Each of the 6 carbons in glucose is oxidized to carbon dioxide by cellular respiration.
- Molecule B is pyruvate.
- Molecule C is acetate (or acetyl-CoA).

The specific steps, enzyme names, and the names and structures of the intermediates of glycolysis and the Krebs cycle will not be tested on the AP Biology exam! Neither will the names of the specific electron carriers of the mitochondrial electron transport chain. You should know ATP synthase, and you should be able to recognize glucose and carbon dioxide.

119. (D) The sun is a source of **photons** (choice A), the **energy needed to energize electrons**. The **energized electrons are removed from chlorophyll and replaced by electrons from water** (generated by **photolysis**). It might appear that the abundance of water would make it a convenient electron donor, but the thermodynamic cost of oxidizing water is high, which is why photolysis is powered by light energy.

120. (A) The electrons end up on **glucose**, carried to the Calvin cycle by NADPH. The C–H bond formed by the reduction of carbon stores energy that can be "released" later by the oxidation of the carbon. **See the picture after answer 123.**

121. (A) The source of high-energy electrons for animals, in other words, mitochondrial oxidation, is glucose. However, fats and occasionally amino acids are also oxidized as fuel.

122. (C) Oxygen is the final electron acceptor at the end of the electron transport chain. Oxygen has a high electronegativity, which makes it a good "electron magnet" at the end of the chain. Once the **oxygen** has been **reduced** by electrons (along with protons), **water is formed.**

$$O_2 + 4\,H^+ + 4\,e^- \rightarrow 2\,H_2O$$

What's the difference between H, H^+, and H^-?

A hydrogen atom, H, is ONE PROTON and ONE ELECTRON. If it loses its electron it becomes a hydrogen ion, H^+.

A hydrogen ion, H^+, is ONE PROTON.

A hydride ion is ONE PROTON and TWO ELECTRONS.

To get oxidized or reduced, an electron must be lost (if oxidized) or gained (if reduced).

In cells, electrons are often lost or gained along with protons, as hydrogen atoms. That is still an oxidation or reduction. For example, the coenzymes NAD^+ and FAD reduced to NADH and $FADH_2$ to carry protons and electrons (as hydrogen atoms) to the electron transport chain in the mitochondria. They are oxidized back to NAD^+ and FAD at the electron transport chain, after they've delivered their hydrogen atoms.

Loss or gain of a hydrogen ion, or proton (H^+) is NOT an oxidation or reduction.

123. To analyze the energetics of organisms, **"follow the electrons."** The processes of photosynthesis and cellular respiration are fundamentally different (see the following table).

- **Because cellular respiration harvests the energy of organic molecules, the process is oxidative. Oxidative processes are generally exergonic.** The electrons will "enter" the process at complex I. It may be useful to imagine that each successive "pass" from one protein to the next is accomplished by each successive electron "acceptor" having a stronger affinity for the electron than the "donor." For example, complex III has a greater electron affinity than complex I, so electrons spontaneously "flow" from complex I to complex III, losing potential energy. The final electron acceptor should have a very high affinity (or electronegativity) for electrons, as it will have to "pull" the electrons away from each previous complex.
- **In photosynthesis, the electrons are energized so they could be used to reduce the carbons in the glucose molecule being synthesized. Generally, reductions are endergonic.** The final electron acceptor must be lower energy than the "electron donor" but high enough to be able to pass the electron off to the organic molecules in the Calvin cycle.

OIL RIG

OXIDATION
Is
LOSS of e^- or hydrogen atoms

REDUCTION
Is
GAIN of e^- or hydrogen atoms

Cellular Respiration		Photosynthesis
The oxidation of glucose is **exergonic**. The energy C–H bond oxidation is stored in the bonds of ATP.	**Energy flow**	The synthesis of glucose is **endergonic.** Fuel for glucose synthesis from ATP synthesized using the energy from light.
The **electrons start on glucose,** which gets OXIDIZED. Electrons are removed from glucose and transported to the ETC via NADH and $FADH_2$. They **end up on water**.	**Electron flow**	The **electrons start on water.** They are removed and energized by light energy, and transported via NADPH to the Calvin cycle where they will be used to reduce the carbon chains. They **end up on glucose.**
Water is produced by the **reduction of oxygen.**	**O_2 & H_2O**	Oxygen is produced by the **oxidation (photolysis) of water.**
Glucose	**Source of electrons for electron transport**	**Water**
High-energy electrons in the bonds of the glucose molecule	**Source of "electron energy"**	**Low-energy electrons from water excited by light energy**
O_2	**Final electron acceptor**	**$NADP^+$**

Follow the Electrons....

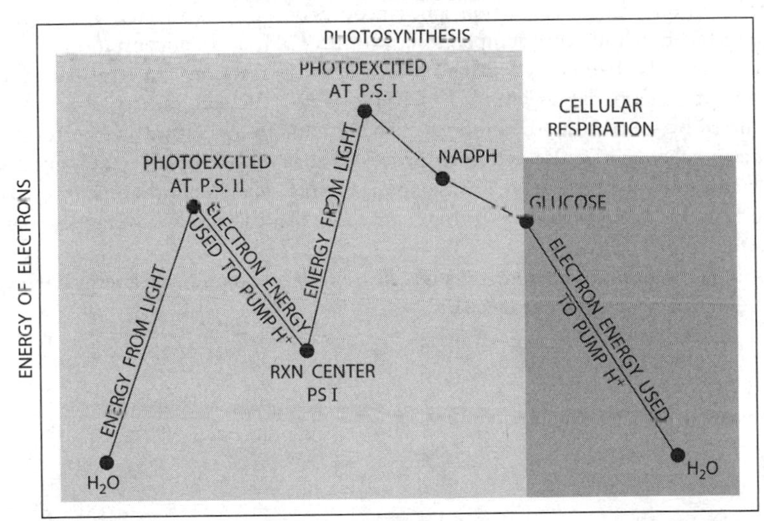

The diagram follows the electrons from a molecule of water, through photosynthesis and into cellular respiration via glucose. The energy changes and time courses are not to scale.

124. (A) Proton gradients store energy. Their energy is potential, like water at the top of a waterfall or dam. Water flows down spontaneously, due to gravity. As it falls, its potential energy is transformed into kinetic energy, heat, and sound energy. If a waterwheel is placed in the water's path, some of the energy of the water will be used to turn the wheel itself, but it could also be transformed to other forms of energy or used to do other work.

When the protons are allowed to diffuse across a membrane through ATP synthase, the energy from the passive flow of protons down their concentration gradient is transformed into the chemical energy of ATP. **ATP synthase is the mitochondria's and chloroplast's version of a waterwheel.**

A great video on YouTube that animates the workings of ATP synthase is called *Gradients (ATP Synthases)*. It was created by the NDSU Virtual Cell Animations Project and is available to view at https://www.youtube.com/watch?v=3y1dO4nNaKY.

Lightbulbs glow (the incandescent kind) because friction and resistance in the filament produce heat (choice B). Although a combustion engine oxidizes fuel to carbon dioxide and water (choice D), it does not do chemiosmosis!

All organisms use chemiosmosis. This fact supports the common ancestry of all organisms and their metabolic pathways.

125. (D) The activation energy (E_A) is independent of the free energy change in an exergonic reaction. **A high E_A is an indication of a stable molecule** (at least under the conditions in which it was measured).

The stability of the products is greater than the reactants in an exergonic reaction. Processes that result in increased stability tend to produce lower energy states.

Why is the hydrolysis of ATP spontaneous but the phosphorylation of ADP is nonspontaneous?

The direction a chemical reaction spontaneously occurs in is affected by several factors (including but not limited to temperature, pressure, pH, and concentration of products and reactants). The free energy change of a process or reaction can be calculated using the Gibbs free energy expression: $\Delta G = \Delta H - T\Delta S$. The spontaneity of a process can be calculated by subtracting the product of the absolute temperature and entropy change from the enthalpy change. **Reactions or processes that release heat ($-\Delta H$, exothermic) and increase entropy ($+\Delta S$) are always spontaneous.** Many endothermic reactions with positive entropy changes (increase entropy) are spontaneous at 37°C (average human body temperature).

"Going downhill" is what makes a reaction spontaneous. Lower energy corresponds to higher stability and greater entropy.

See the box after answer 6 for a strategy for answering questions asked in the *negative* and using the process of elimination.

126. (A) The prokaryotes are the most metabolically diverse of all life. It is not that one type of prokaryote is capable of all types of nutrition, but there is one or more species that can do just about any kind of nutrition.

Choice B is incorrect because animals obtain energy from the oxidation of ingested organic molecules (humans are HETEROtrophs). The high-energy electrons needed for the mitochondrial electron transport are obtained from the ingested (exogenous) organic molecules (**chemo**trophs, see the following chart). An organism that cannot fix carbon (therefore, needs to ingest preformed carbon chains) and obtains high-energy electrons from exogenous molecules (which may or may not need to be organic, depending on the organism) would be a chemoheterotroph.

Plants can synthesize organic compounds from inorganic compounds (they're autotrophs). They obtain low-energy electrons from water but use the energy from visible light to energize them (photoautotrophs) for electron transport in the chloroplast.

The following table compares the four basic types of nutrition. The nutrition of an organism depends on two things:

(1) CAN IT FIX CARBON or DOES IT NEED TO CONSUME PREFORMED ORGANIC MOLECULES?
(2) What is the source of high-energy electrons for electron transport chains? *Does it require low-energy electrons to get energized by light* or **does it consume exogenous compounds that contain high-energy electrons**?

	High-energy electrons from exogenous compounds	*Electrons excited by light*
CAN FIX CARBON	**chemo**AUTOtroph Many species of bacteria	*photo*AUTOtroph Plants, protists with chloroplasts, many species of bacteria
NEEDS TO CONSUME PREFORMED ORGANIC MOLECULES	**chemo**HETEROtroph Animals, protists that lack chloroplasts, many species of bacteria	*photo*HETEROtroph Many species of bacteria

The College Board does not expect you to know the names of the types of nutrition in the table. The purpose of the table is to compare the four ways organisms can harness energy from their environment.

127. ***Chemiosmosis* is the mechanism by which a hydrogen ion gradient is used to drive an energy-requiring process, like the synthesis of ATP.** In mitochondria and chloroplasts, ATP synthase performs this function.

Building a gradient requires energy. If the ultimate goal of chemiosmosis is the synthesis of ATP, the use of ATP to build the proton would, at best, result in no net ATP (and in reality, no energy transformations in the cell are 100% efficient, so you're going to use more energy than you harness). The energy to drive the proton pumps is harnessed from the electrons delivered to the electron transport chain by NADH or $FADH_2$ in the mitochondria and by an electron acceptor in photosystem II in the chloroplast.

In the mitochondria, the movement of electrons from NADH and $FADH_2$ to oxygen, creating water, has a (very) negative free energy change; that is, it is exergonic, spontaneous, and downhill. The potential energy of the electrons is lowered, which means there was a conversion of potential energy to other forms of energy. The change in potential energy, from lower stability glucose to higher stability water is what drives the protons against their concentration gradient. This is a coupled process. A downhill process can drive an uphill process if they are coupled. **Every endergonic process has a price.**

See answer 124 for an explanation of how a proton gradient stores potential energy.

The potential energy of the gradient is raised with each proton that enters the high-concentration compartment and lowered with each proton that leaves. The spontaneous flow of hydrogen ions from high to low concentration drives the endergonic synthesis of ATP.

The exergonic processes are the downhill, driving reactions. They spontaneously happen. The endergonic processes require work. The energy to do the work is harnessed from the reaction it is coupled with. **Coupled processes occur concurrently in the same place. They are physically and temporally linked.** The first two reactions occur at the cytochrome complexes. The oxidation-reduction reactions of the transport of the electrons down the chain provide the energy for proton translocation. The cytochrome complexes are multi-protein complexes composed of electron carriers and a proton pump.

The second two processes are coupled through ATP synthase. The giant, multi-subunit protein complex also translocates protons, but in the direction of their diffusion, harnessing the energy of their diffusion for the synthesis of ATP from ADP and P_i.

128. Mammalian cells contain between 800 and 2,500 mitochondria. Erythrocytes (red blood cells) contain none. The range in the number of mitochondria is from as few as 1 or 2 to as many as a half million!

The mitochondria and chloroplast as obligate cellular symbiotes is an example of co-evolution (**see answer 99**) that is also a **parallel diversification**.

If the AP Biology exam asks for four pieces of evidence, provide only four. They will give you points only for the first four if you provide more, so giving more will only waste time. Choose the four you can explain best.

Because there's a lot of evidence to support the endocytic origin of the mitochondria (and chloroplast). and the AP Biology readers are not grading this answer, more than four are provided here:

(1) Mitochondria (and chloroplast) are similar in size and shape to many prokaryotic cells.
Strength of the evidence: Suggestive, but not definitive "proof."

(2) Endocytosis wraps the engulfed particle within a membrane derived from the plasma membrane of the engulfing cell. Mitochondria have two membranes, inner and outer membranes, which are structurally and biochemically (composition) different. The inner membrane is highly convoluted (into cristae), like that of the plasma membrane of an aerobic prokaryote. The lipid and protein composition of the inner membrane is more similar to a eubacteria than a eukaryote.

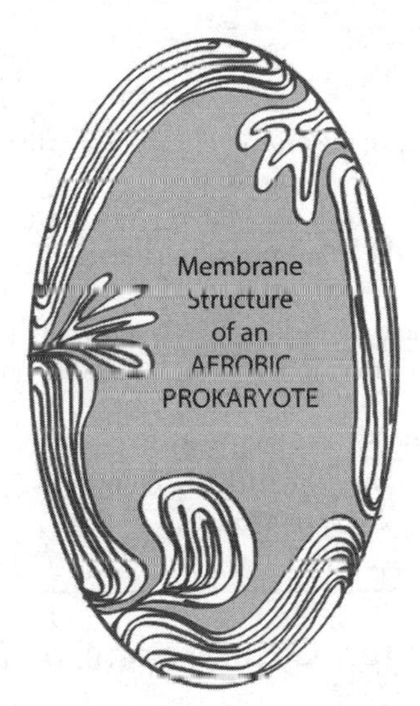

Strength of the evidence: Suggestive, but not definitive "proof."

(3) The mitochondria (and chloroplast) can reproduce semi-autonomously through binary fission (not mitosis). The number of mitochondria in a cell can change throughout the cell's lifetime regardless of whether it has undergone any cell division.
Strength of the evidence: Suggestive, but not definitive "proof."

(4) The mitochondria, like eubacteria, has its own circular chromosome that lacks histones and introns.
Strength of the evidence: Pretty convincing.

(5) Many of the genes from the original mitochondria have been transferred to the nucleus (it's called endosymbiotic gene transfer, and it makes the mitochondria dependent on its host cell).

Strength of the evidence: Pretty convincing if the method for determining the nuclear genes originated from the mitochondria.

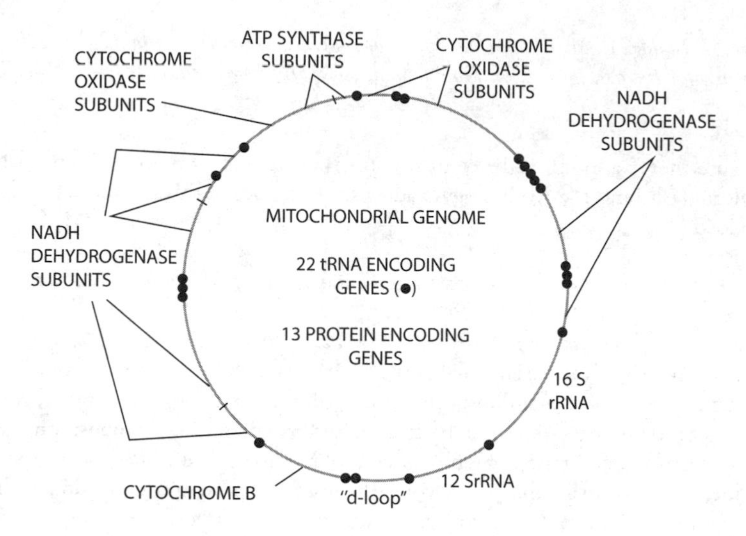

(6) The mitochondrion has its own ribosomes. The large subunit of the mitochondrial ribosome is more similar to the prokaryotic ribosome than the ribosomes in the cytosol.
Strength of the evidence: Pretty convincing.

(7) Antibiotics that work by blocking protein synthesis in eubacteria (like streptomycin) also block protein synthesis in mitochondria (and chloroplasts) but not cytosolic ribosomes.
Antibiotics that work by blocking RNA polymerase in eubacteria block RNA synthesis in mitochondria (and chloroplasts) but not in the eukaryotic nucleus (rifampicin).
Inhibitors of eukaryotic protein synthesis (like the diphtheria toxin) do not inhibit protein synthesis in the mitochondria (or chloroplast).
Strength of the evidence: Pretty convincing.

The Oxygen Catastrophe of 2.3 BYA

Aerobic metabolism (and multicellularity) evolved after the oxygen gas released by cyanobacterial photosynthesis significantly accumulated in the atmosphere. Most organisms could not tolerate the oxidizing atmosphere. This "oxygen catastrophe" caused many prokaryote species to become extinct.

The original endocytosis of the mitochondria occurred between 1.7 and 2 billion years ago and was not followed by its digestion. It's possible that the organelle initially served to detoxify oxygen for its host by converting it to water. Because the reduction of oxygen to water is spontaneous, any cells that could use oxygen to drive electron transport chains would certainly have had an advantage.

The free energy change from the spontaneous, exergonic movement of electrons from organic molecules to oxygen is very large, meaning that a lot of energy is available to couple the synthesis of ATP.

129.

	3	1			

The oxidation of palmitoyl CoA in the mitochondria produces **8 acetyl-CoA** and **7 NADH**.

The complete oxidation of each acetyl-CoA produces **2 CO_2** and **3 NADH** ∴ the oxidation of 8 acetyl-CoA produces **16 CO_2** and **24 NADH.**

7 NADH (1 palmitoyl CoA → 8 acetyl-CoA)
\+ 24 NADH (8 acetyl-CoA → 16 CO_2)
= **31 NADH**

130.

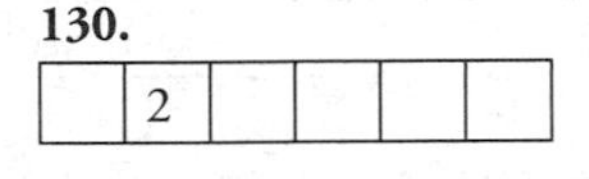

pyruvate → Acetyl-CoA + CO_2

1 glucose →

pyruvate → Acetyl-CoA + CO_2

The splitting of glucose in the cytosol produces **2 pyruvate** molecules from each molecule of glucose. **Each pyruvate molecule produces 1 acetyl-CoA and 1 CO_2 molecule**

131.

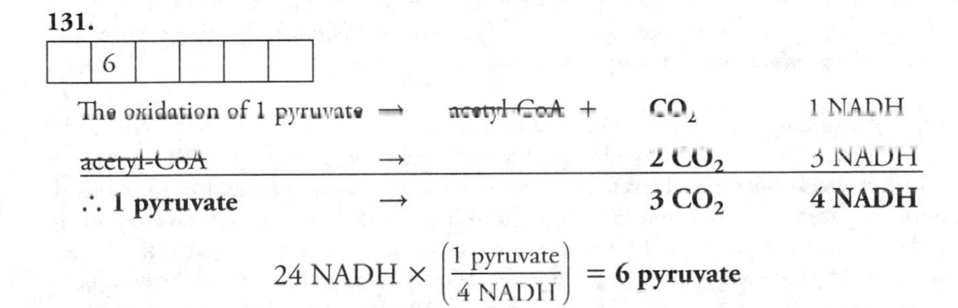

132. (B) Carbon dioxide is an inorganic form of carbon (organic molecules have one or more carbon-hydrogen bonds). The incorporation of carbon dioxide into an organic molecule (or the reduction of carbon with hydrogen) is called **carbon fixation** and is **the defining characteristic of the autotrophs.**

The 3-carbon **product of the Calvin cycle** (glyceraldehyde-3-phosphate or G-3-P, also called phosphoglyceraldehyde or PGAL) can be **used to synthesize sugars, amino acids, lipids, nucleic acids, and all of the other molecules a plant needs** (although they need to take in atoms such as **nitrogen, sulfur, and phosphorus from the soil**).

Choice C is incorrect because light excites, not photons, during light-driven processes in cells. **Photons are particles of light.** Choices A and D describe cellular respiration, not the Calvin cycle.

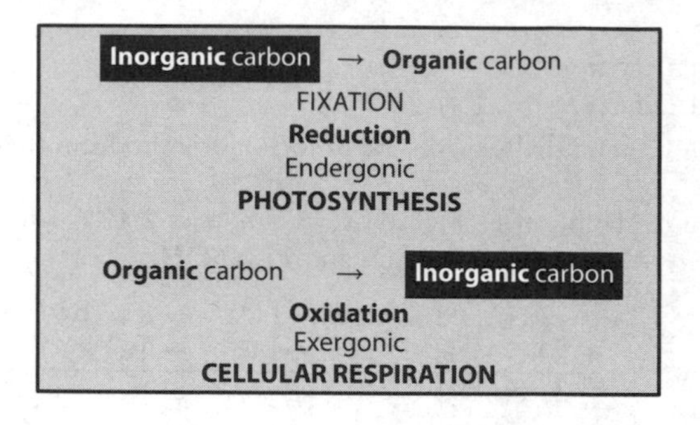

10^{11} tons of carbon fixed globally	
⅔ by terrestrial plants	⅓ by marine microrganisms

133. (A) Synthesis requires energy (it is not spontaneous). The **energy for the synthesis** of organic compounds by the **Calvin cycle** is provided by the **ATP synthesized in the light reactions**. The hydrogen atoms needed for the reduction of carbon in the process are supplied by NADPH (also produced by the light reactions). The *ability to fix carbon* is a defining characteristic of *autotrophs* and the *inability to fix carbon* is a defining feature of *heterotrophs.*

Enzymes are required for practically all of the chemical reactions that occur in cells. Enzymes do not supply the reaction with energy. They **lower the activation energy**, the energy barrier, so that **reactions occur faster**. ***They do not change the energy difference between the product and reactant.***

134. (C) Animals store carbohydrate energy as glycogen; its structure allows it to be rapidly broken down and used for glycolysis. Vertebrates store the majority of their glycogen in their **skeletal muscle and liver.** Skeletal muscle can take up glucose from the blood, oxidize it or store it as glycogen, and remove it from glycogen for oxidation, so it definitely plays a role in the regulation of blood glucose levels (thus choice D is incorrect). Skeletal muscle *can* break down glycogen or it would be deleterious to store it because they couldn't use it (thus choice B is incorrect). In other words, skeletal muscle cells take up glucose out of the blood and store it as glycogen for their own "in-house" use; they never release glucose back into the blood. In contrast, liver cells store glucose as glycogen and also release glucose back into the blood to prevent blood glucose levels from falling.

135. Each ATP is recycled approximately **900 times,** with an acceptable range of 800–1,000.

(1) Calculate the amount of ATP, in moles, that can be made from dietary energy:
At **40% efficiency, 4,680 kJ of dietary energy** is used to synthesize ATP (40% of 11,700 J).

If **50 kJ** of energy is stored by **1 mole of ATP**, 4,680 kJ of energy would fuel the synthesis of 93.6 moles of ATP:

$$4{,}680 \text{ kJ} \times \left(\frac{1 \text{ mole ATP}}{50 \text{ kJ}}\right) = 93.6\text{, round to 90 moles of ATP}$$

(2) Calculate the amount of ATP in the body in terms of moles.

The average 70-kg person contains ~50 grams. One mole of ATP has a mass of 551 grams ∴ **~0.1 moles:**

$$50 \text{ grams ATP} \times \left(\frac{1 \text{ mole ATP}}{551 \text{ grams}}\right) = 0.09, \text{ round up to } \mathbf{0.1 \text{ moles of ATP}}$$

(3) Compare the amount of ATP made from dietary energy to the amount available in the body.

$$\frac{90 \text{ moles of energy to make ATP}}{0.1 \text{ moles of molecules available}} = \text{each molecule is recycled } \mathbf{900 \text{ times}}$$

A more efficient way to solve the problem is to leave out the mole out of the first calculation because 551 grams of ATP (one mole) requires 50 kJ of energy to produce ∴

$$4{,}680 \text{ kJ} \times \frac{551 \text{ g ATP}}{50 \text{ kJ}} = \mathbf{51{,}574 \text{ g ATP}}$$

$$\frac{51{,}574 \text{ g ATP used per day}}{50 \text{ g ATP in the body}} = \mathbf{1{,}031 \text{ times}} \text{ each ATP molecule is recycled}$$

*Important: You don't need to understand exactly what a mole is to answer this question, but if you do this problem will certainly be easier. One of the **7 Science Practices** the College Board expects students to engage in during their AP adventure is **the appropriate use of mathematics**. The AP Biology exam will assess your ability to apply mathematics to **solve problems, analyze data, make predictions, and describe natural phenomenon**. You should be comfortable **quantifying data, estimating answers,** and **describing processes symbolically**. Be ready to **justify your choice of mathematical tools**—they should be appropriate for the situation.*

136. (C) Aquaporins are water channels (∴ passive) expressed by cells in all kinds of organisms including bacteria. For decades, scientists thought that water crossed cell membranes by simple diffusion (by leaking through the lipid bilayer with no help from transport proteins). However, the rate at which water was transported across actual cell membranes (with membrane proteins) far exceeded the rate at which water crossed the lipid bilayer's vesicles synthesized in a laboratory. It was concluded that osmosis occurred through facilitated diffusion and in 1992 the aquaporin proteins were discovered!

Water transport is ALWAYS PASSIVE.

Water always moves from an area of greater "concentration" of water (relative to solute) to an area of "lower concentration" of water (relative to solute). In other words, water moves **from HYPOTONIC** solutions (low, or no, solute concentration) **to HYPERTONIC** solutions (greater solute concentrations).

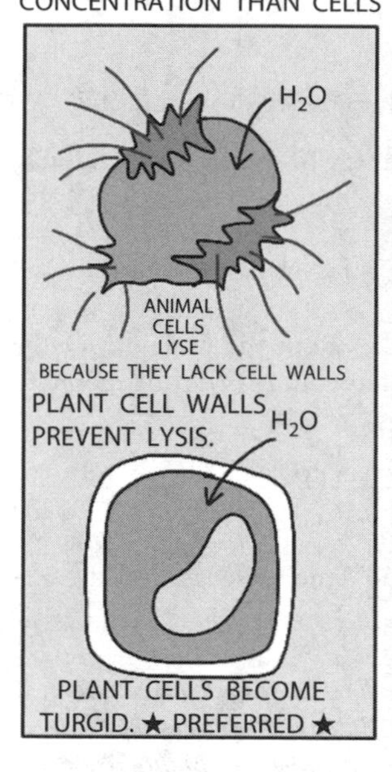

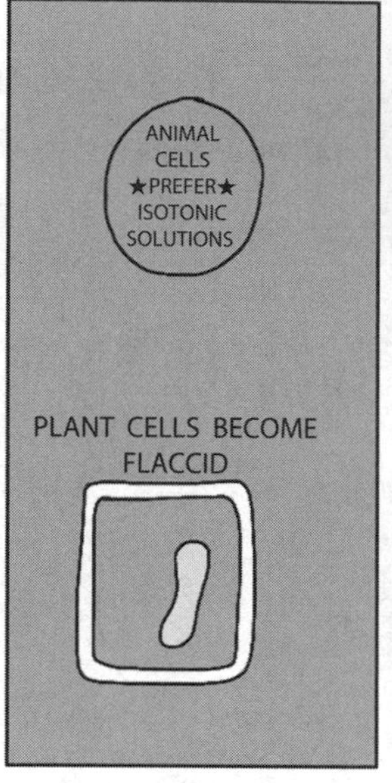

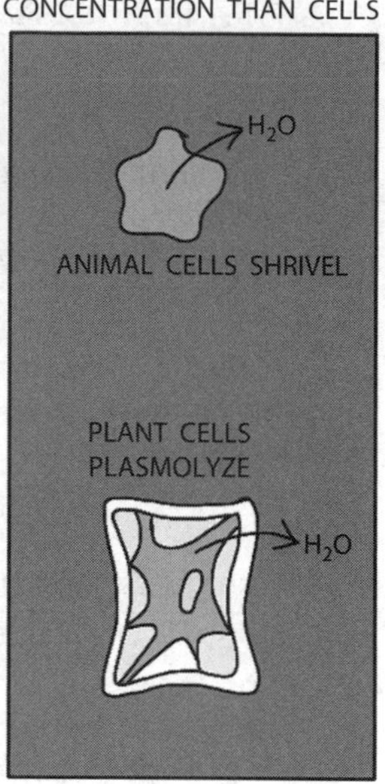

137. **(D) Cells bathed in an isotonic solution**, that is, a solution of the same osmolarity as the intracellular fluid, **will not gain or lose water to the surroundings**.

The cells in diagram A have shriveled because water has moved out of the cells. Water, like solutes, moves from where there is a greater concentration of it to where there is a lesser concentration of it. Cells lose water to a solution because the solution has a higher concentration (greater osmolarity, higher osmotic pressure) than the cells. The cells in diagram B have swelled because they took up water from the environment, which means that the contents of the cell are hypertonic to the contents of the solution.

All cells are permeable to water and it's the movement of water that will most often determine whether a cell lyses or shrivels (or becomes turgid or plasmolyzes, see the figure after answer 136).

138. **(C)** The cells in diagram A have shriveled because water has moved out of the cells. Water, like solutes, moves from where there is a greater concentration of it to where there is a lesser concentration of it. **Cells lose water to a solution because the solution has a higher concentration (greater osmolarity, higher osmotic pressure) than the cells.**

All cells are permeable to water, and it's the movement of water that will most often determine whether a cell lyses or shrivels (or becomes turgid or plasmolyzes; see the figure after answer 136).

139. **(A)** Cell membranes separate the internal environment of the cell from the external environment. **Animal cells do not have cell walls, because animals carefully regulate the composition of the extracellular fluid.**

All cells have membranes regardless of whether or not they have a cell wall. Most unicellular organisms have cell walls, but there are several "protists" such as amoeba, paramecium, and euglena that do not. **Multicellular organisms such as plants and fungus**

(yeast are unicellular) have cell walls to protect cells from hypotonic environments by preventing them from lysing when in dilute solutions. Hypotonic solutions cause plant and fungus cells to become turgid as their cells fill up with as much water as they can physically hold, limited only by the volume the cell wall encapsulates. **Turgor pressure makes cells firm and helps plants and fungi maintain their body structures.**

See the box after answer 6 for a strategy for answering questions asked in the *negative* and using the process of elimination.

140. (A) Graph I represents simple (free) diffusion. **The rate of transport via simple diffusion increases linearly with increasing concentration.**

Graph II represents facilitated diffusion. **The rate of transport via facilitated diffusion increases rapidly with increasing concentration and then "levels out" because the number and type of transporters places a limit on the number of particles that can simultaneously be transported.** When all of the transporters are working as fast as possible, no greater rate of transport can be achieved by increasing concentrations.

Simple and facilitated diffusion are both **passive** and **concentration dependent**, but facilitated diffusion increases quickly with increasing concentration, even at low concentrations and then "maxes out," whereas the rate of simple diffusion increases with increasing concentration at a steady rate. This is **a way to distinguish** between the two types of diffusion.

141. (A) The graph "calibrates" the transporters for 100% transport at 5 mM (millimolar, mmol/L) plasma glucose concentrations, about the average, fasting plasma glucose concentration for nondiabetic humans. **A transporter with a low K_M value (relative to physiological concentrations) functions at a relatively constant rate independent of fluctuations in the concentration of glucose, at least within physiological range.** The rate of glucose transport at 10 mM concentrations would be about 20% greater than that at 5 mM with a 5 mM K_M transporter, and over 50% greater than that at 5 mM with 20 mM K_M transporter.

142. (D) The rate of transport increases significantly with increased concentration (within physiological range) with a transporter with a high K_M. **The K_M value should be large relative to physiologic concentrations,** allowing increased plasma concentrations to increase transport rates, decreasing the time it takes for concentrations to equilibrate across the membrane.

The AP Biology exam does not require specific knowledge of any particular transport protein. However, the basic mechanisms of passive and active transport are considered "essential knowledge." The specific concept of K_M is unlikely to be directly tested on the AP exam; however, you should be comfortable analyzing data that is unfamiliar to you. The AP exam will provide you with the specific details you need for analysis.

143.

	0	.	3	5	

0.35, with an acceptable range of 0.32–0.38. **You can estimate the solute concentration of potato cells by estimating the sucrose concentration at which the potato cores experienced no change in mass** (see the following graph). At concentrations below 0.35 M the cores *gained* mass because water was drawn into the more hypertonic potato cells from the hypotonic solution.

At concentrations above 0.35 M, the potato cells lost mass because water moved from the hypotonic potato cells into the hypertonic sucrose solution.

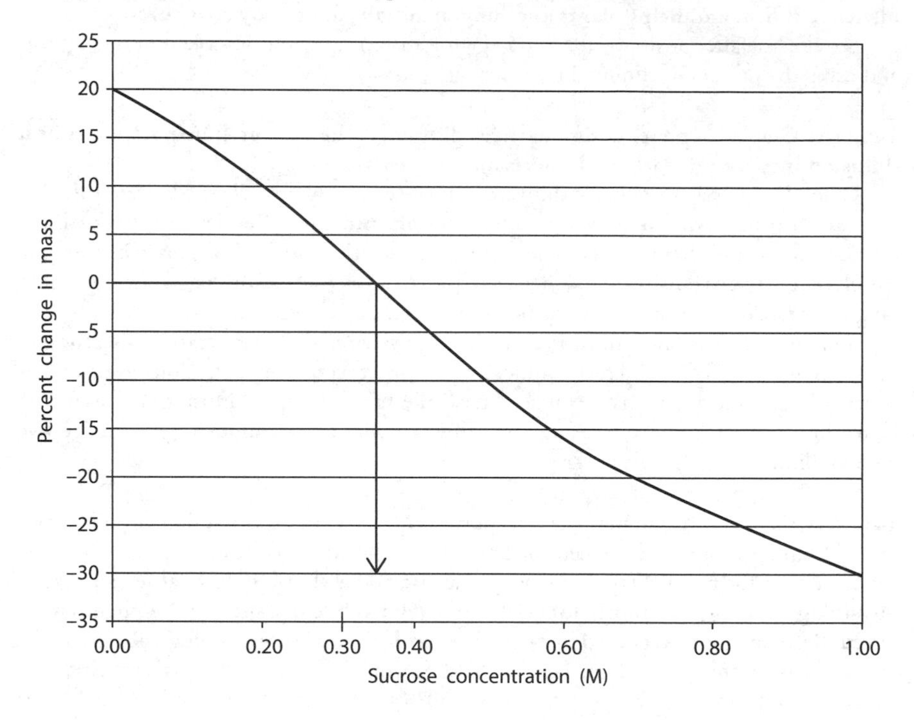

Five data points were used to construct the graph (the gray dots). Because there is a clear, linear relationship between the points **interpolation** is a reliable way to estimate the value of one variable relative to the other. **Interpolation is the estimation or calculation (if you have the formula for the line) of an *x*-value from a *y*-value (or a *y*-value from an *x*-value) by using the "points" on the line that connect the measured data points** (these values were not specifically measured, so technically their values are tentative). There must be data points on each side of the point of interest. **Interpolation is used regularly in many methodologies**, like spectrophotometry (colorimetry), for example, **where a standard curves are constructed.**

Estimating a value outside of the actual data is called **extrapolation** and is typically considered much less reliable because **the relationship is assumed to continue outside the measured range.**

144. The simplest, most straightforward experiment is to separate and weigh the plant parts, dry them, and weigh them again. The parts that lose the greatest percentage of their mass in the drying process contained the most water.

Important aspects of the experimental design include:

- Use **several plants** of the **same species** grown in the **same (or very similar) conditions**.
- Plants should be carefully cleaned after harvesting to remove dirt and contaminants but not in a way that would dessicate (dry out) or hydrate them.

- Plant parts, the roots, stems, and leaves and large parts, should be cut into smaller pieces if needed.

Another way to measure the experiment is a methodology similar to that of question 143. **Use ONE concentration of sucrose** (or another solute that cannot pass through cell membranes). Each root, stem, and leaf sample should be massed immediately prior to immersion into the solution. The container of solution should be large enough so that all of the plant parts can be immersed without contacting the other plant sections. The immersion should be left for several hours or overnight (ideal). The next day, the plant sections should be massed immediately upon removal (and briefly drying by patting off the solution) from the solution. **The plant part that shows the <u>greatest</u> percent change in mass *lost* in a *hypertonic* solution or <u>least</u> percent *gain in mass* in a *hypotonic* solution is the plant part with the greatest percent water by mass.**

145.

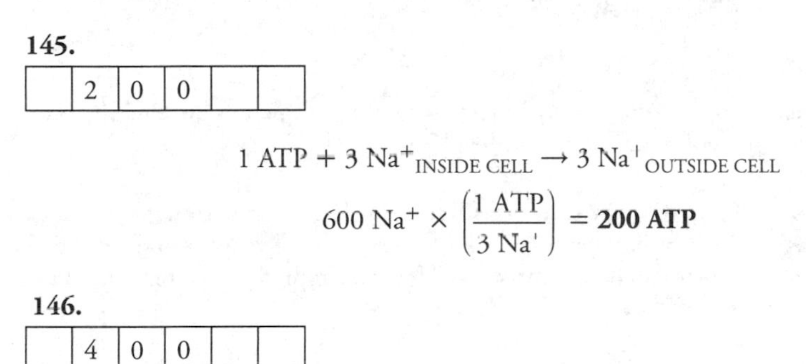

	2	0	0		

$$1 \text{ ATP} + 3 \text{ Na}^+_{\text{INSIDE CELL}} \rightarrow 3 \text{ Na}^+_{\text{OUTSIDE CELL}}$$

$$600 \text{ Na}^+ \times \left(\frac{1 \text{ ATP}}{3 \text{ Na}^+}\right) = \mathbf{200\ ATP}$$

146.

	4	0	0		

$$1 \text{ ATP} + 2 \text{ K}^+_{\text{OUTSIDE CELL}} \rightarrow 2 \text{ K}^+_{\text{INSIDE CELL}}$$

$$200 \text{ ATP} \times \left(\frac{2 \text{ K}^+}{1 \text{ ATP}}\right) = \mathbf{400\ K^+}$$

147. Muscle (including cardiac) cells and neurons are electrically excitable cells. They use the Na^+ and K^+ ion gradients to depolarize their membranes. Active neurons and muscle tissue are expected to have the highest rate of energy expenditure because they are continually "releasing" their gradients through opening channels and then pumping the ions back to the side of the membrane in which they are more concentrated.

Neurons that are firing many action potentials and muscle cells that have a high frequency of contraction are expected to use an even greater fraction of their energy on maintaining Na^+/K^+ gradients. **Up to 70% of the energy expenditure in neurons is spent on maintaining Na^+/K^+ gradients.**

148. The stabilization of the Na^+, K^+-ATPase by ouabain inhibits transport. The increased stability of the pump makes it move to the next position because it has a lower energy. Transitioning to the next configuration would require a greater input of energy to occur (see a picture illustrating the relationship between energy and stability after answer 125).

Ouabain would be very useful in *in vitro* studies of ion transport or the Na^+, K^+-ATPase transporter. It would be helpful in studying cells, like muscle and neurons, which use Na^+ and K^+ flux across the membrane to function. It may be useful *in vivo* to temporarily stop a particular muscle cell or neuron from functioning and then examine the effects. Finally, the study of ouabain would be helpful in finding an antidote to ouabain and ouabain-like compounds.

149. All three molecules share the same basic structure—the steroid nucleus and the 4-carbon oxygen-containing ring, so that is likely the part of the molecule that binds to the Na^+, K-ATPase transporter.

OUABAIN

STROPHANTHIDIN

DIGITOXIGENIN

The sugar attached to ouabain does not affect its binding properties but does increase its water-solubility making it more soluble in the plasma.

150. Animal cells swell and burst by osmosis. If animal cells are, or become, hypertonic to the solution they are bathed in, the result is a large, net influx of water into the cell that increases its volume more than can be accommodated by its membrane. Cells may also burst as a result of cell death.

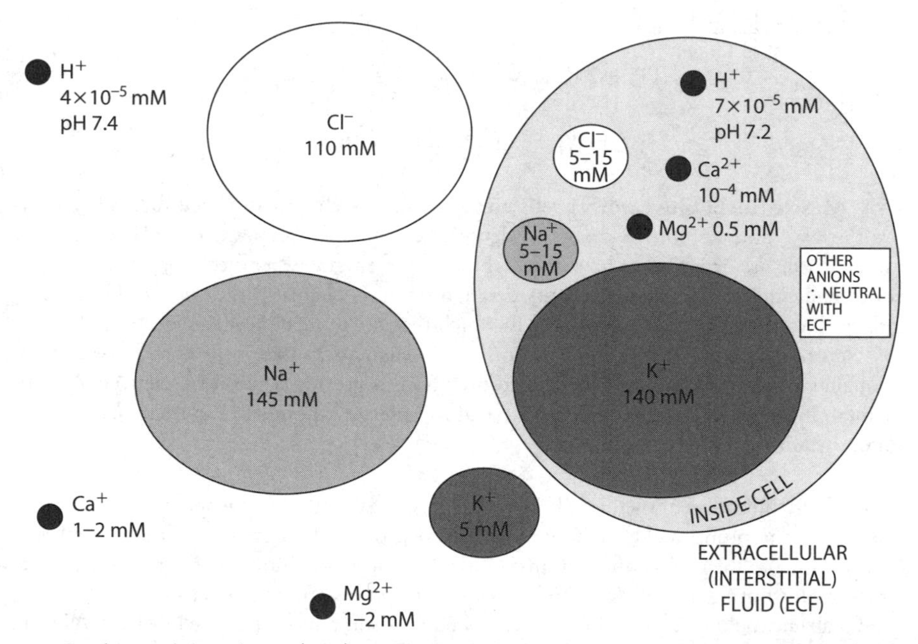

Ouabain inhibits the Na^+, K^+-ATPase ∴ there is little or no active transport of Na^+ and K^+ ions. **With no active transport, K^+ would leak OUT of the cell and Na^+ would leak INTO the cell.** Because the volume of the plasma is much greater than the volume of an individual or even several cells, **almost all of the K+ will leak OUT of the cell, but Na^+ ions will continue to leak INTO the cell until it equilibrates with the solution at 145 mM.**

The Na^+ gradient is necessary for driving the transport of many nutrients into animal cells and plays a critical role in regulating cytosolic pH. Additionally, high intracellular sodium concentrations inhibit many cellular processes, while high intracellular K^+ concentrations are required to activate a variety of cellular processes. Finally, the Na^+, K^+-ATPase transporter (Na^+/K^+ pump) regulates cell volume through its osmotic effects.

You did not have to include all the details of the Na^+/K^+ pump or Na^+ and K^+ concentrations mentioned here. They are provided here to illustrate the importance of the pump and the wide range of its effects.

151.

(1) **Simple diffusion cannot create concentration gradients.** The diffusion of concentration gradients drives many endergonic processes in cells (including the chemiosomotic ATP production common to all forms of life).

(2) **Specificity of transport requires protein transporters.** Without facilitated diffusion, ions and molecules that do not diffuse through the bilayer could not be incorporated into cells or be eliminated by them.

152. (A) The table clearly shows that different NTPs have different functions in the cell. The nitrogenous base differentiates them structurally, which allows specific enzymes to recognize them.

153. (D) A **regulator** for a particular physiological condition will **maintain a fairly constant level** or amount **over time despite the environmental conditions** (within limits, of course). The graph on the left shows that for a particular internal condition, the regulator maintains a steady internal environment relative to the environmental condition. The chimpanzee is a regulator for body temperature in this question because the range of temperature variation is fairly low.

154. (C) The graph on the right shows that the salt concentration of the crustacean body is directly proportional (exactly the same, actually) as the concentration of the water between concentrations of approximately 2%–4%. Above 4% is the lethal range of salt concentration. The crustacean doesn't "need" to match its concentration to that of its environment (choice A, in this case, "need" implies that the change in salt concentration of the crustacean body must match its environment for the benefit of the crustacean).

A conformer is a type of organism that, for the particular variable or variables considered, **does not maintain differences**, or at least not large ones, **from their environment**. At concentrations >4%, the amount of energy required to pump out excess ions is enormous.

155. (C) The table shows that glucose and oxygen are present in **higher amounts in the artery** than in the vein, suggesting that **the tissue is taking these substances out of the blood.** Carbon dioxide, lactic acid, ammonia, and glycerol are present in *higher concentrations in the vein* compared to the artery, suggesting that *the tissue is excreting these substances into the blood.* **The increased ammonia concentration in the vein indicate that amino acids were used for metabolism** (they were deaminated, i.e., their amino group was removed) and the presence of **glycerol indicates that triglycerides have been broken down** (to release 3 fatty acids and a glycerol).

You could use the process of elimination to solve this problem, as well (see the box after answer 6 for more details on when it's best to use the process of elimination). Oxygen is *never* converted to carbon dioxide in cells (choice A). Lactic acid is present in higher amounts in the vein so it must have been *added* to the blood by the cells (not removed and used as an energy source, choice B). Because carbon dioxide is higher in the vein, there must be aerobic respiration occurring in the tissue (choice D).

156. (D) The ***core temperature*** is the temperature of the central nervous system and vital organs of the organism. **The temperature of the core is kept fairly constant, but body temperature is maintained within a range.** In humans, for example, the lowest body temperatures are observed during sleep and upon waking and the highest after a meal or exercise. **In endotherms, metabolism is the source of heat that continuously radiates from the highest to lowest temperatures, typically from the core to the surface and then lost at the surface.**

Body temperature is regulated by the hypothalamus, but the effectors (the skin, sweat glands, respiratory surfaces, and muscles) can be used to regulate heat losses and heat production. Humans, primates, and horses (to a limited extent) can sweat to *dissipate* (not absorb!) heat. Respiratory surfaces are used by most mammals to dissipate heat through panting (increases the rate of evaporative cooling from the lungs). Muscles increase body temperature through shivering (or just contracting).

157. (D) Blood vessels constrict to reduce blood flow and dilate to increase blood flow. The maintenance of the core temperature—the central nervous system and vital organs—is critical so the blood vessel to the surface of the body constricts in the cold to decrease heat losses (the surface is where heat losses occur).

158. (C) Mice are endotherms (mammals), while frogs are ectotherms (amphibians). Endotherms maintain their body temperature within a narrow range despite environmental temperatures (within limits, of course) using the heat generated by metabolism. **Ectotherms maintain their body temperature**, too, but **within a much wider range.** Importantly, **they don't use metabolism to generate heat.** Instead, they **absorb heat from the environment** (and **lose it to the environment**) to regulate their temperature.

The metabolic rate of an ectotherm is a function of its body temperature. A frog with a higher body temperature will have a higher metabolic rate. It will use more energy and therefore require more oxygen. A frog with a lower body temperature will have a slower metabolic rate, use less energy, and require less oxygen. Endotherm metabolic rate is not dependent on temperature. If anything, lower temperatures would *slightly* increase metabolic rate due to shivering (which uses energy). (Question 198 is similar.)

159. (D) The rat is a mammal and therefore an endotherm. **Endotherms require more energy per kilogram of body mass because they use metabolic energy to maintain their body temperature.**

160. (C) A diagram of the **negative feedback cycle** controlling calcium levels is shown on top. When blood calcium levels are too high, the thyroid gland secretes calcitonin which lowers blood calcium levels. When blood calcium levels are too low, the parathyroid glands secrete parathyroid hormone (PTH) which raises blood calcium levels. **All homeostatic**

variables are regulated by negative feedback through antagonistic systems: one that raises the level of the variable and one that lowers it.

Neither diagram addresses energy expenditure relative to calcium levels (choice A is incorrect). The diagram on the right shows the basic mechanism of action of the hormones (and why choice B is incorrect). Calcitonin lowers blood calcium levels by increasing the uptake of calcium from the blood by the bone and increasing the loss of calcium by the kidney. PTH increases blood calcium levels by causing the bones to release calcium into the blood, preventing losses in the urine and increasing uptake of calcium from the diet.

Physiological variables are held relatively constant by negative feedback (see the following diagram). **Negative feedback** works by the determining the **level of a variable relative to a set point**. If the variable is *higher* than the set point, the system responds in a way that will *lower* it. If the variable is *too low* relative to the set point, the system will respond in a way that will *raise it.*

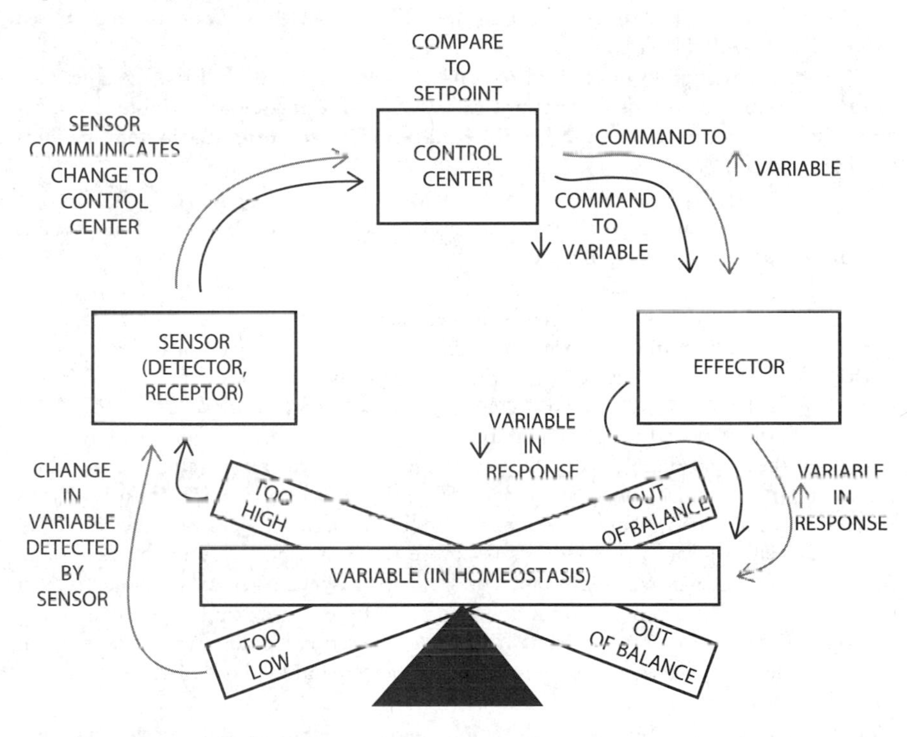

161. (A) Cells can respond to a hormone only if they have receptors for that particular hormone.

162. (B) The figure on the bottom shows the basic mechanism of action of the hormones. Calcitonin lowers blood calcium levels by increasing the uptake of calcium from the blood by the bone and increasing the loss of calcium by the kidney. PTH, on the other hand, increases blood calcium levels by causing the bones to release calcium into the blood, preventing losses in the urine and increasing uptake of calcium from the diet. **If a tissue or organ responds to a hormone, it must have target cells with receptors for that hormone.**

The College Board wants you to know that negative feedback is the mechanism by which dynamic homeostasis is maintained for physiological variables such as body temperature, blood pH, blood Ca^{2+} concentration, and so on. However, NO SPECIFIC BEHAVIORAL OR PHYSIOLOGICAL MECHANISM is required for the AP Biology exam. You should understand the general mechanism of negative feedback and be able to provide several examples of negative feedback for free-response questions.

163. (B) Calcium administration would increase calcium concentrations at first. Because calcium concentrations are controlled by negative feedback, the increased calcium concentration would result in calcitonin secretion, raising calcitonin levels. Calcitonin acts to lower blood calcium levels so as the level of calcitonin rises, the concentration of calcium should stabilize. Calcitonin levels should not reduce calcium levels so much that PTH is released, especially because calcium is administered for the entire hour (choice A). Choice C is incorrect because the release of calcitonin and PTH require levels of calcium to be sensed and compared with the set point.

A high concentration of calcium ions in the solution would kill the rat! There are disturbances to homeostasis that are not "fixable," so **if a particular variable is too far from the set point or far enough from the set point for too long, the system may not be able to recover.**

164. (A) Choice A summarizes the feedback cycle, whereas choices B, C, and D are specific examples of particular aspects of the cycle.

165. (B) The diagram shows a **negative feedback cycle** for the regulation of blood osmolarity. ADH (anti-diuretic hormone) increases the permeability of the collecting ducts in the kidney to water, which results in the removal of water from the filtrate. (This way it is reabsorbed back into the blood. Remember *the filtrate is destined for excretion.*) ADH functions to decrease body fluid osmolarity (so choice C is incorrect).

When the blood osmolarity is too low, the juxtaglomerular apparatus of the kidney secretes renin, which aids in the conversion of angiotensinogen to angiotensin II (ATII). ATII acts on the adrenal glands to promote secretion of aldosterone, which increases Na^+ and water reabsorption in the kidneys (so more water and Na^+ stay in the body) and constricts arterioles, which increases blood pressure. Because renin activates angiotensinogen to ATII, which then promotes the secretion of aldosterone, they cannot be antagonistic hormones (why choice D is incorrect). Choice A is incorrect because renin is secreted in response to low blood pressure or volume, not an increase in blood osmolarity.

166. (D) The basic mechanisms for the regulation of body fluid osmolarity shown are:

HIGH osmolarity:

- Thirst encourages drinking, which dilutes the body fluids.
- Less water is lost in the urine (reabsorbed from the filtrate).

LOW osmolarity:

- Conserve Na^+ in the blood (decrease excretion of Na^+ in the urine).

Blood pressure is regulated, in part, by the regulation of blood volume. Low blood volume increases Na^+ retention to promote water retention ("*where the solute goes, the water flows*") and arteriole constriction to decrease the volume of the circulatory system, which increases the pressure in the vessels.

The volume of the circulatory plumbing, that is, the blood vessels, is not fixed. The dilation and constriction of small arteries and arterioles, and the opening and closing of capillary beds, changes the volume of the blood vessels. Dilation increases the volume, reducing blood pressure, and constriction decreases blood volume, increasing blood pressure.

See the box after answer 6 for a strategy for answering questions asked in the *negative* and using the process of elimination.

167. (D) The volume of the circulatory plumbing, that is, the blood vessels, is not fixed (choice B is incorrect). The dilation and constriction of small arteries and arterioles, and the opening and closing of capillary beds, changes the volume of the circulatory system.

Small, muscular arteries and arterioles can dilate or constrict to direct **blood**, which **will flow along the path of least resistance**. When a blood vessel constricts, its *diameter decreases,* which *increases friction* between the blood and the vessel wall. This *increased resistance decreases blood flow*. **In order for blood flow to be directed, arterioles must dilate and constrict according to the needs of the tissues and organs they serve.** Epinephrine (generally) causes dilation of the blood vessels that supply the skeletal muscles, for example, and constrict the blood vessels that drain the digestive tract.

Capillaries have no smooth muscle in their walls and therefore can't dilate or constrict, but pre-capillary sphincters can contract to prevent blood from entering the capillaries (only about 10% of capillaries have blood flow in them an any given moment).

168. (A) Osmoreceptors are specialized cells that respond to extracellular tonicity (osmolarity). The precise mechanism of osmoreceptors is not known, but the osmoreceptor cells generally increase in volume in dilute extracellular fluid and contract (decrease in volume) with increased extracellular fluid (via osmosis). The change in volume sends a signal to the hypothalamus. Some osmoreceptors rely on extracellular Na^+ concentration to "infer" the hydration state of the organism.

Choice B is incorrect because water receptors have never been discovered (although there are water channels in cells, the aquaporins). Because water is so abundant inside and outside of cells, water receptors are an extremely impractical way to determine the state of water balance in living things. Answer choice C attempted to confuse the mechanism of osmolarity detection with the body's use of pH as an indicator of carbon dioxide levels. The pH of the extracellular fluids is not a reliable way to infer osmolarity. Choice D is incorrect because sodium receptors have never been discovered.

169. (C) Lines that lie to the upper left of the line of equal temperature represent flowers that maintain a higher flower temperature than the environment. *Nelumbo* maintains a temperature above, at (the points that intersect the line of equal temperature) or below (the points on the line that lie to the right of the line of equal temperature). Both *Symplocarpus* and *Philodendron* maintain elevated floral temperatures at all temperatures measured. See the following figure.

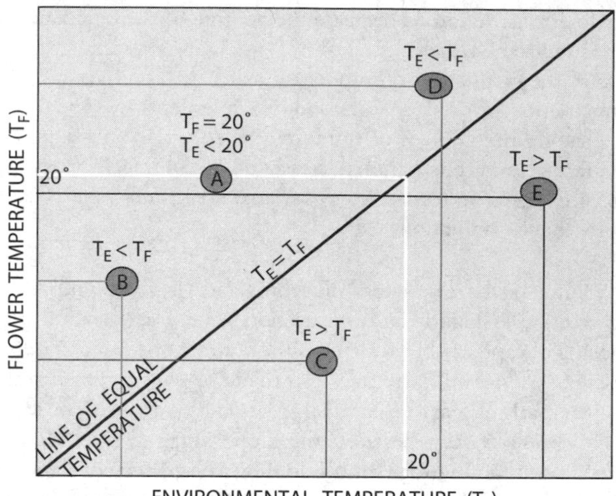

170. (D) One of the characteristics of a regulated physiological variable is that it remains fairly constant despite external perturbations. The most convincing evidence that a plant regulates its temperature is that it can maintain a temperature different from the environment in which it lives.

Choice A *suggests* that temperature regulation may occur as the mitochondria is capable of generating metabolic heat, but it is not convincing evidence. Choice C is incorrect because it indicates the exact *opposite* of regulation: If a plant's temperature directly correlates to the environmental temperature the plant is conforming to the environment.

171. (C) Most organisms have been molded by natural selection to conserve energy, not waste it (which is why it is so easy for most people and domesticated animals to gain weight when given free access to food).

See the box after answer 6 for a strategy for answering questions asked in the *negative* and using the process of elimination.

172. Large flowers have a lower surface-area-to-volume ratio. Heat is exchanged from surfaces so the lower the surface-area-to-volume ratio, the less heat is lost per unit of flower volume, which makes it more efficient to maintain the flower at a higher temperature than the environment.

173. (C) Generally, saturated fatty acids of shorter chain length have the highest melting points. In other words, they are less fluid than lipids with longer fatty acid chains and greater levels of unsaturation (more C=C double bonds). In colder temperatures, increased phospholipid fluidity would prevent plant cell membranes from solidifying. In warmer temperatures, *less fluid* membranes are required to maintain their structural integrity.

At higher temperatures there was an increase in the fatty acid saturation of the membrane phospholipids.

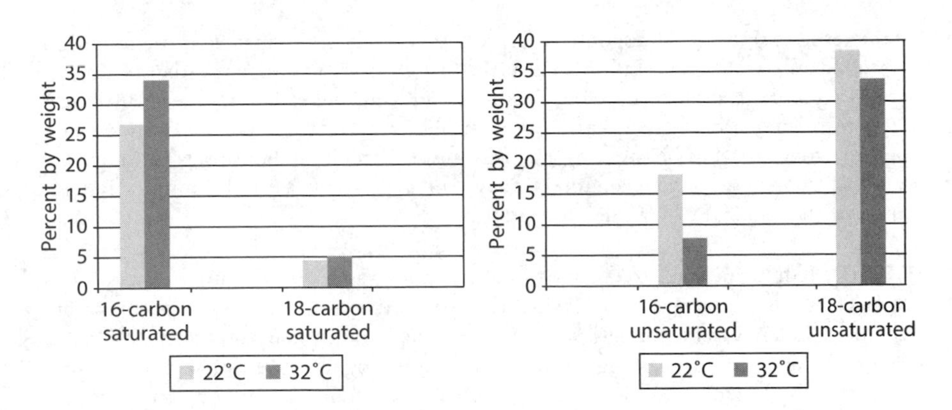

At 22°C, the percentage of 16-carbon fatty acids was 45% and the percentage of 18-carbon fatty acids was 40%. At 32°C, the percentage of 16-carbon fatty acids decreased by about 3% and the percentage of 18-carbon fatty acids decreased by about 1%.

174. (B) Lizards are ectotherms: they regulate their temperature within a fairly wide range through behavioral mechanisms. The maintenance of an elevated body temperature by behavioral means suggests that a fever is *not* a failure of homeostasis. During infections the hypothalamus actually resets the set point for body temperature, which is why the onset of fever is usually accompanied by "the chills"; the elevated set point makes 37°C feel too low. When the set point returns to normal, the fever breaks, often accompanied by sweating, which is the body's way of cooling off. (Question 196 is similar.)

175.

	1	0	0	0	

(1) Calculate antibody production between days 28 and 42:
Day 42: 10^4
Day 28: 1

$$10^4 - 1 = \sim 10^4$$

(2) Calculate antibody production between days 0 and 14:
Day 14: 10
Day 0: 1

$$10 - 1 = 9 \therefore \sim 10$$

(3) Compare:

$$10^4 = 10{,}000$$

$$\frac{10{,}000}{10} = 1{,}000 \therefore \textbf{1,000 fold increase.}$$

$$\text{or, } \frac{10^4}{10^1} = 10^3 = 1{,}000$$

176. The data show the relative concentrations of anti-A and anti-B antibodies as a function of time. The purpose of the second exposure to antigen A on day 28 was to ensure specificity of the (expected) increased response at day 42. The profile of the first exposure to antigen B (from days 28–56) is almost the same as the first exposure to antigen A (day 1–28). The second exposure to antigen A (on day 28) produced a much greater response to antigen A, but not to antigen B (the concentration of antibody A at day 42 is 1,000 times greater than the concentration of antibody B). This means that the response was not a general overproduction of the particular antibody (a type of hyper-vigilance of the immune system), but a specific response to a familiar foreign invader.

177. (D) Fish live in water, so excretion of nitrogenous wastes directly into the water (down a concentration gradient) is more efficient than converting it into a less-toxic form for later excretion. As terrestrial mammals, we store our urine in our urinary bladder for (disposal) when convenient (or for marking territory). Besides having to convert more-toxic forms of nitrogen to less-toxic forms, we store the diluted urea in our bladder and carry it around with us until we urinate, which also requires energy.

178. (D) An essential nutrient is one that must be obtained in the diet. Many molecules are required for cells (or the body) to function, but if it can be made by the organisms that requires it, it is not considered an "essential nutrient."

179. (A) The breakdown of glycogen releases many more glucose molecules per second than the breakdown of starch because there are many more glucose residues (they're *molecules* when they're in the monosaccharide form, but they are called *residues* when they are part of a polymer) available to enzymes. See the following picture.

Animal bodies typically require a lot more energy per kilogram than plant bodies to maintain but animal movement requires an even higher **rate** of energy usage (power).

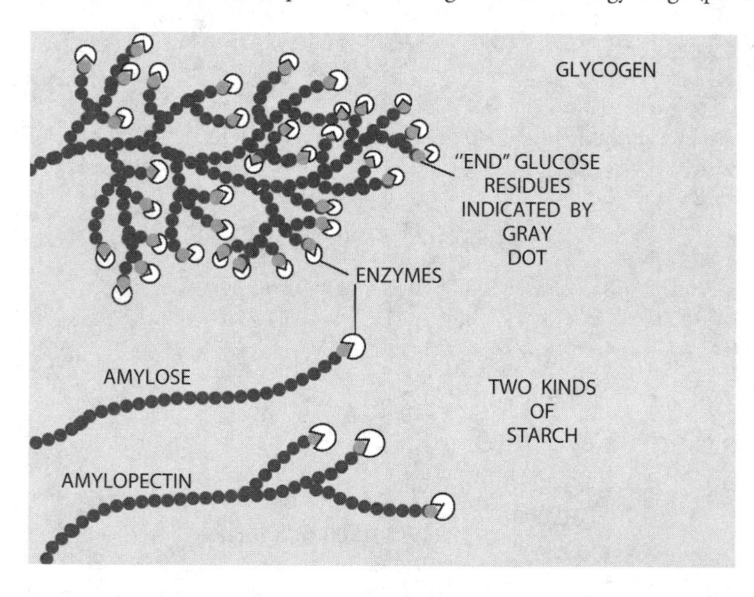

Choice B is incorrect because the storage of glucose in a polysaccharide, whether starch or glycogen, reduces osmotic pressure as compared to many molecules of dissolved sugar. **Osmotic pressure depends on the number of particles of solute**, with no consideration of size or type of particle. A greater number of particles in a given volume results in a higher osmotic pressure, which results in the net movement of water into the system. Choice C is true but does not answer the question, and choice D is incorrect because there's no point in storing glucose for energy if it can't be taken out of storage when needed.

180. (A) Capillaries are the vessels of exchange, and their structure typically suits the particular type of exchange for a particular type of tissue. The fact you needed to know—that capillaries come in a range of leakiness—was provided in the question. What you needed to deduce was that certain tissues require greater or lesser leakiness in order to function or optimally benefit from the blood supply. You also needed to speculate about the specific function of the capillaries of the particular tissue. The lining of the small intestine needs to absorb the products of digestion, so fenestrated capillaries seem to fit that function.

The brain, on the other hand, should have an extra "layer" of protection. **The blood–brain barrier** consists of specialized capillaries and brain cells that prevent the passage of materials from entering the cerebrospinal fluid from the blood. The cells of these capillaries are **connected by tight junctions** that **prevent the passage of materials between the cells.** This means that a substance entering or leaving a capillary must exit *through* a cell. So choice B is incorrect.

Endocrine glands need to secrete substances into the blood, often large protein hormones. Continuous capillaries would not serve this function well (choice C is incorrect). Continuous (un-fenestrated) capillaries are found in skeletal and smooth muscle, connective tissue and lungs.

The glomerulus of the kidney needs to filter blood, which is best accomplished through fenestrated capillaries (choice D is incorrect). Fenestrated capillaries are found in the kidney, villi of the small intestine, and some endocrine glands. Cells with unusually large fenestrations are found in the liver, red bone marrow, spleen, and some endocrine glands.

181. (C) A physiological variable is likely regulated if it remains fairly constant over time.

182. (D) Apoptosis is an important process in vertebrate development (see the following box, "How Cells Die").

How Cells Die

Apoptosis is a form of **programmed cell death.** The program specifies that the DNA fragments and cytoplasm shrink, and membrane changes occur that cause the cell to be phagocytosed rapidly by a neighboring cell or by a phagocyte before any of its contents can be leaked into the intercellular (extracellular) fluid.

- Cell death results without lysis or damage to neighboring cells.
- It is a normal process in multicellular organisms:
  - 50% or more of the nerve cells die soon after they are formed.
  - Billions of cells die in human bone marrow and intestine every hour.
  - Paws and hands are sculpted from spade-like (end plates). Individual cells die to carve out the fingers or digits.

In an adult organism, CELL DIVISION = CELL DEATH ∴ the number of cells stays the same.

During development, DIVISION > DEATH ∴ there is an overall increase in the number of cells.

Necrosis occurs as part of the inflammatory response. It is usually not as "neat" as apoptosis as dead cells spill their contents into the extracellular fluid. This may damage nearby cells and requires "cleanup" by cells of the immune system.

T-cells can assassinate specific cells by perforin or Fas:

- T-cells secrete vesicles containing **perforin** and serine proteases. *Perforin* is a protein that creates pores in cell membranes causing them to lyse and allowing the entry of serine proteases, enzymes that activate apoptosis through the caspase family of proteases.
- T-cell membranes contain a protein called Fas. Binding of the Fas ligand to the Fas receptor (on the cell targeted for death) activates a caspase cascade leading to apoptosis.

183. (B) Cell number is often determined by both cell division and cell death. BOTH of these processes are regulated (choice A is incorrect). Choice C is incorrect because cell division and cell death are the mechanisms by which animal organs regulate their size, not by increasing the individual size of the cells that make up the organ. Muscle cells can be regulated by growth, and neurons can grow to reach their target cells, but phenobarbital works by increased cell division. On a more subtle note, the use of the word *regulate* in choice C is not accurate. Chemicals can be used to regulate, but they are not the actual regulators in this case. The cells that make the chemicals are the regulating agents. Choice D is true but does not support the observations in the question as directly and specifically as choice B.

184. (B) Animal viruses are replicated in animal cells by the host cells. When the immune system recognizes a virus-infected cell, the cell is induced to undergo apoptosis, which kills the host cell (stopping viral replication) and exposes the virus to the cells of the immune system. Preventing cell death would provide the host cell with more time to replicate the virus.

Choice A is *almost* correct. It confuses the processes of cell replication with viral replication. It also requires that the virus is latent, like lysogenic bacteriophages (bacteriophages are viruses that infect bacteria) and retroviruses. Latent viruses can hide their genetic information in the host genome so that each division of the host cell replicates the viral information, too.

185. The requirement for activation is a form of regulation. Caspases, which activate and amplify the apoptosis cascade, should be functional (activated) only when the need for a particular cell to die is certain.

Proteases are enzymes that break down proteins. Proteins are the most abundant class of biomolecule in cells, performing most of the cell's functions. An active protease released by a ribosome could digest the ribosome (which is about 50% protein by mass) and many other crucial proteins in the cell! As protection against unwanted proteolysis, cells synthesize proteases as inactive precursors and secrete them (like the digestive proteases in the stomach and pancreas) or sequester them in compartments, like the lysosome, where the conditions can be maintained to support proteolytic behavior (a low pH and low pH optimum enzymes) without exposing them to the rest of the cell.

Other inactive proteins, like thrombin, are important in blood clotting, an event that needs to happen fast (so the proteins needed should be present and "ready" but not active) but should happen only when needed (so activation is required).

186. (C) The diagram shows that target cells secrete survival factors for developing neurons, which allows the number of neurons to match the number of target cells (choice B is incorrect because target cells "choose" the neurons, not the other way around). Choice A is true in that cell division is energetically expensive, but the increased energy expenditure is the cost, not the benefit, of this strategy. Choice D is incorrect because normal cell division is always regulated.

187. (D) Survival factors are needed for neurons to avoid apoptosis, so excess survival factors would allow more neurons to survive than there are target cells to connect to. Neurons and/or target cells may *eventually* die, but this would not be the *direct* effect; choices A and C could be ultimate consequences, however. (See the box after answer 86 for an explanation of direct versus ultimate causes and consequences.)

188. (B) The control of cell numbers during development depends on the regulation of two processes: cell division and programmed cell death (apoptosis). The cell death that occurs during *C. elegans* development is an example of apoptosis.

189. (A) Each trophic level represents the biomass of all the organisms present in that trophic level. The base of the pyramid represents the primary producers, and theoretically, it provides all the organic compounds and energy for the trophic levels above it. The assimilation of biomass from one trophic level to the next is not perfect or complete. **On average, only about 10% of the biomass of one trophic level is represented in the one above it.** Part of this loss is due to the fact that heterotrophs consume a significant mass of organic nutrients for oxidation. Much of the organic matter "taken" from the pyramid's base will be oxidized to carbon dioxide by the consumers in higher trophic levels.

See the box after answer 6 for a strategy for answering questions asked in the *negative* and using the process of elimination.

190. (B) Producers don't *feed* so choice D can be eliminated. **Omnivores**, able to eat plants and animals, can feed at the level of primary consumer (by eating plants) and secondary (or tertiary) consumer. **Detritivores** and **decomposers** can eat dead things from almost any **trophic level**. Although many **scavengers** are carnivores (so they don't eat plants), the animals they feed on may be primary, secondary, or tertiary consumers. Some decomposers, like certain fungi and bacteria, can eat practically anything (that's dead). "Secondary consumer" is a trophic level. By definition, it is the trophic level in which an animal feeds on another animal (the primary consumer).

Herbivore **Omnivore** **Carnivore**	**Primary consumer** **Secondary consumer** **Tertiary consumer**
Determined by dentition and digestive physiology	Describe an animal's nutritional role in an ecosystem

191. (C) Energy is never "lost" but when it gets converted from one form to another, it typically decreases in usefulness. When people say "energy is lost," they typically mean that energy has been converted to heat and the heat left the system (heat moves from a substance of higher temperature to a substance of lower temperature until thermal equilibrium is reached). Heat is a product of practically all of the energy transformations in your body, and although heat is great when you're cold, it is not very useful for doing work (it's a highly entropic form of energy), particularly in living things.

The **"10% rule" of trophic efficiency** states that 10% of the energy (and biomass, see answer 189) of a trophic level is transferred to the trophic level above it. (This does not apply to photosynthetic primary producers who typically convert only 1% of usable energy from the sun into primary productivity.) Nutrients (like carbon, nitrogen, phosphorus, etc.) are recycled, as the biogeochemical cycles demonstrate. In fact, the majority of the atoms on Earth have been here for 4.5 billion years (the age of the Earth). The word *accumulate* in choice D was an attempt to confuse nutrients with toxins in the process of **biological magnification**, the increased concentration of toxins observed in organisms near and at the top of the food chain.

192. (C) The albino plant is not photosynthetic, so it requires organic molecules from the green plant of the same species to which it is grown in close proximity. Although choice B may seem appealing, the definition of consumer doesn't really apply here, especially given the more accurate description in choice C.

Choice A is incorrect because it may be the albino's lack of photosynthetic ability that prevented it from producing seeds. There is not enough information in the question to determine whether the albino, given the proper nutrition, can produce pollen, eggs, or both.

193. (D) *Stomata* are the openings on the underside (mostly) of leaves where CO_2 enters and H_2O and O_2 exit the plant. Closing stomata results in not only decreased uptake of CO_2 but also an increased O_2 concentration inside the leaf, due to its inability to escape and decreased transpiration, too.

194. (C) Gases are always transported *passively* from areas of high partial pressure (gas concentration) of that specific gas to areas of low partial pressure (of the same gas), so choice A is incorrect. **Gases must be dissolved before they are transported across membranes,**

which is why gas exchange surfaces must be moist (and they would dry out very quickly if they weren't!).

Choice B incorrectly describes **countercurrent gas exchange,** a structural feature of fish gills but not in lungs. **Closed circulatory systems are only found in vertebrates, annelids, and cephalopods;** but all animals are aerobic so they all need to respire.

195. (B) ***Systole*** is the (shorter) phase of the cardiac cycle in which blood is being pumped by the heart. ***Diastole*** is the phase of the cardiac cycle in which the heart is filling with blood. The average of the systole and diastole stages is a weighted average, but not weighted by importance (choice D) but with time.

For example, if the systolic blood pressure was 120 mm Hg for 0.2 seconds and the diastole was 80 mm Hg for 0.8 seconds:

$$120 \text{ mm Hg} \times 0.2 \text{ sec} = \mathbf{24}$$
$$80 \text{ mm Hg} \times 0.8 \text{ sec} = \mathbf{64}$$
$$\mathbf{24} + \mathbf{64} = 88 = \mathbf{\sim 90}$$

You don't have to know a lot about the measurement of blood pressure to answer this question correctly. Choice A is clearly not correct, because it is inconsistent with the information in the question. Be careful of terms like "more important" in answer choices (like choice D). The word *important* can be very biased. It needs qualification to be used accurately in biology. For example, your heart is more important than your brain when it comes to pumping blood, but your heart is *not* more important than your brain in terms of organs you need to live! You *did* need to know that blood pressure is measured *only* in arteries (veins have very little blood pressure).

196. (B) Choices A, C, and D are all true statements about body temperature regulation during a fever. Choice B, which is also true, only demonstrates that a variable that is normally regulated can be affected by an external factor. For example, injecting a starving mammal with epinephrine will further lower its blood glucose levels (of course, it needs to be a reasonable dose—too much will kill it). Over time, the body regains control over the levels of blood glucose levels. Just because the body responds to the exogenous (from an external source) epinephrine doesn't mean blood glucose levels are not a variable maintained by homeostasis. (Question 174 is similar.)

197. (B) Insulin is mainly secreted after a meal, which would increase blood glucose levels due to the absorption of glucose from the carbohydrate-rich mean. Blood glucose is maintained within the limits of approximately 70–130 mg/d (4.5–7 mM). When a quantity in the body is maintained within fairly narrow (or very narrow) limits, it is accomplished through **negative feedback.**

198. (B) The bird is an **endotherm** and the snake is an **ectotherm**. The endotherm will maintain its body temperature despite external changes in temperatures (up to a point, of course), so choices C and D can be eliminated. The ectotherm will conform somewhat to the environmental temperature (particularly if there's no way to extract from it, by sunning on a warm rock, for example, or lose heat to it, by sitting in a cool patch of grass in the shade, for example).

Since the environmental temperature decreased, the snake's temperature will decrease. The body temperature of an ectotherm affects its metabolic rate (so if you must wrestle with an alligator, do it on a cold winter day in Alaska!). A decreased body temperature decreases its resting metabolic rate (∴ its oxygen consumption). Increasing its body temperature increases its resting metabolic rate and oxygen consumption.

Moderate temperature changes in the environment do not affect the resting metabolic rate of an endotherm, because its temperature is maintained by the heat generated by its metabolism. Long stays in the cold are likely to be managed with shivering and other behaviors that generate metabolic heat, which increase oxygen consumption (the extra work done by the muscles requires ATP).

The average body temperature of a bird is 40°C (~105°F)!

199. (D) Depending on the partial pressure (concentration) of carbon dioxide in the inhaled air, increased CO_2 in the respired air could raise blood CO_2 by increased absorption and may even inhibit the diffusion of CO_2 out of the blood. A high CO_2 concentration in the blood would cause **an increase in the H+ concentration, which would lower blood pH**. Elevated carbon dioxide levels are detected by the pons and medulla of the brain as a decreased blood pH, but you didn't need to know that specifically to correctly answer this question. Elevated carbon dioxide levels mean that breathing is not occurring at a high enough rate to eliminate it, so the breathing rate would increase.

200. (A) Exchange occurs at surfaces so cells with the highest surface-area-to-volume ratio have the greatest capacity for transport. The volume of cells determines the requirements for exchange. In this case, it is by creating a waste product that needs to be eliminated. **A larger volume of cytoplasm creates more waste.** It will also **consume more substances that need to be acquired from the extracellular fluid**. Either way, more volume means greater transport requirements, but a larger volume means a lower surface-area-to-volume ratio.

Surface area = $4\pi r^2$
Sphere with radius 1 = 12.6 units^2
Sphere with radius 2 = 50.2 units^2

Volume = $4/3\pi r^3$
Sphere with radius 1 = 4.2 units^3
Sphere with radius 2 = 33.5 units^3

The sphere with radius 1 has a surface-area-to-volume ratio of 12.6/4.2 = 3.
The sphere with radius 2 has a surface-area-to-volume ratio of 50.2/33.5 = 1.5.

The diameter of the larger cell (the radius = 2 so the diameter = 4) is 2 times that of the smaller cell (diameter = 2) but the surface-area-to volume ratio is 1/2.

Do the math! *The AP Biology exam expects you to use mathematics appropriately.*
Use it or lose it: *Practice handling numbers and manipulating data.*

201. (D) Cell D has the largest diameter so any molecule traversing the cell at its maximum width, the center, has to travel 200 μm. One of the pressures put on cell size is to **maximize surface-area**-to-volume ratio. The surface area should be as large as possible to maximize transport capacity, but the volume should be small. **Minimize volume.** Smaller volumes typically require less "stuff," and the rate of intracellular transport is greater because the particles don't have to travel as far.

202. (A) Sphere A has the smallest mass AND the largest surface-to-volume ratio, so it will definitely melt the fastest.

For two quantities of ice of equal mass but different shapes—for example, 100 grams of ice in one big cube or 100 grams of ice in tiny pellets—the different surface-area-to-volume ratios will cause them to melt at different rates. The greater surface-area-to-volume ratio of the small pellets increases their rate of heat gain from the environment that increases their rate of melting.

203. (B) Internal compartmentalization allows the cell to perform many different processes concurrently yet separately. Each compartment can have its own environment. For example, the low pH lysosome performs digestion, while the smooth endoplasmic reticulum is synthesizes lipids and transcription is occurs in the nucleus.

Choice A is incorrect because most organelles are NOT capable of semi-autonomous replication. The mitochondria and chloroplasts are exceptions (see answer 128 for an explanation of the endosymbiotic hypothesis). Choice C misstates the endosymbiotic hypothesis. Although the mitochondria and chloroplast can be thought of as prokaryotes living in cooperation with the eukaryotic cell, all the other membrane-bound compartments of the cell are not prokaryotes.

Although **compartmentalization confers many advantages to cells**, it does not allow them to reproduce faster. Prokaryotic cells reproduce many times faster than eukaryotes (some species can divide every 20 minutes under ideal conditions)! Of course, there are many factors that contribute to their shorter reproduction time. A significant factor is their DNA replication rate. **Prokaryotic DNA polymerase works about 10 times faster than the eukaryotic DNA polymerase AND bacteria have about $^{1}/_{1,000}{}^{th}$ the amount of DNA.**

204. Properties of exchange surfaces include:

(1) Thin exchange surface
(2) Large surface area, almost always highly folded or convoluted
(3) Moist surface
(4) Structural proteins such as channels, carrier proteins, pumps, and protein-lined pores
(5) In animals with a closed circulatory system, they are highly vascularized

Examples of exchange structures:

Structure	How its structure supports its function
Root hairs	High surface area allows for maximum water absorption. Many fine, thin hairs help keep the plant anchored in the soil.
Alveoli	High surface area, highly vascularized, moist interior helps dissolve gases.
Villi	High surface area and highly vascularlized
Microvilli	High surface area with many membrane-bound transporters and enzymes
Gills	High surface area and highly vascularlized
Capillaries	Thin walled and tube shaped ∴ has a very high surface area relative to its volume. Some capillaries have fenestrations (see answer 180).
Cell membranes	High surface area relative to cell volume, many transport proteins and receptors. Structural proteins are required for endocytosis and exocytosis.

205. (C) The diagram illustrates the process of **clonal selection** in which circulating B-cells are activated *directly* by an antigen. Once the specific B-cell (or cells) with receptors for the invading antigen are activated by antigen binding, the cell (or cells) proliferate into two populations: **Plasma cells,** which do not have antigen receptors on their surfaces but secrete soluble antibodies specific to the invading antigen, and **memory cells** that have the same antigen receptor as the cell they were derived from.

Choosing a lottery ticket (choice A) is too random to accurately reflect the workings of the immune system. Although choice B shows a selection process by fit, it does not demonstrate that any replication occurs once the selection is made. Choice D does demonstrate the persistence of the immune system, but incorrectly characterizes it as being really annoying!

B-cells can also be activated through helper T-cells that have been activated by communication with an antigen-presenting cell, such as a macrophage.

206. (B) The adaptive immune system learns and remembers. (Learning is a change in behavior as a result of experience.) Choice A is incorrect because "nonspecific recognition" is contradictory. **All recognition requires some level of specificity.**

Choice C would make having an immune system *less advantageous* to the host because the parasite would be better at evading it.

Choice D is incorrect because natural selection, by definition, occurs on the population level. However, it is important to understand that clonal selection and the "division vs. apoptosis" decisions during development are forms of selection, *natural selection* is the specific mechanism by which species change over time and it acts on a population level.

207. (A) In the process of **phagocytosis**, a cell engulfs a large particle or another cell by enclosing it within a membrane derived from the plasma membrane of the phagocytic cell. The engulfed particle (or cell) becomes trapped within a vesicle that fuses with a lysosome, digesting the contents. The products of digestion can be recycled within the cell or get eliminated.

208. (D) According to the diagram, the plant must lose water to gain carbon dioxide. Water is present at much greater partial pressures in the leaf than carbon dioxide is present in the air so the plant incurs great water losses to the atmosphere to obtain carbon dioxide from the atmosphere.

The process of elimination works particularly well with this question:

- Choice A is incorrect because the diagram shows the loss of water across animal lung and plant leaf surfaces.
- Choice B is incorrect because the process of cellular respiration is the same in plants and animals—the same quantity of ATP are made by the oxidation of comparable compounds.
- Choice C is incorrect because plants use carbon dioxide in the Calvin cycle for the synthesis of organic compounds, which requires ATP and reduced NADPH. Respiration uses oxygen and oxidizes organic compounds into carbon dioxide, and the energy "loss" is stored in the bonds of ATP.

209. (D) There is nothing in the data to support choice A or B. The number 23,000 in the table represents the water-to-air ratio of the mass of the medium versus volume of oxygen, but the actual minimum number of kilograms passed over the gills (assuming 100% efficiency) is 143 kg. Choice C is incorrect because it assumes a causal relationship between the high heat capacity of water and its ability to hold oxygen.

210. (A) Breathing in and out is easy for us because our respiratory medium is literally as light as air! Moving large quantities of water in and out of our lungs would be very energetically expensive, further increasing our oxygen requirements. Fish, on the other hand, move their respiratory surface (gills) through water and use *countercurrent exchange* to dramatically improve the efficiency of diffusion.

Obtaining oxygen and fuel for cellular respiration requires energy.

Choice D is an incorrect description of countercurrent exchange. If the direction of blood flow is opposite to the direction of water flow over gills, the direction of blood flow must be parallel to the direction of swimming (see the following diagram). It is not necessary to work that out in order to answer this question correctly. If you are familiar with countercurrent exchange, you should recognize that choice D does not accurately explain it. Choice A makes much more sense.

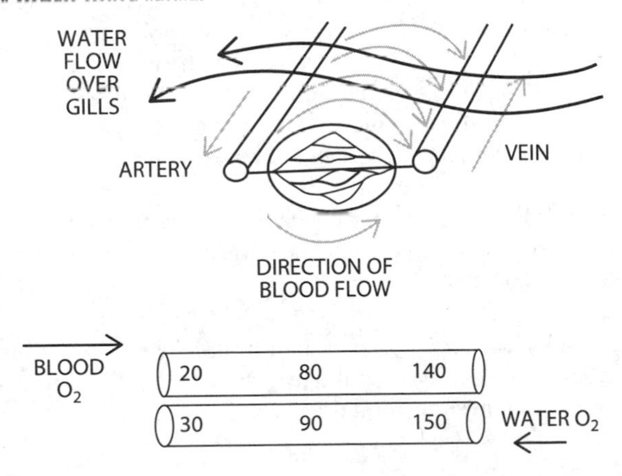

211. (C) The similar oxygen requirements of birds and bats during flight is the best evidence that the respiratory adaptations of birds are not necessary prerequisites for flight. Choice A supports the hypothesis, but choice C provides the best evidence. Choices B and D don't directly compare birds with flying mammals and don't specifically consider the particular requirements of flight.

212. (A) Population-density-dependent factors (density-dependent factors) regulate population size by negative feedback-like mechanisms resulting in a fairly stable population size over time. Larger populations undergo a reduction in the number of individuals over generations, and smaller populations undergo increases. The small deviations in population size hover around a mean (average).

For example, **density-dependent factors include resources such as space, water, food, nutrients, accumulation of wastes, and opportunity for disease to spread**. As population size increases, these become limiting. As population size decreases, resources become more available or less limiting.

Density-independent factors like natural disasters and weather don't typically stabilize populations. They usually result in a severe and random reduction in population size. They are typically too unpredictable to function as a regulator of population size.

The steep growth typically seen at the beginning of a population curve is **exponential** (1930–1950). Only the steepest portion of the curve is exponential. As soon as growth begins to slow, the growth becomes logistic. The increases from 1980–1990 and 2000–2005 are not exponential. The ceiling of 45,000 individuals is strong evidence of limiting factors.

213. (D) A decrease in the average life span of the hares would cause a sustained decrease in the population unless a compensatory reproductive strategy occurred simultaneously. Choices A and B describe normal **predator–prey population cycles** (an increased prey population feeds more predators who then have more offspring, which results in more hungry predators hunting prey, which decreases the prey population, and so on). A sustained increase in **primary productivity** would increase the carrying capacity of the environment. Finally, a rapid increase in the hare population without an increase in primary productivity could decimate the food supply, causing a rapid reduction in the population later.

See the box after answer 6 for a strategy for answering questions asked in the *negative* and using the process of elimination.

214. (B) At 0°C, seawater contains ~78% as much oxygen as freshwater, but at 30°C, it contains over 80%. So although the O_2 concentration decreases with increased temperature in both fresh and seawater, it actually decreases *more* in seawater. Oxygen availability can be a limiting factor in aquatic ecosystems. For example, eutrophic lakes have too much photosynthesis yet not enough oxygen. However, oligotrophic lakes have plenty of oxygen but not enough nutrients to support much life. Importantly, oxygen must stay dissolved in water to be of use to the aerobic organisms in the water.

Choice D is incorrect because most fish live in warm or cold water (although many species migrate). Fish do not typically choose where to live based only on oxygen concentration but, of course, if faced with low oxygen waters, most fish would seek out more comfortable environments.

215. (C) The nonbiological definition of a quorum is "the minimum number of people that must be present at a meeting to make its proceedings valid." In bacteria, quorum sensing is mechanism of monitoring local cell density.

Many bacterial species secrete small molecules that can be sensed by other bacteria. *By detecting the concentration of the chemical in their environment, bacteria can monitor local cell density.* Choices A, B, and D are incorrect because the low-density bacteria culture that seeded the new culture could emit light for a short period of time.

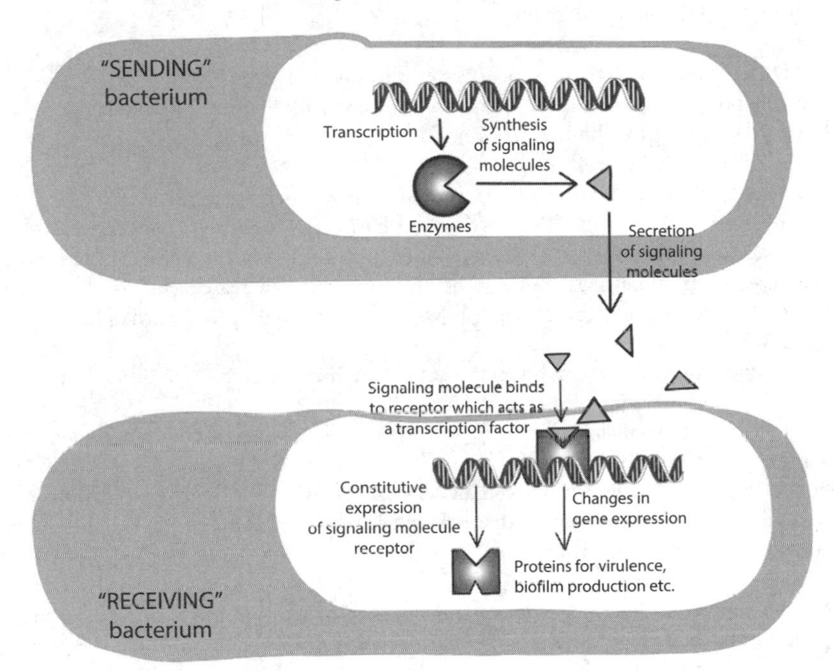

216. (A) Cell signaling is not just for the multicellular! Quorum sensing is required for the successful coordination of populations (and even communities) of bacteria and other single-celled organisms. By sensing the concentration of chemicals secreted by others in their environment, bacteria can determine local cell density. A high density of bacteria create a higher concentration of a particular chemical until a critical level is reached and a response is triggered in the population of a community.

Quorum sensing provides a decentralized mechanism for decision-making, coordinated behavior, and gene expression in these populations and communities of cells. Examples include biofilm formation (a *biofilm* is a group of microorganisms that stick to each other and to a surface by the secretion of a sticky substance) and the production of virulence factors. Please see diagram above.

217. (C) The rate of energy output is about 5 kJ/min over the course of the day but rises to over 35 kJ/min for short periods of time. Energy intake is definitely not constant throughout the study. Most of the time energy intake is zero but it was as high as 150 kJ/min (~35 kcal/min) at approximately 14 hours.

Choice D is incorrect because the data are taken from one individual (what's usual when you have only one day's worth of data from one person?).

218. (A) With rare exceptions of organisms, like some bacteria that can survive for long periods of time in "suspended animation," **cells are always doing metabolic work,** which includes maintaining concentration gradients, transcription, and translation.

219. **(B)** Choice B explains *how*. Choice D is only true because of the existence of cells that store energy-rich molecules for release in times of zero energy input (choice B). Many cells do have small stores of lipid and glycogen, but not all of them. Although choice C is true, it does not explain how multicellular animals can live for long periods of times (maybe weeks or even months) without energy intake.

220. **(D)** Essential nutrients *must* be ingested through the diet.

See the box after answer 6 for a strategy for answering questions asked in the *negative* and using the process of elimination.

221. **(A)** **Allosteric regulation** of enzymes allows the specific conditions in the cell to modulate an enzyme on a moment-to-moment basis. The higher the concentration of an inhibitor, for example, the greater the level of enzyme inhibition. As levels of the inhibitor decline, the inhibition lessens, allowing the reaction rate to increase again. In this way, the immediate environment of the enzyme allows it to be "turned up" or "turned down" like a dimmer switch.

Hormones can cause the activation or inactivation of enzymes through **phosphorylation**, an example of "whole body" regulation (choice B).

Tissue-specific expression of enzyme isoforms is also a form of whole body regulation through differentiation of cell types (choice C). For example, the expression of hexokinase, the first enzyme of glycolysis, occurs in all cells except the liver. The liver isoform is called glucokinase, which has different properties and is regulated differently. The difference is subtle but significant.

222. Two significant determinants of the number of trophic levels an ecosystem can support are the **net primary production** and **ecological efficiency**. The coastal marine ecosystem and the tropical forest ecosystem both have the same net primary production (8,000 kcal/m^2/year), but the ecological efficiency of the open ocean ecosystem (20%) is much greater than the tropical rainforest ecosystem (5%). **The *ecological efficiency* is a measure of how much energy is transferred from one trophic level to the next.** The greater the efficiency, the greater the number of trophic levels that can be supported for a given rate of net primary production.

223. **(C)** As the concentration of blood glucose decreases in the first 10 days, the rate of glucose usage decreases overall to conserve glucose. The levels of ketones and fatty acids increase. Many cells can use these compounds as fuel, conserving glucose for obligate glucose-using cells like the brain and red blood cells.

Organisms use feedback to maintain homeostasis. Glucagon levels rise in response to the initial drop in blood glucose levels and stay high as blood glucose concentrations stabilize at about 4 mM. The concentration of insulin, a hormone secreted in response to elevated blood glucose levels, drops and stays low.

224. **(B)** Curve I represents dietary glucose, the carbohydrate contained in the last meal the men ate. Curve II represents glycogen breakdown, which would be expected to start a few hours after eating. Much of the initial glucose in the meal was stored as glycogen to maintain blood glucose levels during the periods between meals (and also serves the purpose of removing glucose from the blood when it's abundant, to prevent blood glucose concentrations from rising too high). Glycogen stores can maintain glucose concentrations for about 12 hours. Curve III represents gluconeogenesis, the synthesis of glucose from nonglucose molecules, which becomes a significant source of glucose as glycogen becomes depleted.

The AP Biology exam does not expect you to have learned this level of detail regarding the regulation of blood glucose levels. The information and data provided, combined with a general knowledge of homeostasis, are enough to figure out this problem.

225.

	0				

***Water potential* is the tendency of a cell or solution to take up water.** There is net movement of water (by osmosis) to a system of lower water potential from a system of higher water potential. If there is no net movement of water, the water potential is zero.

226. (D) Diving mammals like whales can stay submerged for several hours. Larger diving mammals can stay submerged for longer than smaller ones because their oxygen consumption relative to their body size is lower.

Choice A is incorrect because even though the specific (per kg) rate of oxygen consumption in larger mammals is much smaller, their large body size still requires more oxygen than a small one. For example:

- A **shrew** has a body mass of approximately 5 grams (0.005 kg).
  At a rate of $\frac{7.4 \text{ L } O_2}{\text{kg hr}}$ it consumes **0.037 L O_2 per hour.**
- A **mouse** has a body mass of 25 grams (0.025 kg)
  At a rate of $\frac{1.65 \text{ L } O_2}{\text{kg hr}}$ it consumes **0.041 L O_2 per hour.**
- An **elephant** has a body mass of 3,833 kg.
  At a rate of $\frac{0.07 \text{ L } O_2}{\text{kg hr}}$ it consumes **268 L O_2 per hour.**

Choice B is incorrect because the graph shows that horses and humans have different rates of oxygen consumption per kilogram of tissue. So even if they were the same mass, the horse would still consume more oxygen. Choice C incorrectly characterizes the relationship between body size and oxygen consumption.

227. Small body size in mammals imposes several physiological challenges:

- *Temperature regulation*: smaller mammals have a higher surface-area-to-volume ratio, increasing heat losses and creating greater energy demands to maintain body temperature.
- In small mammals, the *energy and oxygen requirements* are quite large relative to body size, creating greater energy demands.
- The higher oxygen consumption of smaller mammals requires greater *blood flow*. For example, 1 gram of shrew tissue consumes 100 times as much oxygen as 1 gram of

elephant tissue, which requires 100 times greater blood flow in the shrew compared with the elephant.

228.

	1	2			

12, with an acceptable range between 9 and 12. According to the first (top) graph, humans consume approximately 1 liter of O_2 per kilogram per hour ∴ a 0.5 kg rate consumes approximately 0.5 liters per hour.

$$\frac{1 \text{ liter } O_2}{\text{hr}} \times 24 \text{ hrs} \times 0.5 \text{ kg} = 12 \text{ litres } O_2$$

229.

	3	0			

30, with an acceptable range of 24–32. According to the first (top) graph, rats consume approximately 0.5 liters of oxygen per kilogram per hour ∴ a 60 kg human would require 30 liters of oxygen in 1 hour. The tolerance in the answer range allows an interpretation of as little as 0.4 liters to as great as 0.6 liters per kilogram per hour.

$$60 \text{ kg} \times \frac{0.5 \text{ L}O_2}{\text{kg hour}} = \frac{\mathbf{30 \text{ liters } O_2}}{\mathbf{hour}}$$

230. Both graphs express the rate of oxygen consumption per kilogram of body mass as a function of total body mass. Body mass is given on the x-axis using a logarithmic scale. The oxygen consumption, given on the y-axis, is presented on a linear scale on the top graph. The oxygen consumption is presented on a logarithmic scale on the bottom graph, giving the relationship between the two variables an *apparently* linear relationship.

Lines that generate a "best-fit" of data points are called **regression lines**. Most types of data can be graphically represented using a linear scale or a logarithmic scale.

- **Logarithmic scales are typically used when there is a broad range of values.** Logarithmic scales are based on multiplication rather than addition.
- Over a broad range of values, on a logarithmic scale, general trends in the data become apparent.

Mathematical representations of data and regression lines allow predictions to be made about one variable given the other using the general form of the equation for a line, $y = mx + b$, where b is the slope (−0.25) or whatever equation describes the regression line.

*One of the **7 Science Practices** the College Board expects students to engage in during their AP adventure is **the appropriate use of mathematics.** The AP Biology exam will assess your ability to apply mathematics to **solve problems, analyze data, make predictions, and describe natural phenomenon.***

231. (B) Short-night plants flower when they get *less* than a maximum amount of darkness (critical night length sets a *maximum*, compare trials 1 and 2). They *do not flower* if their night length is interrupted by *exposure to far-red light* (trials 4 and 6). Exposure to red light does not affect night-length.

Long-night plants flower when they get *more* than a minimum amount of darkness (critical night length set a *minimum*, compare trials 2 and 1). They *do not flower* if their night length is interrupted by *exposure to red light* (trials 3 and 5). Exposure to far-red light does not affect night length.

232. (C) Comparing trials 1 and 2 shows that *the relative amounts of dark and light affect flowering but do not distinguish between long days and short nights.* Trials 2 and 3 show that, with almost the same day and night length, with the exception of the flash of red light, the *flowering can be affected by an interruption of the night length.*

Trials 1 and 3 do not accomplish this as specifically, because the day length is different between the two trials AND the night length has been interrupted. So which variable, if not both, actually affected flowering cannot be determined by comparing these trials.

Trials 2 and 4 show that interrupting the night with a red flash followed by a far-red flash does not have an effect on flowering, so these trials alone do not support the hypothesis.

233. (A) Choice B suggests that tobacco is a long-night (short day) plant, but does not provide specific evidence that plants detect night length precisely (at least not as precisely as choice A). That information alone cannot be used to determine whether the plant is detecting the day length or the night length. Choice C, like B, is evidence that plants detect photoperiod but does not specifically address whether the day length, night length, or both are detected by the plant.

234. (D) The cells that compose the aggregates are genetically different. The formation of the stalk is necessary for the production of the spores, but disadvantageous to the genomes that produce them because they are not represented in the next generation. Cells that ignore the DIF 1 signal produce spores, leaving stalk formation to those cells that did not ignore the signal. This strategy appears to support cheating; that is, cells that *do not* ignore the signal to form a stalk are expected to be selected *again* while genotypes that *do* ignore the signal get to reproduce and leave behind genotypes that ignore the DIF-1 signal. Eventually, dim A^- genotypes would be expected to greatly outnumber dim A+ genotypes (about one-fifth of the *D. discoideum* cells die without reproducing to produce the supporting stalk).

235. (C) The data show that the number of dim A^- cells is reduced to approximately half in the fruiting body. However, there was *no difference* in the number of spores formed in the dim A^-/AX4 aggregates compared with the wild type AX4/AX4 aggregates.

This demonstrates that the dim A^- mutants can make *more* spores than the AX4 cells when present in the aggregates. It appears that cells that ignore the DIF-1 signal have greater representation in the pre-spore population, but they are ultimately excluded from the final spore production.

236. (A) The diagram shows that the two mating types secrete and respond to different signals: Mating type a secretes a "circle" but has receptors for a "triangle," and mating type α secretes a "triangle" but has receptors for a "circle." There is no indication that they will form a multicellular organism (choice A), that they can respond to their own mating type (choice

C), or that they can *only* respond to chemical signals from the environment (choice D). In all organisms, **both internal and external signals regulate a variety of physiological responses** that synchronize with environmental cycles and cues.

237. (D) The diagram illustrates that chemical signaling is the mechanism by which yeast cells identify their mates demonstrating that **cell signaling is common to yeast as well as multicellular organisms (and bacteria**, too!) See answers 215 and 216.

238. (D) Choice D illustrates an increase in the synthesis of an enzyme, the longest-term regulation. The synthesis of an enzyme takes more time than allosteric modulation or phosphorylation. Importantly, **increased synthesis of an enzyme results in an elevated concentration of the enzyme in the cell** until protein degradation (proteolysis) lowers it, so **the effects can be long lasting**.

Choice A illustrates allosteric regulation, which is local and allows enzyme regulation to occur on a moment-to-moment basis based on the concentration of a particular metabolite of the pathway in which the enzyme is a part. **Allosteric regulation allows enzymes to respond to the immediate conditions in the cell and works constantly to adjust levels of metabolites in the cell.**

Choice B shows an enzyme that is regulated both allosterically and by phosphorylation. **Phosphorylation of enzymes is often mediated by water-soluble (nonsteroid) hormones, which usually work by triggering an amplification cascade, resulting in the phosphorylation of many proteins in the cell**. Some of the target proteins are activated by phosphorylation and others are inactivated by phosphorylation. *The process takes longer than allosteric regulation*. Choice C illustrates an enzyme regulated only through phosphorylation.

239. (C) Tryptophan is one of the 20 amino acids required for protein synthesis. Many species of bacteria can make tryptophan if it's not available from the environment. To synthesize tryptophan, the bacteria make the enzymes that perform the synthesis reactions. There are five proteins coded for in the operon, *trp E* through *trp A*.

Protein synthesis is expensive. For each peptide bond formed, 4 GTP (similar to ATP) are required (see answer 152 for a list of the main functions of the four nucleoside triphosphates, including ATP). If tryptophan is available from the environment, it is advantageous to conserve energy by inhibiting the production of enzymes required for its synthesis and use the exogenous tryptophan instead. The cell is capable of sensing the presence of tryptophan through the *trp* repressor protein. When tryptophan is bound to the repressor, the shape of the repressor changes, making it capable of binding to the operator and inhibiting transcription. When intracellular tryptophan levels decline, the repressor protein no longer has tryptophan bound to it and can no longer bind to the operator.

It is important to understand that **the repressor is activated by the presence of tryptophan in the cell. The source of tryptophan is irrelevant. This allows the system to respond to intracellular conditions as well as external conditions.** In other words, if the level of tryptophan increases through excess synthesis relative to consumption, enough free, cytosolic tryptophan will be available to bind to the repressor and inhibit the production of more tryptophan-producing enzymes.

The College Board wants you to understand that organisms respond to changes in their environment through a variety of behavioral and physiological mechanisms. You don't need to memorize a specific mechanism.

240. (C) The *trp* repressor protein binds to the operator and represses transcription only when tryptophan is bound to it. *The binding of tryptophan to the* trp *repressor protein induces a shape change that makes the DNA binding site on the protein accessible for binding to the operator.* When intracellular tryptophan levels are low, the repressor protein does not bind to tryptophan. Without tryptophan binding, the DNA binding site of the repressor protein is not accessible to binding to the operator. When the repressor protein is *not* bound to the operator, transcription occurs.

241. (B) The use of radiolabeled tryptophan allows the scientist to distinguish tryptophan synthesized by the bacteria (or produced from the digestion of a protein). The presence of the molecules bound to the protein inside the cell means the molecules must have been transported across the membrane, from the outside of the cell to the inside to the cell. The large size and polarity of tryptophan requires a transport protein in the membrane. The following diagram compares tryptophan, lactose, and water. Highly polar groups are shaded gray.

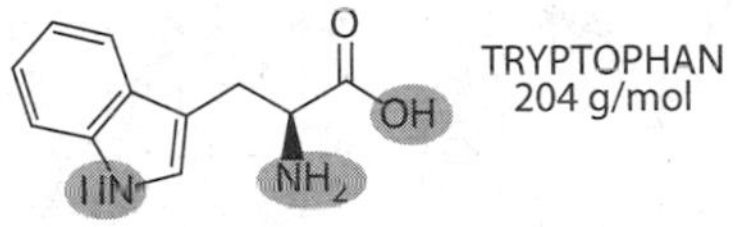

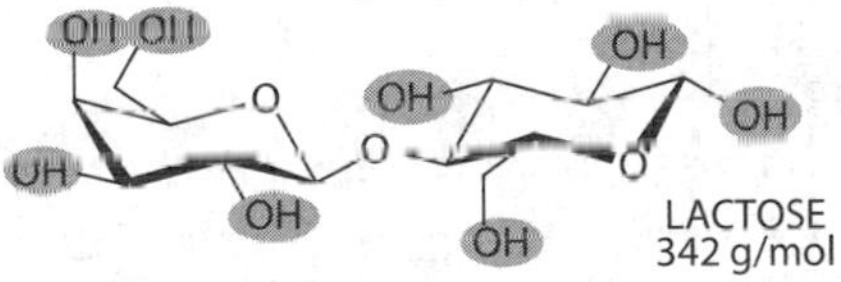

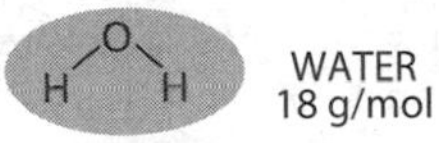

TRYPTOPHAN
204 g/mol

LACTOSE
342 g/mol

WATER
18 g/mol

You don't have to know the structure of tryptophan to answer this question correctly. Answer choices A and C are incorrect because the purpose of the radiolabel is to distinguish the tryptophan provided by the scientist from the tryptophan produced by the bacteria. Although the presence of radioactivity in the cell could certainly change gene transcription, the specific binding of radioactive tryptophan to the *trp* repressor is not evidence that radioactivity changed gene transcription (the question states the answer choice "*must* be true"). Choice D is incorrect because the diagram clearly shows that tryptophan and lactose have the opposite mechanism of action (the presence of tryptophan *represses* transcription of the genes of the *trp* operon, whereas the presence of lactose *induces* transcription of the genes of the *lac* operon).

242. (B) The degradation of proteins does not provide the energy to synthesize other proteins. Many processes that "release energy" in the cell are not coupled to processes that require energy, so the energy "released" by these processes is never harnessed to do work.

In order for cells to respond to the needs of the organism, they must be able to alter their chemical composition. Proteins do most of the work of the cell, so changing the relative amounts of the different types of proteins allows the cell to change what it does and how fast it can do it.

Although proteins are energetically expensive to synthesize, the regulatory control over cellular function through the combined efforts of the appropriate collection of proteins to meet cellular requirements and the ability to keep amino acids in the most useful proteins is apparently worth the price!

See the box after answer 6 for a strategy for answering questions asked in the *negative* and using the process of elimination.

243. (B) Disruptions to ecosystems impact the dynamic homeostasis (balance) of the ecosystem. In this particular case, humans caused the disruption by introducing the perch.

Eutrophication (or hypertrophication) is the enrichment (or over-enrichment) of water by nutrients. It results in excessive bacterial and/or algae growth, often called "blooms." When these organisms die, the oxidative breakdown of the detritus leads to an oxygen shortage (see answer 114). There is no indication that the extinction of other fish was due to predation. True, Nile perch are predators. But similar disruption to an ecosystem could be caused by a primary consumer (herbivore) that outcompetes native species. Although the mass extinctions on Earth have been succeeded by diversifications, a local extinction event is not necessarily followed by a diversification.

244. (D) *Phytochrome detects different wavelengths of red light, red and far-red* (far-red has a longer wavelength than red light). The quality of day light changes over the course of the day and throughout the year, so plants use phytochrome to detect changes in light. **Flowering, seed germination, and phototropism are all dependent on photoperiod.** ***Apical dominance*** is the suppression of the growth of axillary buds by the apical bud of a shoot. It is mediated by hormones and is not dependent on 1photoperiod.

See the box after answer 6 for a strategy for answering questions asked in the *negative* and using the process of elimination.

245. (D) The rapid conversion of P_R to P_{FR} resets the plant's biological clock each morning, regardless of the length of the night. Although the plant's circadian rhythm is innate, the plant *can* learn about its environment, specifically about its photoperiod, through phytochrome.

246. Negative feedback maintains homeostasis, while **positive feedback** amplifies a physiological variable to result in a specific culminating event that, once accomplished, resets the variable back to pre-event levels.

Negative Feedback	Positive Feedback
Maintains levels of a physiological parameter within a range	Amplifies a physiological parameter
Mechanism by which homeostasis is maintained	Allows an "all or none" process to occur
Mechanisms to increase or decrease physiological parameter needed	Amplifies a response
Examples include: the maintenance of blood glucose concentration, temperature regulation, and sodium/potassium gradients across cell membranes	Examples include: depolarization during action potential, labor, LH surge preceding ovulation, blood clotting, and orgasm

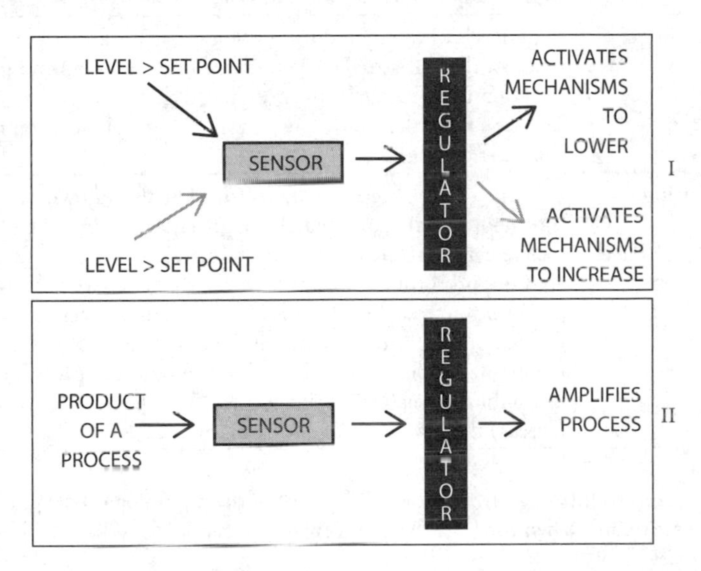

247. The general mechanism of positive feedback is shown in the preceding diagram: *The product of a process amplifies the process,* increasing the product that further amplifies the process **feedback,** which amplifies a physiological variable and results in a specific culminating event that, once accomplished, resets the variable back to pre-event levels.

248. Graph B represents positive feedback. The level of the physiological parameter deviates far from the homeostatic set point before eventually returning to it (where the set point equals the baseline).

Variable	Process
Lactation in mammals	Suckling initiates nerve impulses in the stretch receptors of the nipples to the hypothalamus, increasing release of prolactin-releasing hormone from the anterior pituitary.
Onset of labor during childbirth	Contraction of uterine muscle forces the baby's head or body through the cervix. Stretch receptors in the cervix send nerve impulses to the hypothalamus, causing the release of oxytocin, which stimulates uterine contractions.
Ripening of fruit	Ethylene initiates fruit development at low concentrations. Ethylene also stimulates the production of more ethylene, which increases its concentration to further the effects of ripening as well as the production of even more ethylene.
Ovulation in humans	High levels of estrogen during the last part of the pre-ovulatory phase of the menstrual cycle increase luteinizing hormone (LH) secretion from the anterior pituitary, which increases estrogen secretion from the ovaries, which increases LH secretion from the anterior pituitary.
Blood clotting	Damaged vessel release tissue factor (TF, also known as thromboplastin) causes platelet aggregation at site. The platelets secrete chemicals that attract more platelets. TF also initiates activates prothrombinase, an enzyme that converts inactive prothrombin to active thrombin. Thrombin accelerates the formation of prothrombinase, increasing the production of thrombin and further increases the formation of prothrombinase. Thrombin also activates platelets to release substances that increase the formation of prothrombinase.

Graph A represents negative feedback. The level of the physiological parameter hovers around the set point. When the level rises, it decreases to set point. When the level falls, it increases to set point.

249. (D) The purpose of the pigment was to track the cells transplanted from the dorsal lip of the gastrula. The results show that the transplanted cells induced the development of a second notochord and neural tube, structures that were identified by their pigment.

The purpose of the experiment was to determine the fate of specific cells in the pigmented gastrula. There is no evidence in the diagram showing the resulting embryo was viable (capable of living on its own, choice A). Although not explicitly stated, the pigment was chosen because it was inert (nonreactive, choice B).

Although the pigment will get diluted with each cell division as the pigment molecules are distributed into daughter cells, the results of the experiment did *not* show that pigmented cells differentiate but that daughter cells *do not contain pigment* (choice C).

Names of the specific stages of embryonic development are beyond the scope of the AP Biology course and exam.

Chapter 3: Information

250. (D) All living cells, whether they are prokaryotic or eukaryotic, have **double-stranded DNA** as their repository of information. **DNA contains genes and the regulatory sequences** that control the expression of those genes. The **genes encode** the amino acid sequence information of **polypeptides** and the nucleotide sequences of **ribosomal, transfer, and other types of RNA molecules** that do not get translated. DNA regulatory sequences include promoters, enhancers, and operators.

*There are many types of RNA, but you only need to know a few of them for the AP Biology exam—**mRNA, tRNA, rRNA, and sRNA.***

251. (D) Exons, regions of the DNA that encode proteins, rRNA, and tRNA, make up only 1.5% of the human genome. Introns and regulatory sequences (e.g., promoters, enhancers and operators) make up 24% of the genome.

252. (C) Complementary base pairing of nucleotides is the critical structural feature of nucleic acids that imparts their information function—their ability to replicate with high fidelity and take on complex three-dimensional shapes like in tRNA, rRNA, and sRNA. Base pairing is what **allows nucleic acids to serve as the repository of information** because it can be replicated with high fidelity and copied by RNA polymerase to create molecules of mRNA that can be translated with the help of tRNA (whose anti-codon base-pairs with codons on mRNA and whose structural specificity is accomplished through complementary base pairing). The functionality of rRNA, which includes catalysis of the peptide bond between amino acids during protein synthesis, is also a result of structural properties manifested by the molecule by complementary base pairing.

In 1952, Erwin Chargaff found that *the number of adenine bases was equal to the number of thymine bases and the number of guanine bases was equal to the number of cytosine bases* in the DNA of a particular organism. This information was critical in helping Watson and Crick deduce the structure of DNA.

253. (A) Retroviruses contain a reverse transcription enzyme that synthesizes a DNA molecule from an RNA template. *Retroviruses acquire an average of one point mutation in each replication cycle* because the viral **reverse transcriptase** can't fix misincorporation errors (it has **no proofreading ability**).

254. (C) Gene expression commonly refers to the processes of transcription AND translation, though technically, a gene has been expressed once it's been transcribed. **In a multicellular organism, ALL cells have the same genome** (with some exceptions) because they were produced by **mitosis, the type of cell division that produces a genetic clone of the replicating cell.**

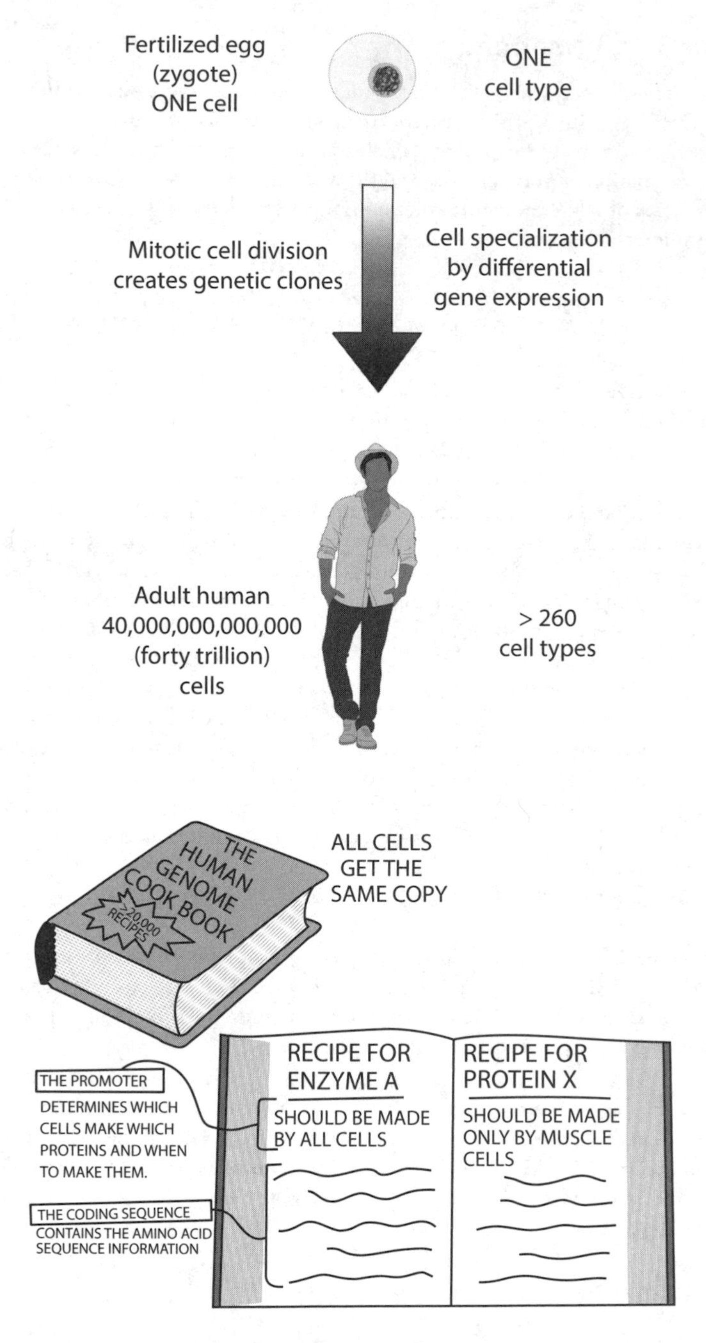

The **regulatory sequences of DNA are made of nucleotides,** just like the genes. When a newly formed cell gets a copy of the genome during mitosis, it receives a nearly identical copy of genes and regulatory sequences to the parent cells (there may be a few base-pair substitution mutations from uncorrected DNA polymerase mistakes).

255. (B) The mechanisms of transcription and translation are extremely similar in ALL known organisms. The genetic code is the "Rosetta stone" of life. It is, for all intents and purposes, universal to life on Earth (but may be, ironically, limited to life on Earth).

For example, **a yeast cell can express an animal gene** because the yeast cell will transcribe the animal gene into the same mRNA molecule as the animal cell does. The yeast ribosomes will synthesize the same protein based on the mRNA sequence as the animal cell. **The specific triplet codon encodes the same amino acid in every known organism.**

Although DNA replication is very similar in yeast and animals (they're both eukaryotes), DNA replication isn't specifically required for a yeast cell to express an animal gene (choice C).

You do NOT have to memorize the genetic code.

256. (D) The radioactive atom tagged the phage DNA or proteins so they could be traced separately throughout the experiment. In one trial, the nucleic acids were tagged with ^{32}P. In a separate trial, the proteins were tagged with ^{35}S. The phages were allowed to infect the bacteria for a short time (known prior to the experiment) before the solution was gently blended and the solution was centrifuged to form a *pellet of bacteria* at the bottom of the tube. The solution above the pellet is called the *supernatant.*

The ^{32}P-radioactivity was detected in the pellet (the bacteria cells), indicating **the phage DNA gained entry into the bacteria cell.** The ^{35}S-radioactivity was detected in the supernatant. This demonstrated that the phage capsids, the proteins, were "left behind" on the bacteria cell wall.

257. (C) The radioactivity tagged the phage DNA and proteins so they could be traced separately throughout the experiment. Proteins and nucleic acids both contain carbon, hydrogen, oxygen, and nitrogen (CHON), but **proteins contain sulfur** (in the amino acids cysteine and methionine), while **nucleic acids contain phosphorus** (in their 5' phosphate groups). Importantly, *no amino acids contain phosphorus and no nucleotides contain sulfur.* The difference in elemental composition between the two types of macromolecules allowed them to be tagged and traced in isolation.

258. (D) DNA and RNA polymerases can only synthesize their respective nucleic acids from the 5' end to the 3' end. In other words, **they can only attach the 5' phosphate end of the nucleotide being added to the 3' hydroxyl of the nucleotide that's already been incorporated** (so they can't add the 3'OH of a nucleotide to be added to the 5'PO_4 end of a nucleotide that has already been incorporated). Because the two strands of DNA in the double helix are oriented antiparallel to each other, building a strand from the 5' to the 3' end requires copying the 3' $\rightarrow$ 5' strand, which means the new strands gets built on the template strands in opposite directions. (See the following diagram.)

Transcription of mRNA from DNA is also 5' $\rightarrow$ 3' so **the 3' end of the DNA contains the codon for the 5' end of the RNA**. Translation of mRNA starts at the 5' end so the codon for first amino acid in the polypeptide (the N-terminus) is located at the 3' end of the gene that coded for it.

You do not have to know all the steps and all of the particular enzymes and proteins involved in the processes of DNA replication and transcription.
BUT YOU SHOULD KNOW:
DNA polymerase, RNA polymerase, helicase, and topoisomerase.

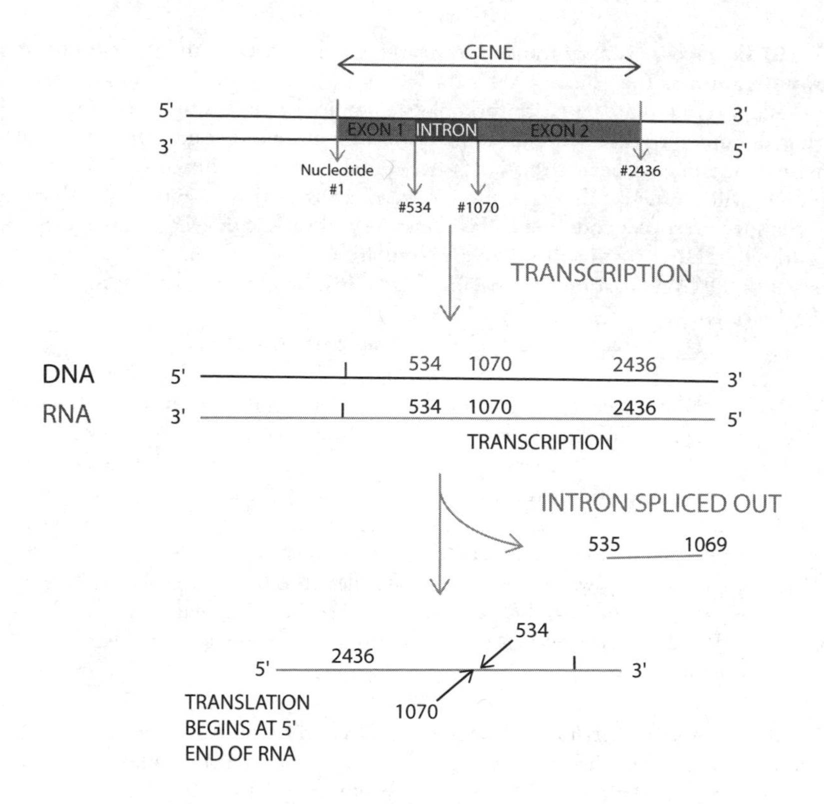

259. (C) The initial transformation of S-cells into R-cells occurred because **S-cells carry molecules that contain heritable information.** The nature of the molecule could not be determined from the transformation results alone (why choice A is incorrect).

Although a random mutation could be beneficial to bacteria, the pathogenicity of the bacteria has no clear benefit or disadvantage in this experiment (choice B). Choice D is incorrect because no antibiotics were used in the experiment, so we can't know if the S-cells are actually resistant to them.

260. The S-cells of *Streptococcus pneumoniae* (pneumococcus) are usually surrounded by a gummy polysaccharide capsule with a **s**mooth, glistening appearance when grown on a solid culture medium. *The capsule confers virulence to the bacteria by preventing them from being engulfed by macrophages and neutrophils in animals.* R-cells lack capsules and produce **r**ough-looking colonies. After many generations of cultivation on an artificial medium, some S-cells lose the ability to make the capsule and become harmless.

Uptake of foreign DNA provides the R-cells (including the S-cells that were randomly transformed into R-cells by losing their ability to make the capsule) **with the genes to make the capsule.** Bacteria that can make the capsule can evade the phagocytes.

Injection of a single S-cell into a mouse will kill it within about 24 hours, whereas an injection of over 100 million (1×10^8) R-cells is harmless. The R-cells lack the capsule and are therefore completely defenseless against the phagocytes.

The word *transformation* is used in a few different but related ways.

When the S-cells were transformed into R-cells, it was through ***random mutation*** in the genes that code for the proteins of capsule formation. The cells were still *transformed,* but it was not by the uptake of foreign DNA.

The word *transformed* is also used when somatic cells are transformed into **cancer** cells.

A cell is transformed when its genetic information changes and produces a change in the cell's phenotype. The *mechanism* of the genetic change is not the defining characteristic of transformation.

261. The transforming factor is DNA.

262. (D) The results of this experiment, originally performed by Griffith in 1928, showed that **R-cells were transformed into S-cells,** but **the identity of the transforming factor was not determined** until several years later.

The mice in group 4 died when given an injection of heat-killed S-cells and live R cells, so choice A is incorrect. Choice B is incorrect because live R-cells and live S-cells were never administered together in the experiment. **Dead cells cannot be brought back to life** (at least yet), as choice C suggests.

263. (B) Proteases digest proteins, and **nucleases digest nucleic acids.** If the transforming factor were a protein, the mixture digested with proteases would lose its ability to transform because its transforming factor would have been destroyed. The mice would live because the R-cells would not transform into S-cells. **If the transforming factor were a nucleic acid, the mixture digested with nucleases would lose its ability to transform because its transforming factor would have been destroyed.** Again, **the mice would live** because the R-cells would not transform into S-cells. Because the transforming factor is DNA, the mixture digested with protease would retain its transforming ability. The R-cells would be transformed into pathogenic S-cells and the mice would die ☹.

264. (B) The data in the table were taken from Chargaff, who discovered **two rules** that helped lead to the elucidation of the structure of DNA. The first rule is that **the number of cytosine bases equals the number of guanine bases and the number of adenine bases equals the number of thymine bases.** This was tantamount to the discovery of the base-pairing rules.

Although choice A is true, the data in the table do not provide much, if any, evidence to support it. The *E. coli* data suggest that base pairing is possible between all the bases but also supports that the number of each kind of base is the same for all four bases. For example, if there were equal numbers of each base, A = 25%, T = 25%, C = 25%, and G = 25%, the ratio of any two bases is 1.

Choice D is incorrect because the data in the table show that the A:G ratio is 0.40 for avian tubercle bacillus, which means that there are 4 adenine bases for every 10 guanine bases, *not* that A:G base pairs make up 40% of the bases.

The "formula" in choice B can be derived as follows:

There are only four bases $\therefore$ A + T + C + G = 1
Assume AT and GC base pairing $\therefore$ A = T and C = G $\therefore$ 2 A + 2 C = 1
2 A = 1 − 2 C

$$A = \frac{(1-2C)}{2}$$

Chargaff's second rule: **The relative amounts of the bases varies from species to species.** For the variety of *E. coli* in the table A:C = G:C, but that is not true for all species.

265. (A) All the cells of an adult animal are descendants of the zygote (fertilized egg). They are genetic clones because they were produced by mitotic cell division. What makes somatic cells different from each other is **differential gene expression. Out of the tens of thousands of genes present in each human cell, for example, only a fraction get expressed** (transcribed and, if applicable, translated).

A significant number of **developmental genes** are expressed only during animal development in the process of cell differentiation (specialization), and then they are inactivated ("turned off") for the rest of the life of the organism. **Some genes**, like those that code for the enzymes of glycolysis or the Krebs cycle**, are expressed throughout an organism's entire life.**

266. (B) The process described is **cloning.**

267. (A) The nucleus contains the **genome** and therefore determines the phenotype of the young mouse. However, the **mitochondria,** which **contain a small amount of DNA**, came **from the egg cell donor** (mouse 2). The mitochondrial genome does not appear to make a significant of a contribution to phenotypic diversity unless there is a harmful mutation that disrupted cellular respiration.

268. (C) Extra-nuclear genes are those present outside the nucleus, like those of the mitochondria (and chloroplast). The mitochondria (and chloroplast) are present in the egg and are inherited with the egg. They were once free-living bacteria, so they have small, circular genomes of their own.

269. *Horizontal gene transfer* is the process by which a piece of DNA can be transferred from the genome of one cell to another cell, even if the cells are different species. The donor and recipient cells can be in the same generation, and the gene transfer is not the result of cell division. Horizontal gene transfer is **a common mode of generating genetic variation in bacteria.** (***Vertical gene transfer* of genetic information is the transfer of genetic information from parent to progeny**, whether on the organismal level or the cellular level.)

- ***Transformation*** is the process in which pure DNA is taken up by bacteria (across the cell surface) and incorporated into their genome.

- ***Conjugation*** is the process by which the DNA in one bacterium is copied and transferred to another bacterium through a conjugation tube (pilus).
- ***Transduction*** is the viral-mediated transfer of DNA from one bacterium to another bacterium during infection.

If the AP Biology exam asks for two examples, provide only two examples! You do not have to know the specific details of the mechanisms of horizontal gene transmission in bacteria.

One benefit to the cell that receives the genetic information is that the cell gains genetic information it did not previously have and it may be beneficial, such as a gene for antibiotic resistance.

Another benefit not directly beneficial to a particular bacterium is the general ability to generate genetic variation that allows natural selection to act on a population. A population of bacteria unable to adapt to a changing environment would clearly be at a disadvantage to a population that could.

One reason horizontal gene transfer doesn't appear to have a significant role in multicellular organisms is that **a multicellular organism requires the coordinated efforts of all the cells**. Horizontal gene transfer would likely result in the **transformation of only a few cells** in a multicellular organism, and that may cause **the non-transformed cells to lose their ability to communicate and coordinate with the transformed cells** and may cause **the defensive cells to perceive the transformed cells as "non-self."**

Answers 270–273 refer to the following table:

Wild-type	The wild-type produces β-galactosidase (β-gal) only in the presence of lactose, but it takes a few minutes to get production going.
Mutant 1	Mutant 1 does not produce β-gal under any of the conditions present in the experiment.
Mutant 2	Mutant 2 produces β-gal whether lactose is present or not. The high rate of β-gal produced does not change significantly throughout the experiment.
Mutant 3	Mutant 3 produces β-gal whether lactose is present or not. The low rate of β-gal produced does not change significantly throughout the experiment, although there is a slight increase at 15 minutes, 10 minutes after the addition of lactose.

270. (A) The data for the wild-type suggests that β-gal expression in inducible. The presence of lactose induces the synthesis of β-gal, the enzyme that breaks down lactose. The fact that mutant 1 cannot make β-gal in the presence of lactose suggests that either the protein is not made (or it is made in a non-functional form). If a mutant *never* expresses a gene, it is likely that there is a mutation in the gene sequence or in the regulatory region (the promoter/operator) of the gene. (Please see preceding table.)

Since the latter is not an answer option, the former must be true. Choice A is a type of mutation that would prevent a protein from being produced in a cell.

Choice B is incorrect because the bacteria were grown in a glucose medium before lactose was added.

Choice C is true, but it is does not explain why mutant 1 did not produce β-gal; it states the consequence of the inability of mutant 1 to express β-gal in the presence of lactose (mutant 1 can't break down lactose because it doesn't express β-gal).

There is no indication in the data that choice D is true.

271. (A) A good way to **test for a cause-and-effect relationship** between lactose and β-gal gene transcription is to **measure the amount of β-gal mRNA before and after the addition of lactose.** Neither isolating nor sequencing the β-gal gene (choices B and D) will help determine whether β-gal transcription is activated by the presence of lactose (but isolating the gene was a preliminary step in the synthesis of the complementary probe used to hybridize with the β-gal mRNA). Inhibiting *all* RNA synthesis doesn't tell us anything about the β-gal synthesis specifically. It may also kill the bacteria since some gene transcription is occurring all the time (but which genes are transcribed depends on many factors).

Transcription can be:

- **inducible** (expressed is response to an environmental signal)
- **repressible** (expression stops in response to an environmental signal)
- **constitutive** (expressed all the time)

272. (A) Bacteria need to produce β-gal only when lactose is present. The human intestine does not secrete lactose, a disaccharide of glucose and galactose (choice B). Lactose is a sugar found mainly in milk. The small intestine of humans (that are not lactose intolerant) contains lact*ase*, the enzyme that breaks down lact*ose*. Choices C and D are not true.

273. (C) β-gal gene expression is inducible in the wild-type, so **constant (constitutive) expression in mutant 2 suggests there is a mutation in the regulatory portion of the gene** (the promoter or operator). If the experiment was done properly, there wouldn't be "minute" amounts of lactose present (choice A). Choice B is incorrect because a mutation in the amino acid coding portion of the gene would produce a normal (if it were a **silent mutation**) or a non-functional protein (or no protein at all if the mutation introduced an early stop codon), but an altered amino acid sequence would *not* cause the gene to be transcribed constantly (constitutively). Choice D doesn't make sense, because glucose is a product of lactose hydrolysis and therefore the continued production of β-gal would result in lots of glucose (if lactose was present).

274. (D) The graph plots the rate of β-gal production *relative to the wild-type*, so wild-type expression appears unchanging, even though it may not be unchanging (it's probably changing). Strain 1 produces little to no β-gal in the absence of lactose (like the wild-type), but while the expression of β-gal increases in the wild-type after exposure to lactose, the level of β-gal expression does not change. The line shows decreased β-gal production relative to the wild-type because the wild-type *increased* production. But since **the data are plotted**

as *relative to wild-type*, there appears to be no change in the wild-type, when in reality, the wild-type increased its rate of β-gal production while strain 1 did not.

Strain 2 shows a high rate of β-gal production relative to the wild-type when lactose is absent. In the first 5–10 minutes, the expression rate of β-gal may be decreasing in strain 2 or increasing in the wild-type. It is not possible to determine with certainty, but if we suppose the wild-type displays the regular expression pattern, then it appears that strain 2 was decreasing its rate of β-gal production until lactose was added. After the addition of lactose, the rate of β-gal synthesis remained slightly lower than the rate of β-gal production in the wild type.

The following graph shows the "absolute" rates of β-gal production. Notice that at around 10 minutes, the rates of β-gal production are the same in both strain 2 and the wild-type.

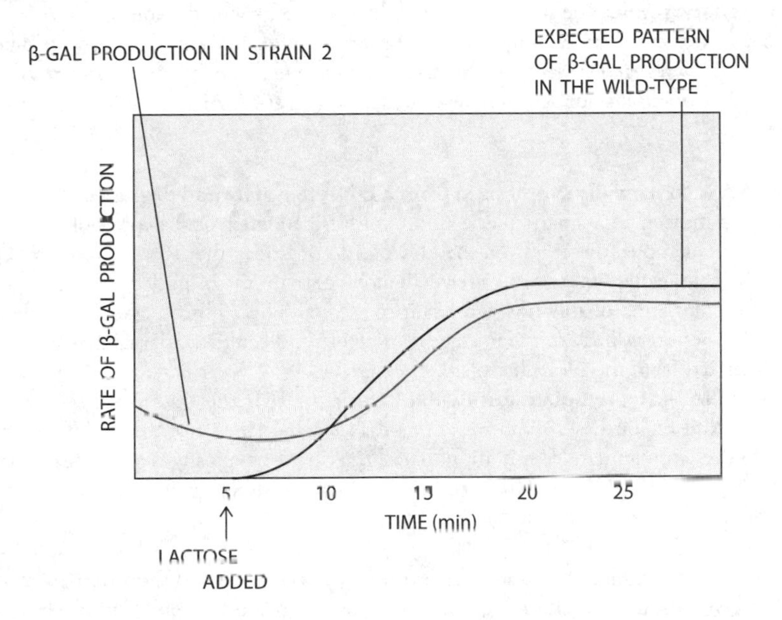

275. (D) Choice A is incorrect because the average lungfish weighs less than 50 pounds and has the largest genome of all the organisms listed in the table: 130,000,000 kilobase pairs (one set of chromosomes contains 130,000,000,000 base pairs!).

Organism complexity is difficult to qualify, but if we assume humans are the most complex (which is questionable), we still don't have the largest genome or the greatest number of genes (so choice B is incorrect). Choice C is incorrect because *Aeropyrum pernix* has 2,630 genes in a genome of 669 kilobase pairs.

276. (B) The amino acid sequence and therefore the gene sequence of the **histone H4 protein has changed the least** in the billion years shown in the table, **which indicates a strong selective pressure against changes in the sequence**. It is arguable that the data show that histones are more important than fibrinopeptides, hemoglobin, or cytochrome c (although on a cellular level, histones are extremely important!). Their conserved sequence *could* simply mean that cells are very sensitive to changes in histone sequence.

Fibrinopeptides, hemoglobin, and cytochrome c can tolerate sequence changes more than histone H4 (so choice A is incorrect). Hemoglobin (extracellular oxygen carrier protein) is a common protein used to date the time of divergence between species (choice C), but the fact that a greater number of species do not express hemoglobin makes it less than ideal. In addition, for long periods of time since divergence, a protein that has undergone less sequence changes may be preferable. *Genetic drift* is a change in allele frequencies due to random processes, so there is no relationship between genetic drift and the fibrinopeptide sequence differences.

Fibrinopeptides are peptides that are cleaved from fibrinogen during blood clotting and *cytochrome c* is an electron carrier in the mitochondrial electron transport (respiratory) chain.

277. (A) DNA polymerase has proofreading ability, but not all mistakes are corrected and mutations result. The first sentence in the question states that despite a high rate of misincorporated nucleotides, only one mistake per 10^9 ends up being passed on to daughter cells. *If DNA polymerase corrected all mistakes, there would be no (or much less) variation* and natural selection could not act to allow organisms to adapt (or it would have much less to work with).

278. RNA is continually synthesized from a DNA template and degraded. Many copies of the RNA molecule for a particular gene are transcribed when gene expression is activated, but they are all copied directly from the DNA and not from other RNA molecules. Out of many RNA molecules, only one or a few will likely contain the mutation. Mistakes in a few RNA molecules are probably not significant to have an effect if not passed on to daughter cells, and the effect is likely transient since those RNA molecules are degraded and new ones are synthesized from the DNA template.

The DNA is the template for all RNA molecules. If the nucleotide sequence in a protein-coding region has a mutation, every RNA transcribed from that DNA sequence will have the complementary mutation. The DNA is not typically degraded, particularly in dividing cells. All progeny cells would receive the mutated sequence, an incorrect copy, of the gene.

279. (D) Only a plate that contains ampicillin will have only ampicillin-resistant bacteria growing on it. Plate II does not have bacterial growth, but plate IV has several colonies, each of which contains the bacteria that are resistant to ampicillin, which is why they can survive, grow, and reproduce on a plate that contains ampicillin.

280.

(1) Plate I: **Most of the bacteria survived the plating and incubation.**

There is *no ampicillin* on the plate, so there is *no selective pressure to be ampicillin-resistant*. The bacteria plated on this plate were *not exposed to the plasmid*, though they were *subject to the transformation procedure*. Plate III did not inhibit the growth of any bacteria. **Plates I and III have a lawn of bacteria**. The entire surface of the plate is covered by them. The bacteria do not grow in colonies with large spaces of no or dead bacteria between them. It is **likely that none of the bacteria are ampicillin-resistant** on plates I and III.

(2) Plate II: **Almost none of the bacteria survived the plating and incubation.**

The bacteria plated on this plate were *not exposed to the plasmid*, though they were *subject to the transformation procedure*. There was *ampicillin on the plate*, but there were no bacteria that could live in the presence of it, so *they were killed when they were plated*. Any that did survive were not able to reproduce enough to form visible colonies on plates.

(3) Plate III: **Most of the bacteria survived the plating and incubation.**

These bacteria *underwent the transformation procedure with plasmid*. However, there was *no ampicillin* on the plate to put a selective advantage on those who had the plasmid or a pressure against those that didn't have it. **Plates I and III** have a **lawn** of bacteria; they did not inhibit the growth of any bacteria.

(4) Plate IV: **Most of the bacteria are ampicillin-resistant.**

These bacteria were *transformed with a plasmid* and then plated on an *ampicillin-containing plate* to **select against the bacteria that did not have the plasmid and to select for those that did have the plasmid.** In other words, the purpose of this treatment is to kill all the bacteria that did *not* absorb the plasmid during the transformation procedure.

281. (D) Any bacteria that contain the intact plasmid should be able to produce the proteins on the plasmid. Often an inducer is added. An inducer activates a transcription factor inside the cell. The promoter is specifically designed to activate the gene when the inducer is present. This way the scientist chooses when to express the gene.

282. (D) Ampicillin is a selecting agent. Its purpose is to allow the bacteria that were transformed to survive the plating, then live and reproduce on it. It must be able to use the nutrients supplied on the plate and live in the presence of ampicillin, and it may be induced to express other genes on the plasmid based on the molecules supplied in the media. Since the plasmid contains several genes, selection for one selects for them all. The bacteria that did not pick up the plasmid—the vast majority of them—are killed by ampicillin.

283. Prokaryotic gene expression: Prokaryotes have fewer genes than eukaryotes and they are **all unicellular, so no mechanism for coordinated, differential gene expression between cells during development is necessary.** The prokaryotic genome is mostly contained in **one, large circular chromosome,** although they **may also carry smaller, circular plasmid** with genes of non-essential functions. **Genes of related function are linked together under the control of one promoter in operons.** The promoter may contain an operator, a **simple on/off switch for transcription.** There may one to a few switching signals (*lac* operon glucose and lactose).

Eukaryotic gene expression: Eukaryotes have many more genes than prokaryotes. Many eukaryotes are **multicellular**, so **there must be regulation of the differential gene expression in cells to create tissues and organs during development**. Gene expression in a single cell can change dramatically over the course of differentiation, so the development of multicellular eukaryotes **requires many master regulatory genes and transcription factors to coordinate the process.**

Eukaryotes have several, linear chromosomes, each containing many genes. **Each gene is under the regulatory control of its own promoter.** Promoters contain *response elements* for specific gene regulatory proteins that are required for the expression of a particular gene.

Enhancers are another type of regulatory sequence that are typically situated at a distance from the gene whose transcription they enhance. One consequence of this arrangement is that one promoter may be under the regulatory control of many enhancers.

Another requirement for eukaryotic gene transcription **is the assembling of several general transcription factors at the promoter to initiate the binding of RNA polymerase.** The transcription factor assembly can speed up or slow down transcription initiation, so there is *great control over the rate of transcription.*

Histone proteins organize and assist regulation of gene transcription by the coordination of DNA packing into **chromatin**. This provides *additional regulatory opportunities* by allowing some parts of the DNA to be accessible to transcription, while ensuring other segments of DNA are not accessible. This is another example of how the structural arrangement of a molecule supports or inhibits one or more of its functions.

284. (C) Protein synthesis is energetically expensive. For each peptide bond formed, 4 GTP (similar to ATP) are required (see question 152 for a list of the main functions of the four nucleoside triphosphates, including ATP). **If tryptophan is available from the environment, it is advantageous to conserve energy by inhibiting the production of enzymes required for its synthesis.**

Bacteria are capable of sensing the presence of tryptophan through the *trp* repressor protein. When tryptophan is bound to the repressor, the shape of the repressor changes, which makes the DNA binding site (where the repressor would bind to the operator) inaccessible to binding DNA (*the protein is incapable of binding to the operator and inhibiting transcription when tryptophan is bound*).

When intracellular tryptophan levels decline, the repressor protein no longer has tryptophan bound, so it can no longer bind to the operator.

(See answer 239 for more information about the *trp* operon and see the following box.)

Operon Regulation: *trp* versus *lac*

***trp* is repressible**	***lac* is inducible**
Gene expression is the default mode. Tryptophan represses it.	**The default mode is no gene expression.** Lactose induces it.
The enzymes to make tryptophan are synthesized unless inhibited.	The enzymes to metabolize lactose are not synthesized unless activated.
Tryptophan is an amino acid that is always needed by the cell for protein synthesis.	Lactose is an optional energy source that may be only intermittently available.
The presence of tryptophan turns OFF the production of the enzymes for tryptophan synthesis to conserve energy for the bacteria when tryptophan is freely available and to keep the cell's amino acids in useful proteins.	The presence of lactose turns ON the production of the enzymes for its metabolism. The energy investment in the production of the proteins is likely to have a good return in the uptake of sugar.

285. (C) Most proteins are flexible structures. They change shape when substances (ligands) **bind to them and when they are phosphorylated** (one or more phosphate groups are added to an amino acid side chain). Two examples are the binding of tryptophan to the repressor which changing the shape of the DNA binding site and the allosteric modulation of enzymes.

In order to respond to the environment, an organism needs to be able to detect and interpret its environment. The bacteria senses the tryptophan because it has a protein that has a binding site that is complementary to tryptophan and other chemical features that allow it to **specifically bind to tryptophan when it is present**, even in the presence of many other molecules.

But then *that* information has to be passed on to a part of the cell that can respond to the information—in this case, the DNA. **Proteins relay information to the other molecules through their binding site availabilities/conformations**. One protein usually has several binding sites, so the particular composition of its environment causes one or more particular binding sites to fill up, which influences the other availability of other binding sites (for other molecules). For example, tryptophan binding makes the DNA binding site more accessible. The absence of tryptophan makes it more accessible. If the *trp* repressor binds DNA, no transcription of the *trp* genes occurs. **Which binding site, tryptophan or DNA, is occupied in the *trp* repressor determines the activity of the entire pathway.**

The binding of tryptophan activates the repressor molecule by inducing a shape change—it's like they are shaking hands. *Each hand is complementary to the other, but they both conform around the other when contact is made.* When tryptophan levels decline as it gets used for protein synthesis, the concentration in the cell is not high enough to keep the *trp* repressor proteins binding sites occupied. So the *trp* repressor loses its ability to bind to and DNA transcription can proceed.

> The **function** of a protein is **determined by its structure.**
> The **structure** of a protein is determined by its **amino acid sequence.**
> The **amino acid sequence** of a protein is determined by the **nucleotide sequence of the gene** that codes for it.

286. (B) The purpose of repression is to synthesize the required molecules unless it becomes available. Gene expression (and protein synthesis) is the default mode of operation, but it can be inhibited. The *implication* is that the product of the pathway is always needed. The purpose of **induction** (the mode of regulation used in expression of the *lac* operon genes and during development) is to perform a process or make a molecule **only when needed**. The *implication* is that the molecule is only useful under certain conditions. See the box before answer 285 for a comparison of *lac* and *trp* operon regulation.

287. (D) Bacteria replicate by binary fission, not meiosis. They have one, large circular chromosome so they are haploid; they lack homologous pairs of chromosomes.

288. (A) The expression of the genes controlled by the *trp* operon promoter (via the operator) is turned off by the presence of tryptophan, so the GFP gene would be expressed in the absence of tryptophan.

289. (C) The colonies that appear green under UV light were **transformed by the plasmid and are expressing the GFP gene.** In a laboratory transformation, *only a small percentage of transformations will occur.* So few bacteria would be transformed, but they would be plated with many, many more that were not, and there is no selective pressure for having the plasmid (so you would need a microscope to see them). Choice D is incorrect because a random mutation in the DNA that occurred during the experiment is unlikely to cause it to glow green under UV light as a result.

290. (C) Transduction occurs as a result of mistakes or inefficiencies in the packaging of the viral genome that occurs inside of bacteria cells. Occasionally and likely accidentally, a phage packages a piece of the bacterium's DNA (which the viral enzymes often cut into pieces) into one of the capsids. When the viral particles with the bacteria DNA infect a different bacterium, new genes are introduced into the infected bacteria. Some of those genes are viral, and some of them are from the previous bacterial. *The volume of a viral capsid is very tiny.* There is not much room, if any, for extra DNA, so if bacterial DNA is packaged in the capsid, the viral genome is likely incomplete or absent.

291. (B) Tay-Sachs is caused by any of several different mutations in the *HEXA* gene, which encodes one subunit of the β-hexosaminidase A enzyme. This enzyme is located in the lysosomes of cells in the brain and spinal cord. Its substrate is a ganglioside, a large molecule composed of lipid and sugar. The accumulation of this substance is toxic to cells.

Diploid organisms, like humans, **have two copies of each gene** (except for XY males who have only one copy of the genes on the X and Y chromosomes). **In Tay-Sachs heterozygotes, one allele is mutated and β-hexosaminidase A cannot be made from the information contained within it. The other allele is normal and can provide the information to synthesize β-hexosaminidase A.** *Enough of the enzyme can be synthesized from this single copy of the functional allele* that the accumulation of β-hexosaminidase A is not a threat. Choice A is incorrect because heterozygotes show the dominant phenotype.

292. (C) Of course, the most complete and convincing interpretation involves all four dishes; however, *by the end of the experiment, the two dishes had the same ratio of yellow to green seedlings.* The condition of the first seven days was irrelevant.

In dish A, *21 seeds* germinated by day 7, *1 green* and *20 yellow*.
By day 14, all 28 had germinated in a 3:1 ratio, 21 green and 7 yellow.

Because the number of yellow seedling decreased with time, the yellow color could be due to developmental stage or lack of light. The color of the last 7 seeds to germinate can't be determined from the data provided, but the number of additional seeds that germinated was 7, the exact number that had germinated after day 7. This observation supports the hypothesis that yellow color in the first 7 days is a developmental stage.

However, dish B had light the whole 14 days. All the seeds had germinated by day 7: 20 of them were green and 8 of them were yellow. Nothing changed by day 14. **At the end of the two-week period, both dishes had a 3:1 ratio of green to yellow seedlings.**

293. (B) *Thalassoma bifasciatum* is an example of a form of phenotypic plasticity called a **polyphenism—a trait for which multiple, discrete phenotypes can develop from a single genotype based on the environment.** This is in contrast a *reaction norm*, which describes *the range of phenotypic expression of a single genotype over a range of environmental conditions*. The critical difference is that *the reaction norm describes a continuous range of potential phenotypes* in which the environment the genome encounters determines the phenotype. Height and weight traits in humans are subject to the norm of reactions. **Polyphenisms are a discontinuous expression of the genotype.**

Examples of polyphenisms include:

- sex determination in *Thalassoma bifasciatum*
- the two phenotypes of the migratory locust *Schistocerca gregaria.* One is short-winged, one color, and solitary. The other is long-winged, multicolored, bright, and gregarious.
- caste system in some social insects
- seasonal pigmentation changes in many insects
- temperature-dependent sex determination in some reptiles

Thalassoma bifasciatum is an example of population-dependent sex determination. It is also ***reversible* sex determination**. *The benefit of reversible, population-dependent sex determination in a population is that there is always a mating couple when two animals of the same species are present.*

294. (A) The genetic diversity as well as the biomass of the planet has generally increased over time. The number of organisms and the diversity of organisms has also increased. New alleles are created through gene duplication events, where the duplicated genes can undergo mutation to create new alleles.

295. (C) The endocrine and nervous systems work together to coordinate the trillions of cells in your body to maintain homeostasis, respond to stimuli, and in some organisms, move, think, solve problems, and create! But they work differently—the endocrine system uses the bloodstream to circulate hormones from different glands throughout the body, whereas the nervous system (often) has a central processor and many extensions throughout the body.

In general, signals from the nervous system travel faster and produce faster but shorter-lasting effects in target cells.

Endocrine signals (hormones) travel more slowly and produce longer-lasting effects on target cells.

296. (A) Regulation of a multicellular organism occurs at many levels. Inside the cell, **allosteric** regulation of enzymes allows cells to adjust the concentrations of molecules within the cell according to cellular needs on a moment-to-moment basis. **Phosphorylation** of enzymes by protein kinase A occurs intracellularly, but the elevated cAMP levels that activate protein kinase A are the result of **hormonal** activation, which occurs on a body-wide level. Choices C and D are also regulatory processes that involve the coordinated effort of different tissues.

297. Each cell in the body of a multicellular organism is programmed to respond to specific combinations of signaling molecules. However, *different cell types can respond differently to the same signal (or signals). Cells vary in what receptors they possess* and thus what signals they will respond to.

Two different cell types with the same receptor may bind to the same molecule but respond differently. **It is the intracellular machinery that effects the response to the signal the cell received through its receptor.** This allows one chemical to have a multitude of effects in the body, depending on the cells, their receptors, and the intracellular machinery that is attached to them.

For example, acetylcholine binds to (nicotinic) acetylcholine receptors at the neuromuscular junction. Acetylcholine binding to nicotinic receptors opens sodium channels and promotes muscle contraction. Acetylcholine binding to muscarinic receptors on heart muscle cells opens potassium ion channels and slows the rate of heart muscle contraction. Adrenaline (epinephrine) induces glycogen breakdown in muscle, and increased rate and force of contractions of the heart. In fat cells, it promotes lipolysis.

298. Multicellular organisms must have the ability to differentially regulate gene expression in genetically identical cells to create tissues and organs during development. From development throughout life, multicellular organisms **require a complex system of coordination and communication**. These processes require several different kinds of molecules that can be detected and responded to in predictable ways. Cells must also adhere to one another. They must be able to monitor the body and coordinate over time and space.

> ***Multicellularity* means that some cells, but not others, will get to reproduce. Cells must cooperate, rather than compete, for their mutual benefit.**

299. Signal transduction is a complex process!

- **Receptors** capture specific chemical signals.
- **Intracellular response molecules** like G-proteins and protein kinases are required to effect an intracellular response from the signal.
- In some signaling pathways, a mechanism for signal **amplification** is required.
- **Enzymes** are needed to synthesize the signaling molecules.
- **Gene sequences** that code for all the necessary proteins must be present and transcribed.
- If the signal works by activating gene transcription, the **transcription factor** binds to the signaling molecule and then the **response element** in the promoter of the gene that is expressed in the presence of the signaling molecule.

300. (A) A short half-life is a key characteristic of a regulatory molecule. The protein with the shortest half-life in the graphs has a half-life on 1 minute. When production decreases 10-fold, the relative concentration (relative to its initial concentration in the cell) drops to nearly its lowest levels within 5 minutes and drops to its lowest level by 8–10 minutes. *The rapid decrease in concentration allows a quick response from the molecule that responds to that molecule.* When synthesis rates increase 10-fold, the concentration rises to near maximum levels in 4–5 minutes, allowing for a quick response. If the half-life of the molecule were long, the process would take longer to activate and then be activated for a longer time.

301. (D) Many endocrine glands secrete hormones that are distributed throughout the body by the cardiovascular system. These signaling molecules are called *circulating* *hormones.* **Only target cells bear receptors for a particular hormone. In most cases the proximity of the endocrine gland to the target tissue is irrelevant, since the hormone travels to most places in the body.**

302. (D) While glands are a component of many animal signaling systems, they are *not* a requirement of chemical signaling. **Cellular life requires the capacity for chemical communication.**

See answers 215 and 216 for an example involving **quorum sensing in bacteria** and answers 236 and 237 for an example of **mating type communication in yeast.**

See the box after answer 6 for a strategy for answering questions asked in the *negative* and using the process of elimination.

303. (B) Simple diffusion means *directly across the bilayer* without a transport protein. Channels provide specificity and can increase the rate of diffusion. **Two important molecules that rely on simple diffusion for transport across cell membranes are O_2 and CO_2.**

Function	Transport mechanism
Maintenance of Na^+/K^+ gradient	Active transport by Na^+, K^+-ATPase (pump)
Na^+ inflow and K^+ outflow	Voltage- or ligand-gated facilitated diffusion by Na^+ and K^+ channels
Ca^+ inflow from extracellular fluid into axon terminal	Voltage-gated facilitated diffusion by Ca^{2+} channel
Maintain low intracellular Ca^{2+} concentration	Ca^{2+}-ATPase (pump)
Neutrotransmitter release	Exocytosis
Removal of neurotransmitter from synapse/neuromuscular junction	Degradation of neurotransmitter on post-synaptic surface Pre-synaptic cell reuptake of neurotransmitter or the products of its breakdown

304. (B) Endocrine signaling plays a critical role in the basic metabolic functioning of the organism, in particular, maintaining the composition of the extracellular fluids, which maintains each cell with the proper environment to access nutrients and oxygen and eliminate wastes. Choice A is incorrect because *movement is, for the most part, coordinated by the nervous system.* Choice C is true only of some specific cases of endocrine signaling, like epinephrine. An important exception is *plants* and *sessile animals.* **The endocrine system works mainly though negative feedback,** not positive feedback (choice D). In addition, cells technically don't "exchange" hormones.

305. (D) Structure A is a *receptor.* *Receptors get activated by the binding of a specific chemical signal,* often referred to as their *ligand.* The binding of the ligand induces a *shape change* in the receptor protein. **Cell surface receptors are transmembrane proteins,** so they extend through the membrane and into the intracellular environment where **they activate and inhibit certain processes through a cascade of events that their binding-induced shape change triggers.**

306. (B) Structure B is a vesicle awaiting exocytosis. In a cartoon drawing of a cell, however, it could be a lysosome or an endocytosed vesicle (a nucleus will usually contain something "hairy-looking" to indicate chromatin), so to answer this question correctly requires that you *see the cell in context.* The diagram shows the cell exocytosing the vesicles into a capillary.

307. (C) Structure A is a cell-surface receptor. Non-steroid, also called water-soluble, hormones like proteins, peptides, and amino acid derivatives are ligands for cell surface receptors. They act by inducing a shape change in the receptor upon binding that in turn activates a cascade of specific events inside the cell.

Structure C is a steroid-hormone receptor that binds to steroid hormones intracellularly. The binding of the hormone induces a shape change in the receptor that exposes the DNA binding site. With the DNA binding site exposed, the hormone-receptor complex is able to bind to a specific response element in the promoter of the genes whose transcription is sensitive to, or regulated by, the presence of the hormone. **The ligand-receptor complex act as a transcription factor.** It binds to promoters with specific DNA sequences (called *response elements*) and activates the expression of those genes. For example, there is a response element for the estrogen receptor. If estrogen is present, it binds to the receptor, inducing a shape change that eventually results in the ligand-receptor complex binding to the response element in the promoter of estrogen-sensitive (or responsive) genes, the genes whose expression is affected by the presence of estrogen.

308. (B) Figure II shows two neurons, one exocytosing neurotransmitter into the synapse (the pre-synaptic neuron), while the other (post-synaptic neuron) may respond to the chemical (not electrical!) signal. **Neuron communication is one-way due to the polarity of the action potential . . . It only moves from dendrite → axon → axon terminals.**

309. (C) Choices A and D are true, but the experimental design relied on these assumptions in order to **test the hypothesis that the heart rate was slowed by a chemical factor released by the vagus nerve.** Technically, the vagus nerve releases acetylcholine to exert its effect on the heart, but that Loewi could activate the nerve by electrical impulse was also a requirement of the experimental design.

At the time of the experiment, Loewi wrote in 1953, it seemed unlikely that "if a nervous impulse released a transmitting agent, it would do so in sufficient quantity to influence the effector organ, in my case the heart, but indeed in such an excess that it could partly escape into the fluid which filled the heart, and could therefore be detected." But he tried it anyway and *discovered chemical signaling in the nervous system!*

310. The illustration of the **heart** shows that **acetylcholine receptor binding causes K^+ channels to open.** The table shows that K^+ ions are more concentrated inside of cells, so opening K^+ channels would cause K^+ to diffuse out of the cell. **K^+ outflux increases the polarity of the membrane by lowering the concentration of positive ions inside the cell, making the inside of the cell "more negative" than at rest.** The graph of the action potential shows that in order for an action potential to occur, the threshold of about –50 mV must be achieved. **K^+ outflow causes the resting potential to be more negative, making it harder to reach threshold.**

The type of acetylcholine receptor in the heart is called *muscarinic.* It is coupled to *inhibitory* G-proteins that open K^+ channels and inhibit adenylyl cyclase (so no cAMP is made). **Acetylcholine binding to muscarinic receptors makes membranes more resistant to depolarization.**

The illustration of the **skeletal muscle** shows that **acetylcholine binding to receptors opens Na^+ channels,** which the table shows is more concentrated outside the cell. **Opening Na^+ channels would cause an influx of Na^+ ions inside the cell, making the inside of the cell more positive, moving toward the threshold potential.** If enough Na^+ ions flow into the cell, threshold potential will be achieved and the membrane will become depolarized (action potential).

The type of acetylcholine receptor in skeletal muscle and other locations (except the heart) is called *nicotinic.* **Acetylcholine binding to nicotinic receptors increases the likelihood of membrane depolarization.**

The opposite response can be achieved by the same molecule, acetylcholine, because there are, in this specific case, **two different kinds of acetylcholine receptors**. They both bind acetylcholine, but the heart version is linked to a K^+ channel, whereas the muscle version is linked to a Na^+ channel. This is common in cell signaling—different cell types have different responses to the same situation, so their intracellular response to the same signal is different.

The receptors have the same basic binding site to recognize the chemical signal, but the receptor may be "hooked up" to different intracellular "plumbing" that causes the effect.

You do not need to know the details and specific mechanisms of any particular receptor system.

311. (A) Acetylcholine initiates the muscle contraction by depolarizing the muscle membrane. The increased breakdown of acetylcholine would decrease its concentration in the neuromuscular junction, decreasing or eliminating acetylcholine binding to the receptor on the muscle cell membrane.

312. (A) Atropine directly increases the heart rate. **Atropine and curare are selective receptor antagonists.** *Selective* means that the molecule binds to a *specific subtype* of receptor. A molecule that binds to *all* acetylcholine receptors has a different effect than one than binds to only one type of acetylcholine receptor. *Compounds of this nature have been pivotal to the study of receptor functions and mechanisms.*

Acetylcholine binds to muscarinic receptors in the heart to decrease heart rate. Atropine prevents acetylcholine binding and is expected to have no effect on heart rate or *to increase heart rate because it would inhibit the inhibition of heart rate.* Choice B would only work in atropine bound to nicotinic receptors in skeletal muscle (and prevented acetylcholine binding while not activating the receptor); however, atropine selectively binds to muscarinic receptors.

Acetylcholine binds to nicotinic receptors in skeletal muscle to promote contraction. Curare *irreversibly inhibits* the nicotinic acetylcholine receptor in skeletal muscle, which causes paralysis.

Pyridostigmine is an antidote to the curare poison. It is a *reversible inhibitor of acetylcholinesterase, the enzyme that breaks down acetylcholine in the synapse.* Inhibiting the enzyme allows the concentration of acetylcholine in the synapse to increase so a greater concentration of curare is required to have an adverse effect.

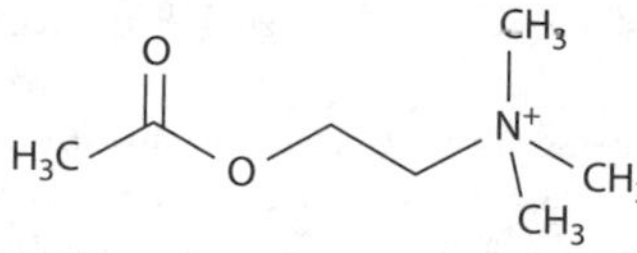

ACETYLCHOLINE

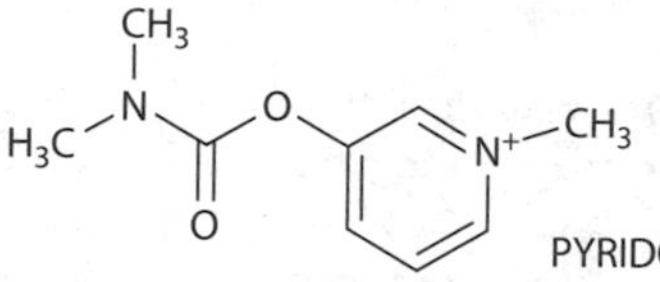

PYRIDOSTIGMINE

313. (D) In the diagram, the cells on the left appear to be the same (by their similar shape and lack of differentiation), but the cells on the right are different from each other or are doing different processes. The lettered shapes indicate the molecules that determine the fate of the cell. The arrow shows the name of the process that was induced by the molecules.

Molecules A, B, and C are present for the first three cells. The first cell receives only signals A, B, and C and survives but does not divide or differentiate. **The cell at the bottom receives no chemical signals and dies.** It is likely that molecules A, B, and C are required for cell survival. The presence of D and E promote cell division in the second cell. Molecules F and G induce differentiation in the third cell.

Choice D is incorrect for two reasons, it is not an interpretation of the diagram, but more importantly, it's not true. Apoptosis is a necessary part of homeostasis for multicellular organisms (including plants and fungus).

The AP Biology exam will assess your ability to construct and interpret diagrams representing various biological processes. Diagrams are, in a sense, symbolic representations of models.

314. (C) The identity of the numbered molecules is as follows:

(I) cortisol
(II) estradiol
(III) testosterone
(IV) vitamin D_3

The **steroid nucleus** is common to all the structures except IV, vitamin D, whose structure suggests that the molecule was made from a molecule with a fused ring structure.

The identity of the molecules in the answer choices is as follows:

(A) uric acid (a product of purine catabolism)
(B) retinoic acid (a form of vitamin A)
(C) **cholesterol (an important steroid!)**
(D) phenylalanine (an amino acid)

315. (D) Molecules I, II, and V are nearly identical. They may appear different because they are oriented differently in the diagram (see the following diagram). If you find the side of the ring with the two hydroxyl groups (OH groups), you'll notice molecule I has them facing left, molecule II has them oriented straight up, and molecule V has them facing left. The substituents (the atoms or group of atoms that occupy a specific position on a molecule) are bonded to the same carbon in the ring structure, but they differ slightly.

The identity of the molecules in the diagram are as follows:

(I) Epinephrine (adrenaline)
(II) Norepinephrine (noradrenaline)
(III) Acetylcholine
(IV) Seratonin (a neurotransmitter synthesized from the amino acid tryptophan)
(V) Dopamine (a neurotransmitter synthesized from the amino acid tyrosine)

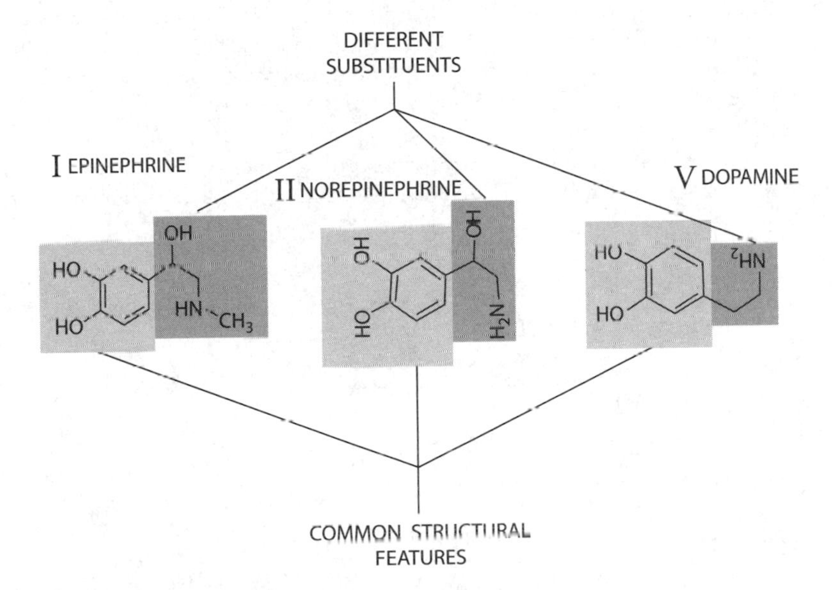

316. (B) Stimulating three different parts of the brain triggered three different responses, so answer choice A is a logical conclusion (and therefore incorrect). The patient's report that she felt something brush against her face when a region of the somatosensory cortex was stimulated is evidence that the brain generates sensations felt in the body. This is definitely true. Some amputees continue to feel sensations in their amputated limbs (a phenomenon called *phantom limb syndrome*). There are also many different molecules that act as pain relievers but act only on the brain, not at the source of pain or inflammation. The question states that the motor cortex and somatosensory cortex are located (nearby) in the brain (so choice D is a logical conclusion and therefore incorrect).

Choice B is not a logical conclusion, because anticipation still requires information and stimuli. The brain responded differently to one type of stimulation in three different regions.

The process of elimination and the "negative question strategy" are both particularly useful for answering a question like this (see the box below answer 6).

317. (D) The observation that electrical stimulation of the brain triggered laughter in one person does not suggest that humans have pre-programmed responses to all stimuli. Clearly, humans *do not* have pre-programmed responses to all stimuli, but *on a more theoretical level, the statement is too broad to make sense of one piece of data.*

There is a rare neurological disorder called *aphonogelia,* which is the inability to laugh audibly.

318. (B) Exons, regions of the DNA that code for protein, rRNA, and tRNA, make up only 1.5% of the human genome. Introns and regulatory sequences make up 24% of the genome.

The "one gene–one enzyme hypothesis" was proposed by Beadle and Tatum in 1941 (before Avery!). This hypothesis has not withstood the test of time due to the discovery of **alternate mRNA splicing** in eukaryotes (and some Archae). **Many genes are made up of multiple exons that when spliced together differently code for different polypeptides,** but B is still true. A gene can code for a polypeptide. Genes exist in alternate forms called *alleles,* not exons and introns. Depending on the reason an allele is recessive, it may be expressed and just not observed in the phenotype unless the dominant allele is absent. Finally, **genes only make up about 1.5% of the human genome.**

319. The general function of a neuron is rapid communication between cells, sometimes over very long distances. An individual neuron may process information from many sources.

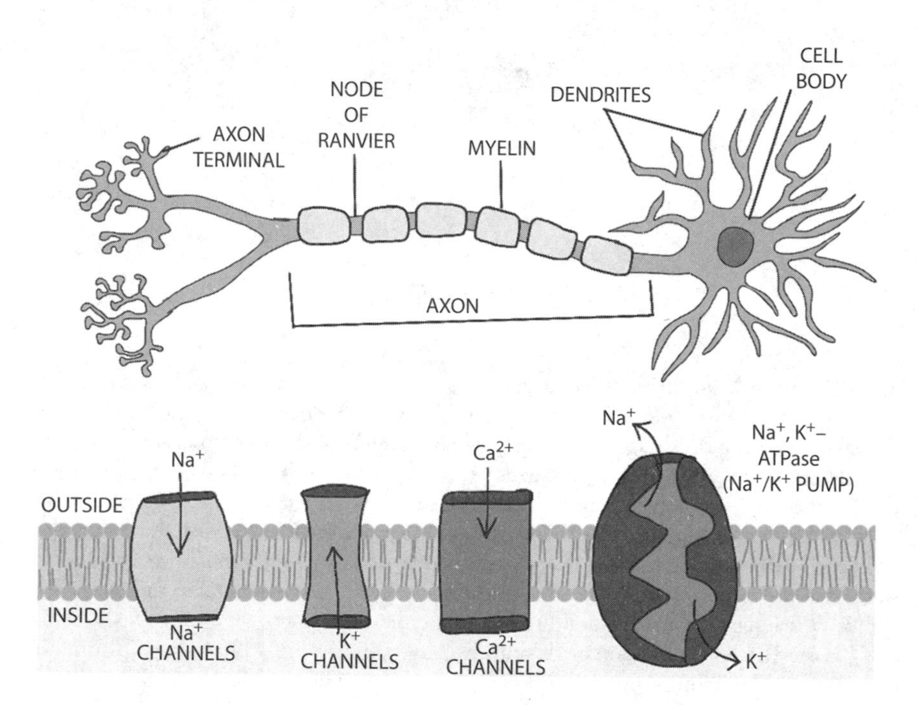

Structure	Description	Function it supports	Specifics about structure that supports its function
Dendrites	• Several long, thin tube-like projections out of the cell body	• Information intake	• Several long dendrites allow the cell to receive information from many cells at the same time. • The distances of these cells can vary because within a particular neuron, the lengths of its dendrites can vary.
Axon	• (Usually) one long, thin, tube-like projection from the cell body • May branch at the ends. • Leads to axon terminals. • May be myelinated (vertebrates) or have a relatively large diameter (invertebrates).	• Rapid signal transmission over long distances	• The electrical conduction of an action potential along the membrane occurs very rapidly due its long, thin shape. • Myelin and large diameter both increase the speed of an action potential down the axon.
Projects from axon and axon terminals	• Branching from axon • Vesicles containing neurotransmitters "wait" in the terminal.	• Transmission of signal to other cells	• Branches allows one cell to form connections with many cells. • Pre-loaded vesicles allow for rapid exocytosis of neurotransmitter when triggered.
Na^+ and K^+ channels and pump	• Transport proteins in the cell membrane	• Pump: maintain Na^+/K^+ gradient at rest and recovery from action potential. • Channels: generation of action potential	• Binding sites specifically recognize Na^+ or K^+ ions as a specific ratio: 3 Na^+ are exchanged for 2 K^+.
Ca^{2+} channels	• Protein channel in the membrane	• Trigger exocytosis of neurotransmitter.	• Ca^{2+} gradient allows Ca^{2+} to flow into cells when channel is open. • Channel is voltage gated so action potential opens it. • Specific to Ca^{2+}

320. (A) Signaling can be *between cells* or occur *within cells,* as part of the signaling cascade that executes the response. Choice A is not a mechanism by which a molecule that interferes with cell signaling would work.

The process of elimination and the "negative question strategy" are both particularly useful for answering a question like this (see the box following answer 6).

321. Enzymes of opposite function can be controlled in a variety of ways. Glycogen synthase and glycogen phosphorylase, the enzymes that synthesize and break down glycogen, respectively, are regulated (in part) by phosphorylation and dephosphorylation.

	Phosphorylated state	Dephosphorylated state
Glycogen phosphorylase	*Active*	**Inactive**
Glycogen synthase	**Inactive**	*Active*

Epinephrine triggers the signaling cascade that results in the phosphorylation of the two enzymes. *Epinephrine is secreted in times of stress and low blood glucose.* The phosphorylation of the enzymes "turns on" glycogen breakdown and "turns off" glycogen synthesis.

When the stress is over or blood glucose concentrations rise, the phosphorylated state of the enzymes is not maintained, so the enzymes are in the dephosphorylated state. This activates the synthesis of glycogen and inhibits the breakdown.

322. (B) Epinephrine triggers a phosphorylation cascade in the cell that results in the phosphorylation of both glycogen synthase and glycogen phosphorylase. However, **glycogen synthase is inactivated by phosphorylation,** and **glycogen phosphorylase is activated by phosphorylation.**

See answer 321.

323. (A) The only enzymes that are necessary to consider are *glycogen synthase kinase* and *glycogen synthase.*

The diagram clearly shows that glycogen synthase (GS) is inactive when phosphorylated. The question states that when glycogen synthase kinase (GSK) is phosphorylated, it is inactivated. The active form of glycogen synthase kinase phosphorylates and *inactivates glycogen synthase,* so **when insulin binding triggers a different phosphorylation cascade that phosphorylates glycogen synthase kinase and *inactivates* it, glycogen synthase remains *unphosphorylated and active*, so glycogen synthesis can occur.**

Let me restate that more simply. We know from the question that insulin causes a phosphorylation event, and in the cells mentioned, that phosphorylation event leads to the inactivation of the kinase (GSK) that phosphorylates glycogen synthase, GS. That means that GS will not get phosphorylated. Phosphorylated GS in inactive, so unphosphorylated GS must be active, which means that the phosphorylation by GSK inhibits GS, so GSK is an inhibitor. Phew! That was tricky! *Sometimes things with long names can prove extra confusing. Use nicknames or abbreviations and draw a diagram or make a flow chart to simplify!*

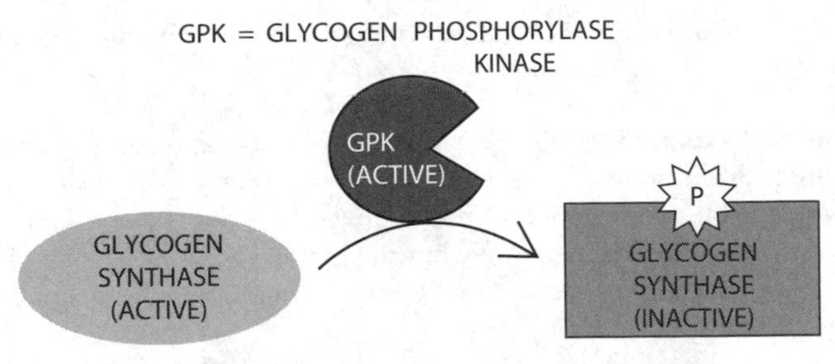

GPK IS ACTIVE WHEN DE(UN)PHOSPHORYLATED
∴ GLYCOGEN SYNTHASE GETS PHOSPHORYLATED
WHICH INACTIVATES IT.

GPK IS INACTIVE WHEN PHOSPHORYLATED SO
GLYCOGEN SYNTHASE REMAINS ACTIVE.

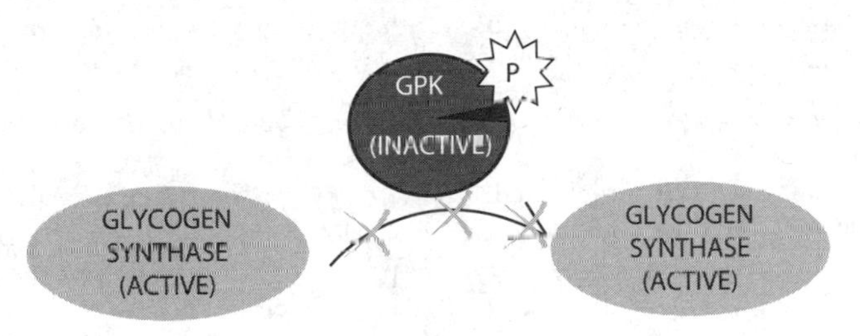

This makes sense—**insulin lowers blood sugar by promoting the uptake of glucose by cells, particularly liver and muscle cells.** Once glucose is inside these cells, insulin promotes its storage into glycogen (and its oxidation via cellular respiration).

324. (A) The diagram shows that cAMP binds to a regulatory subunit of PKA and changes its shape in a way that activates it. *Adenylyl cyclase converts ATP into cAMP*; it is *not* activated by cAMP (choice B). **ATP is often the phosphate donor from phosphorylation events.** The diagram does not show that, nor does it show that cAMP is a phosphate donor (which it's NOT, choices C and D).

325.

	1	0	0		

$$\frac{10^8 \text{ glucose molecules released}}{10^6 \text{ phosphorylase enzymes activated}} = \frac{10^2 \text{ glucose molecules}}{\text{kinase}}$$

326. Fusing the cells combines their cytoplasms. In the cell cycle, S phase follows the G_1 phase, so the fusion of S phase cells and G_1 phase cells indicates that the G_1 nucleus was triggered to begin DNA replication by something in the S cell cytoplasm.

The fusion of M cells with G_1 cells also triggers mitosis in G_1 cells despite the fact that DNA replication had not yet occurred.

The hypothesis: **Cytoplasmic factors induce later stages in the cell cycle to occur.**

You do not have to know any particular growth factor or cyclin-CdK pair that regulates the cell cycle.

327. (A) Small cells have a larger surface-area-to-volume ratio compared to large cells. For two things of the same shape, the larger of the two has the greater surface area as well as a greater volume, but the surface-area-to-volume ratio is not as large. As a cell grows, its surface area increases as well as its volume, but the surface area grows more slowly than the volume (remember the units of area are m^2 whereas the unit of volume is m^3).

328. (C) Remember that you are looking for the incorrect answer in this question (refer to box under answer 6 for a negative question strategy). **Mitosis creates genetic clones** of the parent cells, so recombination does not occur (choice B). Mitotic cell division occurs in both haploid and diploid cells (so choice C is incorrect). Its function is growth, healing, cell replacement, and normal turnover (choice D). Fungi are haploid, and mitosis is how they achieve massive growth (the largest organism on the planet is a fungus!).

Meiotic cell division occurs only in diploid cells (or cells with multiple sets of chromosomes, like many plants) because it functions to separate homologous pairs of chromosomes. If a haploid genome were cut in half, half the genes would be missing.

You do not have to memorize the names and phases of mitosis (or meiosis).

329. (D) Eukaryotic cell division consists of two separate functions: division of the genome (**mitosis**) and division of the cytoplasm (**cytokinesis**). Cells spend a lot of energy replicating the genome in the S-phase of interphase. The distribution of the two copies of the genome to daughter cells in mitosis must be as perfect as possible to ensure both cells have a complete copy of the genome. **Organelle segregation and division of the cytoplasm occur during cytokinesis.** Cytokinesis may not be symmetrical, and "accurate" is hard to qualify for the cell. DNA replication occurs during interphase.

330. (B) Meiosis allows sexually reproducing species to maintain their chromosome number while generating genetic diversity. Meiosis separates homologous pairs of chromosomes so fertilization can provide a new set of alleles.

Meiosis separates homologous pairs of chromosomes so *haploid cells*, which contain only one set of chromosomes, *cannot undergo meiosis* since there are no pairs to separate. *Meiosis creates genetic variation by shuffling alleles between homologous chromosomes* to create recombinant chromosomes during crossing over. *Meiosis also creates variation by the independent assortment of homologous chromosomes*, which randomly assorts one of each homologous pair of chromosomes into daughter cells, creating over 8 million different gametes in human egg and sperm (2^{23} combinations = 8,388,608).

331. (C) The benefit of having two sets of chromosomes to the individual is that harmful, recessive alleles can be masked if a copy of the dominant allele is present.

Choices A and D are both true but neither describes an advantage of being diploid. Choice B is not true—*haploid organisms can produce gametes by mitosis*. Fertilization produces a diploid zygote that undergoes meiosis to produce four haploid spores, each of which has the potential to grow into a multicellular, haploid organism.

332. (C) Bacteria don't have a nucleus, so they do not undergo meiosis or mitosis. The bacterial genome consists of one, large circular "chromosome," so they don't cross over or do independent assortment. Finally, the closest thing to bacterial sex is conjugation. Meiosis and fertilization, the defining processes of sexual reproduction, are not part of bacterial reproduction.

Mushrooms are the reproductive structures (fruiting bodies) of some fungi. Fungi are typically haploid, but they can fuse haploid gametes (produced by mitosis) to form a diploid zygote that undergoes meiosis to produce haploid spores that grow into multicellular fungus. Many "protists" can reproduce sexually and produce gamete plants that reproduce sexually. Trees are plants and are capable of sexual reproduction. Many people are allergic to pollen, which is basically plant sperm, the product of meiotic cell division.

You do not have to know the details of the sexual reproduction cycle in plants and animals.

333. (C) Tetrads form only in meiosis so that crossing over between homologous chromosomes can occur. Each tetrad contains the homologous pair of chromosomes with their sisters (the result of DNA replication). Choice A is true of **prophase II** (however, **reductive division**—the separation of homologous chromosomes—occurs in **meiosis I**). **Kinetochores** and **spindles** are present in meiotic and mitotic cell divisions, and the chromosomes always condense during nuclear division (mitotic or meiotic).

334–339 Hedgehog binding to Ptc activates Smo and allows the Gli protein to enter the nucleus, intact, where it acts as a transcriptional activator of Hedgehog responsive genes.

334. (B) The three proteins are the result of three different genes, but **they are expressed in different cell types** because the developmental effects of their signaling abnormalities (or lack of signaling) produce different phenotypes. The timing of the expression cannot be determined from just the observations in the question.

Alleles are different "flavors" of a particular gene.

- A *gene* is a nucleotide sequence at a particular location on a chromosome.
- The *allele* is the specific nucleotide sequence at that location.

A gene can code for:

- one or more polypeptides (if there is alternate RNA splicing of exons)
- tRNA
- rRNA
- other untranslated, functional RNA molecules

335. (C) Mutations that activate the pathway ectopically cause cancers. An effective treatment would inactivate Hh signaling in those cells.

Hh signaling results in the transcription of Hh responsive genes, so one possible mechanism is the repression of Hh responsive genes. Hh triggers Smo activation, so a molecule that can block it would inhibit Hh signaling. There are several steps after the receptor (downstream) that would produce a similar effect. For example, **a molecule that activates the PKA/SLIMB system would cleave Gli and inhibit transcription.**

Choice A is incorrect because a molecule that activates the Ptc receptor (the Hh receptor) will have the same effect as Hh. A molecule similar to Hh could act as an antagonist, a molecule that binds to the same receptor but does not activate the signaling pathway. But it does prevent binding by the molecule that *does* activate the signaling pathway.

Choice B is incorrect because increased CBP binding is associated with an activated Hh signaling pathway.

Choice D is incorrect because the phosphorylated state of Cos 2 and Fused are associated with an activated Hh signaling pathway. The Gli protein does not get cleaved, so it acts as a transcriptional activator of the Hh responsive genes.

It is difficult, if not impossible, to predict the ultimate effect on the pathway when changing even one small part of a pathway. The more interactions that are understood, the better the chance of prediction, but there are a lot of "moving parts" in the process. Generally, a factor that could bind to the Ptc receptor without activating it, while preventing Hh from binding to it, is expected to prevent signaling. But until the experiment is done and the data are analyzed, it's just a hypothesis.

336. (B) If cholesterol is not linked to SHH, SHH cannot function normally.

337. (A) Proteins recognize genes through **response elements** (a type of regulatory sequence in the DNA, usually residing in the promoter, to which specific transcription factors bind). Genes whose expression is either activated or inhibited by the presence of a particular compound are responsive (or sensitive) to that substance. Hh responsive genes are turned on by Hh binding to Ptc. The Gli protein in its *uncleaved* state is the transcription factor that *induces* gene expression, but in its *cleaved* state is also the *transcription repressor.*

The binding of two molecules requires complementarity between their surfaces. Typically there is shape and charge/polarity complementarity (positive changes on one surface are attracted to negative charges on the complementary surface).

Answer B is incorrect because promoters that contain a nucleotide sequence that is recognized by a transcription factor are considered responsive to that factor (or the substance the factor binds). In other words, if the Gli protein recognized the promoter, that gene and promoter pair are Hedgehog responsive. Choice C confuses the Gli protein with the processivity factor (the sliding clamp) of DNA polymerase. Choice D is incorrect because TATA boxes are present in many promoters. Binding to the TATA box would not confer enough specificity to the process.

338. (D) Morphogen gradients result in differentiation of cells according to their spatial orientation. Morphogens *provide spatial information in their concentration gradients.* A "**field**" is produced, with **the strength of the field proportional to the concentration of the morphogen.**

339. (C) The fitness of type I plants declines immediately in the presence of even light herbivory. The decreases in fitness decreases linearly with the increase in herbivory intensity. Type II plants do not experience a reduction in fitness when herbivory levels are low, but their decline in fitness occurs at the same rate as type I once their resistance threshold has been exceeded. The type III plants actually increase fitness with herbivory until a critical point, but that point is at a very high intensity of herbivory relative to plants of type I and II.

340. (C) The mechanisms of how choices A, B, and D are responsible for the shape of the curve as not as important in answering this question as the fact that choice C cannot be responsible for the shape of the graph. Type III plants may produce compounds that encourage low to moderate intensities of herbivory as their fitness increases with increased herbivory.

341. (B) In matters of fitness, only reproduction matters.

342. (D) Plants can't move to escape predation, so they rely heavily on the production of secondary compounds to diminish or eliminate herbivory. In this example, **the wasp detects the compound emitted from the plant and follows its increasing concentration closer to the source, which results in the discovery of a caterpillar—a potential incubator for her eggs.** Just like you don't exhale carbon dioxide so mosquitos can find you more easily, all organisms leave behind a "chemical trace" or trail of breadcrumbs. *Any organism that can exploit the ability to detect that trace can use the organism that left it behind to their advantage.*

Choice A doesn't make sense, because the wasp is ultimately successful in laying her eggs, and the eggs hatch (the larva that hatch from the eggs are also wasps, so the chemical should be toxic to them, too, if choice A was correct).

Choices B and C could be true, but neither is the best hypothesis, because there is no evidence that the chemical is toxic to either insect, that the wasp protects the caterpillar from the chemical, or that the wasp uses the chemical as food.

343. (A) The wasp must have a receptor for the volatile chemical to detect its presence. Although the detection mechanism is likely part of the wasp's *olfactory system* (sense of smell—an "old factory" probably *smells*). The olfactory system links to the central nervous system, so a response to a specific, volatile compound can be executed. The wasp is compelled to move toward the plant secreting the chemical (*positive chemotaxis*—the wasp moves toward an increasing concentration of the substance, like sniffing your way to the source of something delicious).

Choice B is incorrect because plants do not have a nervous system. A nervous system is unique to animals (though not all animals have one, like Porifera).

Choice C is incorrect because the chemical is secreted from the wound, or by the leaf, at least. There is nothing to indicate that processing in the root is necessary or that the chemical in the caterpillar saliva activates the secretion of the chemical by that particular mechanism.

344. Each of the 20 amino acids used to make proteins in cells has a specific R-group that determines its chemical function in the protein. The amino acid sequence determines the three-dimensional structure of the protein that determines its function and specificity. Because the overall protein structure is an emergent, complex effect of the chemical properties of hundreds of amino acids, any change in the protein (like ligand binding or phosphorylation) can bring about a change in its folding and therefore the function.

Sickle-cell anemia is a devastating blood disorder that has multiple phenotypic effects and is caused by a single amino acid substitution in a polypeptide of hemoglobin. The effect of substituting some or all of the arginine residues with canavanine can have a large effect on multiple proteins.

The enzyme that adds the amino acid to the tRNA (aminoacyl-tRNA synthetase) clearly mistakes canavanine for arginine. Their chemical structures are nearly identical (see the following figure).

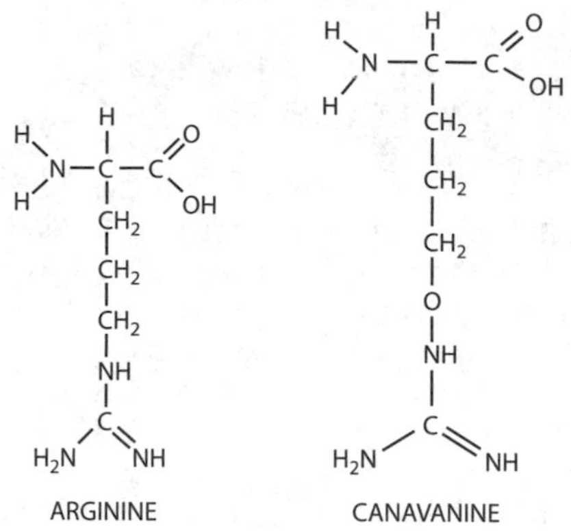

ARGININE CANAVANINE

345. (B) Although the observation that male mice that lack the VNO attempt to mate with practically anything might suggest they have a *stronger* sex drive, there needs to be more quantitative data to confidently state that the drive is stronger. The sex drive is definitely *less specific*, however.

Although ultimately anything that increases the likelihood of sex that will result in viable offspring will contribute to greater variation in the next generation, **the *direct* function of pheromones is, at least in part, to help guide animals to mating with the appropriate partners.**

Choice D may be true, but there is no information given about the age or developmental state of the mice.

346. The pAPBIO plasmid has two Hnd III restriction sites, so digestion with Hnd III cuts the plasmid into two pieces. Digestion at the first site cuts the plasmid into a linear piece of DNA, and the second cut cuts it into two pieces. There are 2,260 base pairs (bp) of coding and regulatory sequences between the two Hnd III sites on one side of the plasmid. The entire plasmid is 5,650 base pairs (given in diagram of plasmid) so there are 5,650 – 2,260 = 3,390 base pairs on the other side of the Hnd III restriction sites.

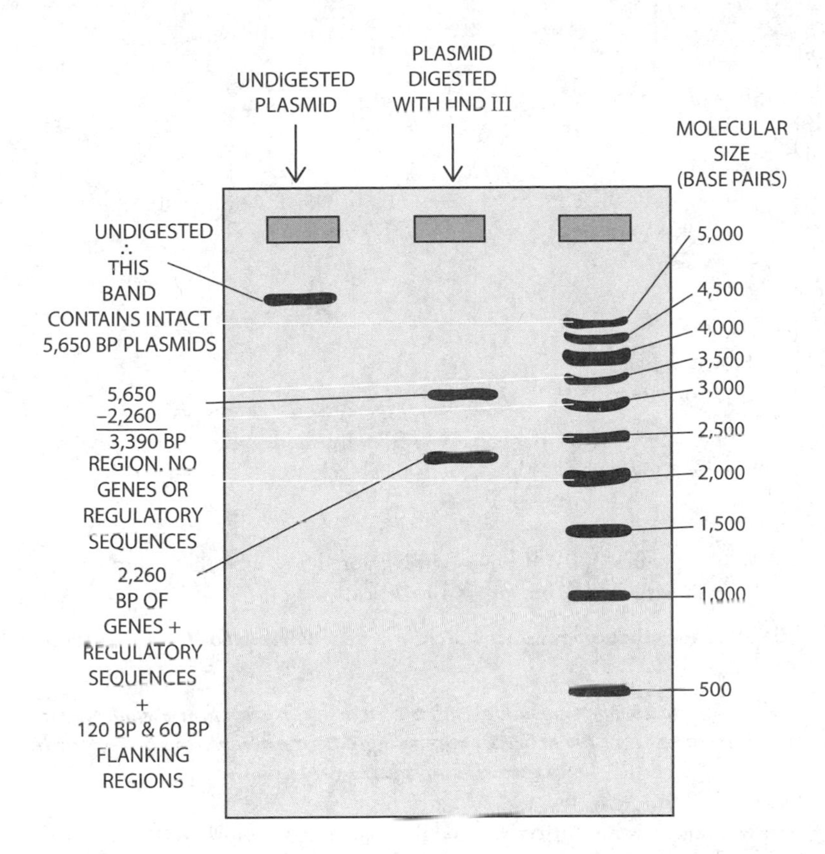

The promoter of the two genes, the amp resistance gene and the *gfp* gene, is a 400 bp sequence. The *amp* gene is 900 bp and the *gfp* gene is 700 bp. There are two pieces of "flanking" DNA (situated on each side or one side). There is a 120 bp sequence between the promoter and one of the Hnd III sites and a 60 bp sequence between the *gfp* gene and the other Hnd III site. *These sequences do not encode amino acid sequence information or regulatory information.*

347. (C) Learning is a change in behavior due to experience. Because the male bird was only exposed to another species' song, he was unable to learn his own species' song. However, *there must be some genetic involvement, because the young bird was unable to learn the song it was exposed to.* Choices A and B do not take into account the inability of the bird to sing its own species' song. Choice D can't be inferred from the observation.

348. (C) If the inheritance pattern was co- or incomplete dominance, there would be different ratios of progeny from each cross.
Assume:

BB = black
Bb = brown
bb = white

Predicted: BB × bb → 100% Bb brown
Table shows black × white → 100% black

Assume:

BB = brown
Bb = black
bb = white

Predicted: Bb × bb → ½ Bb black and ½ bb white
Table shows black × white → 100% black

Assume:

BB = white
Bb = brown or black
bb = black or brown

Predicted: BB × Bb → ½ BB white and ½ Bb colored
Table shows white × black → 100% black

OR

Predicted: BB × bb → ½ Bb colored + ½ bb other color
Table shows white × brown → 100% black

Epistasis **is the phenomenon in which one gene is affected by one or more other genes.**

You don't have to know any details of epistasis (in fact, it is one of the excluded topics), but you should understand that ***the expression of a single trait often relies on the coordinated expression of several genes.***

To answer this question correctly, you only needed to recognize that simple Mendelian genetics doesn't apply in this situation.

349. (A) Hybrids (heterozygotes) crossed with hybrids produce hybrids and true breeders in approximately equal ratios because they **produce two kinds of gametes.**

True breeders only produce one type of gamete.

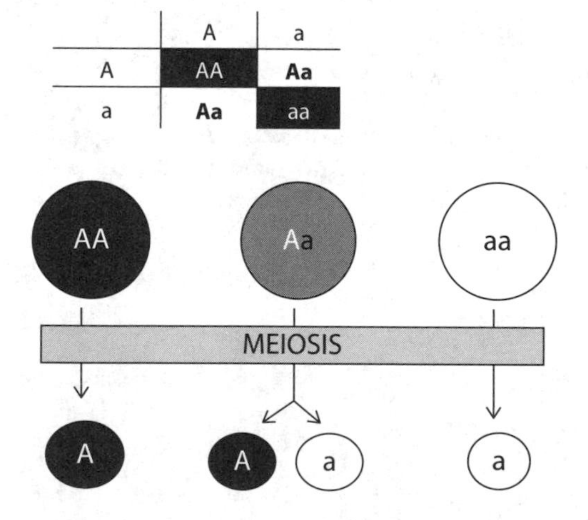

350. (C) If the genotype of an organism with a dominant phenotype is unknown, a test-cross can reveal its genotype. **A test-cross involves crossing the organism of unknown genotype with a homozygote recessive** (their genotype and phenotype for the particular trait are known). If any of the offspring have the recessive phenotype, the dominant parent was heterozygous.

A self-cross can accomplish the same thing, though it may require more offspring to be sure of the results. A hybrid crossed with a hybrid will produce some recessive offspring (~25%) whereas a cross between homozygous recessive plants will only produce homozygous dominant offspring.

351. (D) There are **two alleles** for flower color in the snapdragon plant but **three phenotypes**. You may have heard of this phenomenon as incomplete dominance which gives the impression of "blending" but the fact that pink-flowered plants can produce white an red-flowered plants demonstrates that **the hereditary factors are indeed discrete and do not blend**. The pink-flowered plant is heterozygous so it produces two kinds of gametes with regards to flower color: red and white. When two pink plants are crossed, 50% of their progeny are pink hybrids and the other 50% are true breeding, half for red flowers and half for white flowers.

352.

	9	.	3	2	

100 total progeny:

95 pink observed, 100 pink expected.
5 white observed, 0 expected.
For pink-flowered plants: 95 – 100 = 5

$$\frac{(5^2)}{100} = 0.25$$

For white-flowered plants: 5 – 0 = 5

$$\frac{(5^2)}{0} = \text{undefined}$$

For red-flowered plants: 0 – 0 = 0

Three possible outcomes: pink, red, and white ∴ 2 degrees of freedom

$$\therefore \chi^2 = 9.32$$

353. Round seed shape and yellow color are dominant if a cross of true breeding plants that produce round, yellow seed were crossed with true breeding plants for green, wrinkle seed and all the seeds were round and yellow. **They are also heterozygous for seed color and shape** (∴ 100% are RrYy)

RrYy × RrYy is a dihybrid cross and will produce the 9:3:3:1 ratio.

$^9/_{16}$ round, yellow
$^3/_{16}$ round, green
$^3/_{16}$ wrinkled, yellow
$^1/_{16}$ wrinkled, green

354.

	1	6			

The genotype RrYyPpTt can produce 16 different gametes.

There are four traits, each with two allele options (dominant or recessive).
$2^4 = 16$ (two choices for each trait raised to the number of traits).

355. The tall, purple-flowered plants are all heterozygous because their parents were tall with purple flowers and short with white flowers. There is no way for them to be homozygous dominant for either trait because of their short, white parents. The parents were true breeders because 100% of their offspring were tall with purple flowers. Therefore, there should be a 9:3:3:1 ratio of tall with purple flowers ($^9/_{16}$), tall with white flowers ($^3/_{16}$), short with purple flowers ($^3/_{16}$), and short with white flowers ($^1/_{16}$).

356. The null hypothesis is that there will be a 9:3:3:1 ratio. The genes are not linked. The alleles independently assort. In other words, nothing unexpected is happening and any deviation from the expected is due to chance.

Observed	Expected	(observed − expected)^2	(observed − expected)^2 ÷ expected
220	225	25	**0.11**
84	75	81	**1.10**
66	75	81	**1.10**
30	25	25	**1.00**

$$\Sigma = \mathbf{0.11 + 1.1 + 1.1 + 1 = 3.31}$$

Four possible phenotypes – 1 = **3 degrees of freedom**

$$\chi^2 = \mathbf{3.31}$$

The null hypothesis is supported. In order for the null hypothesis to be rejected, a χ^2 value of at least 7.8 is needed.

357. (C) Induction is the developmental process by which one population of cells influences the development of an adjacent population of cells by close-range interactions with them. Choice C suggests induction occurs to form the lens of the eye with the word *interacts.* Although this does not explicitly illustrate how induction occurs, it is the only choice that provides the slightest hint of an induction mechanism, which is typically accomplished by chemical communication.

358. (D) Learning is a change in behavior as a result of experience. **Trial and error** is a form of learning that, like the name suggests, involves testing various options and assessing the results.

An **instinct** is an inclination toward a particular behavior that is inborn (innate). One extreme of instinct is the **fixed action pattern (FAP),** a series of actions triggered by a specific sign stimulus. Once the series of behaviors has been triggered, each behavior, in the specific order, is performed to completion. The sequence of behaviors will not be prematurely terminated even if the stimulus has been removed. FAPs are not learned and thus cannot be "unlearned" (or even changed). At the other (end) is the urge to survive. The mouse is probably driven to find its way out of the maze to get food or escape (entrapment) but the way it figures out how to get out of the maze is through trial and error.

Classical conditioning is a simple form of **associative learning** (Pavlov's dogs were conditioned to salivate when a bell rang). **Insight** is an understanding of cause and effect that is usually the result of observation and deduction. Mice are generally not credited with the ability to have insight.

359. (B) Bees use visual information to navigate, and it has been shown that they can generate maps of particular areas in their brain. The **waggle dance** uses *both visual and auditory information (buzzing)* to convey information about food sources.

360. (D) The behavioral algorithm is based on odor, not touch. The sweeping of the antenna is most likely to ensure that odor molecules bind to chemoreceptors on the antennae. Even though it is possible that learning occurred, there is no indication of learning in the algorithm and choices A, B, and C must be true in order for the algorithm to work, but learning is not necessary.

361. (B) Pheremones are secreted chemicals for communication between members of the same species. There are many ways ants can follow a trail. No choices included a food source, which is definitely a possibility.

362. (C) Ants use a variety of techniques to navigate and cooperate, but they do *not* appear to learn and memorize anything about their environment.

363. (B) Nerve signals are sent to different areas of the brain, which detect the image and then decipher its meaning. (Remember that different parts of the brain have different functions.) *You don't have to know the details of the visual system for the AP Biology exam* but you should know certain facts that will allow you to answer the question. For example, photons are *not* sent to your brain through your eye (choice A). You should also know that **action potentials are "all or none,"** the same intensity of membrane depolarization occurs whether the stimulus is light or intense, as long as an action potential is generated, they are all the same (not dependent on image content as choice C incorrectly states). The pattern of action potentials, their number, and their rate, which are communicated with the post-synaptic cell, determine the intensity of the response.

The process of elimination and the "negative question strategy" work particularly well with a question like this (see the box following answer 6).

364. (A) A consequence of low intracellular Ca^{2+} levels is the ability to elevate them by allowing Ca^{2+} influx from the blood (and extracellular fluids, which maintain a concentration of 1–2 mM, about 1,000 times greater than inside the cell). *Many cells also have a high concentration of Ca^{2+} ions sequestered in the smooth endoplasmic reticulum* (in skeletal muscle, the smooth ER, called the sarcoplasmic reticulum, is highly specialized for Ca^{2+} flux because of the importance of Ca^{2+} ions in the contraction process).

365. (D) DAG is a breakdown product of PIP_2 by the action of phospholipase C (choice C is incorrect). *The diagram shows an arrow connecting DAG to protein kinase C, indicating there is a physical interaction between them.* There is nothing in the diagram or the given information to indicate that any proteins are phosphorylated (or dephosphorylated) in this cell signaling pathway, at least at this stage.

Chapter 4: Living Systems

366. (B) The diagram shows the assembly of monomers into a polymer by the removal of water, a process called both **condensation** and **dehydration synthesis** (and sometimes just dehydration). The diagram shows monomers getting linked together with **water as a product**.

The monomers and polymers are not identified in any way, so choice C can't be correct. The final structure and function of the polymer are also not clear, so choices A and D can't be correct. *It takes practice to extract the appropriate level of detail from a visual representation.*

367. (D) Photosynthesis is a *synthetic* reaction, but it is *not a polymerization process.* **Polymerization** refers to the process in which many monomers, or single units, are linked together to form a large structure like a *protein, nucleic acid, or complex carbohydrate*. Starch, glycogen, and polypeptides (the product of translation) are all polymers (see the following table).

Monomer		Polymer
Glucose	⇌	Starch, cellulose, or glycogen
N-acetyl glucose	⇌	Chitin
Amino acids (20 different kinds)	⇌	Polypeptides (proteins)
Ribonucleotides (4 different kinds)	⇌	RNA
Deoxyribonucleotides (4 different kinds)	⇌	DNA

You do not have to know the molecular structures of specific amino acids, nucleotides, lipids, or carbohydrate polymers.

368. (D) The word *directly* is the operative term here. **The process is transcription and translation.** Molecule B is just the untwisted version of molecule A. Molecule C is the single-stranded (RNA) product of the transcription of one strand of molecule B, and molecule D is a polypeptide. These processes are required for all the answer choices! However, the question asks *directly*, which narrows the focus to choice D, the only *direct* result of the processes. *Ultimately*, the processes allow A, B, and C to occur.

See the box after answer 86 for a comparison of ultimate and proximate causes and consequences.

369. (A) In biology, **reproduction is the ultimate goal**. Cells and bodies die, but the genetic information they contain continues to get passed on from one generation to the next.

Every living thing alive today has an unbroken lineage of descent from at least 3.8 billion years ago—perhaps longer! Every single ancestor of theirs lived long enough to leave behind offspring who then did the same. And those lineages that resulted in no offspring died out.

The process that results in protein synthesis accomplish the functions in answer choices B, C, and D, but they are more *direct* consequences of the processes.

See the box after answer 86 for an explanation of ultimate and proximate causes and consequences.

370. (D) *A change in nucleotide sequence does not necessarily mean there will be a change in amino acid sequence.* The genetic code is redundant, so there are often 2–4 different codons that code for each amino acid (there are 61 codons that code for 20 amino acids). Mutations that produce no change in amino acid sequence or no change in the phenotype of the organism who harbors them are called **silent mutations.**

However, we will assume the change *does* produce a change in amino acid sequence, and it was the bottom strand of molecule B that was the strand transcribed to produce molecule C. **In this case, the mutation would be nearest the N-terminus of the polypeptide chain. If the top strand of the molecule was transcribed to produce molecule C, then the amino acid change would be closer to the C-terminus.**

371.

	DNA	RNA	Polypeptide
Monomer	• Deoxyribonucleotides contain a deoxyrobose sugar.	• Ribonucleotides contain a ribose sugar.	• Amino acids
Types of monomers	• 4 kinds of nucleotide monomers: A, T, C, G	• 4 kinds of nucleotide monomers: A, U, C, G	• 20 kinds of monomers, each with the same N-C-C backbone, but with different chemical properties due to their different side-chains: polar, non-polar, charged, acidic, basic
Macro-structure	• Long, double-stranded polymer • Cells, usually one or two copies of the genome (though some organisms have multiple copies) • In eukaryotes, complex interactions with histone proteins help regulate gene expression.	• Many kinds of single-stranded molecules (and some double-stranded molecules). • Single-stranded molecules like rRNA and tRNA can take on complex three-dimensional structures to perform their functions.	• Hundreds to thousands of different kinds of proteins in each cell and many copies of each kind

Functional differences	• Information storage	• Information retrieval, regulation of gene expression, catalysis, mRNA splicing in eukaryotes	• Enzymes • Receptors • Transporters • Signaling molecules • Structure (cytoskeleton, extracellular matrix) • Storage and regulation of gene expression (histones and transcription factors) • Antibodies • Ribosomes
Functional similarities and differences		• Polymers are unbranched chains and have directionality. • DNA and RNA are synthesized 5′→ 3′ • Polypeptides are synthesized from the N-terminus to the C-terminus. • Their combined functions provide the cell with functional proteins and RNA molecules to perform all life functions.	

372. All matter is constructed from simpler subunits—atoms, which themselves are made up of simpler units: protons, neutrons, and electrons. We don't have to go any smaller than the three subatomic particles to understand that the number of each of the **three particles** in an atom determines everything about that atom and allows the existence of at least *92 naturally occurring kinds of atoms (the elements)*.

The four DNA bases (and four RNA bases) provide 64 different triplet codons. Twenty amino acids provide enough chemical variety to synthesize millions of different proteins (if not more) that can perform a huge number of functions based on structures and chemical (electrostatic) properties.

The 26-letter alphabet (in English) and 9 digits allow the communication of an almost infinite number of complex and abstract ideas. Reading and writing in *one direction*, for example, left to right in English, is much more efficient than a language that has no specific direction. **Without directionality in information processing, misinterpretations would abound** . . . *pals* or *slap*? *Reviled* or *deliver*? *Repaid* or *diaper*? (*semordnilap* are words that can be read differently, but validly, forward or reverse.) Many small molecules (i.e., not polymers) are "right handed" or "left handed." Amino acids, for example, are left handed (L-enantiomers) and sugars are right handed (D-enantiomers).

373. (A) Because the cells were from the same organism, they would have originally had the same DNA. Even though one strand was randomly removed and replaced, the sequence information is the same. It's equivalent to replication where each strand can serve as a template strand resulting in the production of a ½ new but identical double-stranded DNA molecule (assuming no mutations occur).

374. (D) The protein coding regions of a gene are typically on one of the two strands, but which strand has which protein coding regions varies. The *sense strand* is the strand that has the protein coding information, the sequence that is transcribed, but along a chromosome, each strand contains coding regions. In other words, one of the two strands is *not* the sense strand for all the genes on the chromosome. To simplify, we will assume that one strand of the double-stranded molecule contains half the protein coding regions of the chromosome and the other strand contains the other half.

375. (A) Each amino acid has a particular structural and chemical properties that give a role in the polypeptide (or protein) of which it is a part. Some amino acids that are not part of the proteinogenic amino acids (the "translation 20") are often modified once their precursors are incorporated into the protein. However, this is not the case for **selenocysteine or pyrrolysine.** These amino acids **are two of the rare exceptions to the genetic code.**

376. (D) The function of a polypeptide is determined by its structure. Its structure is determined by its amino acid sequence. Its amino acid sequence is determined by the nucleotide sequence of the gene that codes for it.

The side chains of the amino acids determine much of a polypeptide's folding. Because many three-dimensional protein structures have been solved by X-ray crystallography, the ability of scientist to predict a proteins structure by its amino acid sequence has greatly advanced. Databases of particular structural motifs connect their specific structure with a particular function or binding capability. For example, zinc fingers and leucine zippers are DNA-binding motifs: proteins that have these structures have binding sites for DNA. When a scientist determines that one of these structural motifs is part of a protein, they can be confident it is capable of binding to DNA (and is likely a transcription factor).

- **Nonpolar side chains** may drive proteins to fold into a particular structure based on their "water-avoidance behavior."
- **Polar side chains** may increase the solubility of a protein in water and may influence the ability of an enzyme to participate in chemical reactions.
- **Acidic and basic side chains are ionized at physiological pH** and can participate in electrostatic interactions and the formation of salt bridges (ionic bonds).
- **Histidine**, in particular, plays an important role in proteins that function as buffers.
- **Aspartate and glutamate** can bind metal ions.

377. (A) The process of protein synthesis is represented in the diagram. Polypeptides are synthesized from the N-terminus (+) to the C-terminus (-):

$$(^{+}\text{N-C-C}^{-}) + (^{+}\text{N-C-C}^{-}) \rightarrow {}^{+}\text{N-C-C-N-C-C}^{-} + \text{H}_2\text{O}$$

Polypetides are linear chains of amino acids that fold into their specific shapes due to the sequence of amino acids. **Their synthesis requires energy** (so it is not exergonic), and it produces water by multiple dehydration reactions.

378. (D) The diagram shows **cell-cell recognition**, chitin (the polymer of N-acetylglucose used in many biological **structures**, including the **exoskeleton** of arthropods), and the

cellulose cell wall of plants. The bottom-right picture shows the **synthesis of ATP from the oxidation of glucose.**

379. (C) Eicosanoic acid is synthesized from the 18-carbon linoleic acid, though linoleic acid must be consumed in the diet. **Allosteric activators are usually metabolites of a pathway that accumulate when an enzyme later in the pathway does not work fast enough to convert the metabolite to a product.**

380. (D) Answer choices A, B, and C are all functions of lipids in the body. Joints are places where two bones meet. Some joints are immovable, but movable joints are cushioned and lubricated in several ways. The joint itself is encased in a fluid-filled capsule, but **the composition of synovial (joint) fluid is not lipid.** It is a thick, stringy fluid made of hyaluronic acid (a glycosaminoglycan, a long, unbranched polysaccharide made of repeating disaccharides) and lubricin (a proteoglycan, a protein that has been heavily glycosylated).

The box following answer 6 contains information about the process of elimination and the "negative question strategy" which are particularly useful in answering this type of question.

381. (A) Because the nuclear envelope fragments during prophase and reforms during telophase, the proteins must be able to "find their way" back into the nucleus every time the cell divides. The signal sequences are attached to the amino acid chain during translation so it is not reattached once the protein has left the ribosome.

Choice B is incorrect because daughter cells *do* inherit pieces of organelles and whatever proteins they contain at the time of cytokinesis. Choice C is incorrect because the protein translocated into the nucleus through the nuclear pore, so it had to be small enough to get into the nucleus. Choice D just doesn't make sense.

382. (D) Most membrane-bound organelles can only be synthesized from "pieces" of membrane-bound organelle. During cytokinesis, the endomembrane system (including the golgi and ER) is divided up into daughter cells. These fragments of membrane are used to construct new membrane. The mitochondria and chloroplast are important exceptions. These organelles are capable of semi-autonomous replication within cells (they used to be free-living prokaryotes). Still, organelles come from pre-existing organelles in some form.

383. (A) Models are theoretical frameworks. They can be phylogenies, equations, or analogies, or take on any other form that allows predictions to be made based on data and observations that have already been collected. The continued testing of the model permits adjustments to be made to the model. **No model is perfect,** which is why scientists continue to test and refine them.

384. (B) Because of **alternate RNA splicing**—the process in which two or more mRNA sequences can be constructed from one gene due to different arrangements of the exons—several polypeptides can be synthesized from the information contained in one gene.

385. **(A)** An accumulation of the intermediates prior to the point of inhibition is expected. The inhibitor should specifically inhibit *only* enzyme 3, but that would have inhibited metabolite D from forming, so there would be a very low concentration of metabolite D, the substrate for enzyme 4, and the precursor of metabolite E. Therefore, the concentration of molecule E is also expected to be diminished.

386. **(A)** Mutating the enzyme to make it non-functional (or knocking out the gene so that it isn't expressed) is like inhibiting the enzyme.

Mutating (or knocking out) the enzyme allows the effects of the enzyme to be determined by measuring the loss of functionality in the cell (in this case, the buildup of metabolite D is expected).

The major difference between inhibition and knocking out a protein is that in a traditional knockout, the protein is *never* expressed and/or *never* active in the cell, so the knockout may have serious developmental effects. Alternatively, a conditional knockout is one in which the expression of the particular gene can be blocked at any time after development the experimenter desires.

387. **(B) The enzymes of most metabolic pathways are sequestered together in one compartment of the cell.**

- The enzymes of glycolysis are located in the cytosol.
- The enzymes of the Krebs cycle are in the mitochondrial matrix.
- The enzymes of the Calvin cycle are in the stroma of the chloroplast.

388. **(B)** Recent studies have shown that the mammalian genes involved in lipid metabolism, glycolysis, gluconeogenesis, and other pathways are **regulated in part by circadian rhythm**, an endogenous 24-hour cycle observed in all eukaryotes and in some bacteria.

389. **(B) The activation energy of reaction is the amount of energy required for the reactant (or substrate) to achieve the transition state**. The **transition state** is a short-lived, high-energy conformation that a molecule (or molecules) must "pass through" to form a product. The **activation energy** is determined by **subtracting the energy of the reactants from the highest point on the graph**—the "top of the hill," which represents the energy of the transition state. Reaction B clearly has the largest activation energy.

390. **(A)** The **free energy change** is determined by **subtracting the energy of the reactants from the energy of the products**. If the free energy change is positive, the reaction is **endergonic** and the products have more energy than the reactants. If the free energy change is negative, the reaction is **exergonic** and the products have less energy than the reactants. This kind of process is also called **spontaneous**, indicating it requires no action from the environment to occur. Choice A has the greatest difference in energy between products and reactants.

391. **(A) The free energy change is the difference in energy between the reactants and products.** See answer 390.

392. **(B)** Reactions with a high activation energy take the longest amount of time to occur and require an energetically expensive **conformational change** to transform into their transition state. See answer 389.

393. **(A)** Because the catalyst increases each reaction rate by the same amount, the catalyst doesn't need to be considered. **The reaction with the highest overall rate will still have the highest rate if all reaction rates increased by 100-fold.** Reaction A has the lowest activation energy and would proceed the fastest.

394. **(D) The endomembrane system is composed of the nuclear membrane, the ER, Golgi apparatus, vesicles, and lysosomes. The lipid composition of the nuclear envelope, golgi, lysosome, and plasma membrane are more similar to each other,** each containing a large percentage of phosphatidylcholine, phosphatidylethanolamine, and cholesterol. The mitochondrion has a large percentage of ethanolamine, little cholesterol, no sphingolipids, and a lot of cardiolipin.

395. **(C) The diffusion of CO_2 and O_2 across plasma membranes does not require transport proteins.** The diffusion is "simple," which means the substances diffuse right through the membrane from where it is more concentrated to where it is less concentrated. **Diffusion is always passive and always occurs in the thermodynamically favored direction** (from an area of higher concentration to an area of lower concentration).

396. **(A) One important function of membrane sidedness is signal transduction**. For G-coupled receptors that rely on calcium signaling, **the activation of phospholipase C is required to break down phosphatidylinositol into IP_3 and DAG. DAG activates protein kinase C, while IP_3 opens calcium channels in the endoplasmic reticulum.**

Choice B is true, but the membrane change does not indicate anything about the sidedness of the membrane. Choices C and D are evidence that the membrane leaflets differ in their lipid composition, which suggests that membrane sidedness is important, but it is not evidence that sidedness is important.

397. **(B)** Red blood cells contain a large amount of the hemoglobin protein, so any mutation on the surface of the protein could interfere with interactions between hemoglobin molecules within the cell.

Choice A is incorrect because the base substitution of adenine to thymine would result in the placement of adenine to the complementary strand of DNA during replication. Choice C is incorrect because there are no hydrophobic interactions that occur between the amino acid backbones. Choice D is incorrect because glutamic acid was replaced with valine, not the other way around, so no "acid-base" reaction would occur between the red blood cell membrane and the hemoglobin surface.

398. **(C) Red blood cells are the only cells in the human body that cannot use oxygen.** Fast-twitch muscle fibers can work anaerobically for short periods of time, but otherwise all cells in the body need a constant supply of oxygen to function. Just a few minutes of oxygen deprivation can kill you!

399. Most of the pill bugs will be found on the moist side of the chamber because all known living things require water to survive. Pill bugs that do not choose moist environments when available are less likely to find the appropriate conditions that support life and therefore less likely to survive and leave behind offspring that have similarly poor preferences.

400. There are numerous improvements that could be made, but one suggestion is to insert each pill bug individually, one bug at a time, note where they end up, remove the bug, and repeat with the same pill bugs to see if it makes the same choice repeatedly (and then repeat with other bugs). This eliminates the possibility that the behavior of some pill bugs is affected by the behavior of other pill bugs.

401. One hypothesis is that the pill bugs prefer the light, so most of them will be found in the light compartment after 30 minutes. Your hypothesis could also be that the pill bugs prefer the dark, and so a greater number would be found in the dark compartment after 30 minutes. The null hypothesis is that there is no preference: that half of the pill bugs will be in the dark compartment and half will be in the light compartment after 30 minutes.

402. The null hypothesis (see answer 401) is that there is no preference so 50% of the bugs will end up in each chamber.

$$\text{Dark chamber: } \frac{(\text{observed} - \text{expected})^2}{\text{expected}} = \frac{(85-50)^2}{50} = 4.5$$

$$\text{Light chamber: } \frac{(\text{observed} - \text{expected})^2}{\text{expected}} = \frac{(15-50)^2}{50} = 24.5$$

$$4.5 + 24.5 = \mathbf{29}$$

Two choices − 1 = 1 degree of freedom

The data support the hypothesis that the bugs prefer the dark chamber because the chi-square value exceeds the limits for a P value of 0.05 and 0.01%, which means that the probability that the results differ from the null hypothesis due to chance is much less than 1%. The null hypothesis is rejected.

403. Taxis is directed movement toward or away from a stimulus. Positive taxis is moving toward the stimulus, and **negative taxis** is the movement away from the stimulus. Moths exhibit positive phototaxis whereas mosquitos exhibit negative chemotaxis to certain insect repellants.

Kinesis is the random movement of an organism that continues until the desired environment is reached. Pill bugs are known to exhibit kinesis toward moist environments but you wouldn't need to state that specifically to answer this question correctly.

The most important thing to consider is that *for both taxis and kinesis, the organism requires a mechanism to detect the stimulus.*

404. Most animals must act within their environment in order to survive and reproduce. If an animal does not seek out food or escape predation, it is unlikely to survive and leave behind offspring. Behaviors have a molecular, cellular, and physiological basis.

Behaviors range in complexity, but the simplest ones are (carried out) by inborn circuits in the nervous system, like a fixed-action pattern. Others are complex and depend on learning, like social interactions between humans.

405. A molecule does not have an odor. **The scent of a chemical is a construction of the nervous system.** In order **to detect the molecule, a receptor is required**. The receptor must have a mechanism with which to communicate with the nervous system once the presence of the molecule is detected.

Olfactory signaling is important to many organisms in finding food, finding mates, and detecting potential threats from organisms of the same or different species. The circuitry that relays the chemical information at the receptor to the processing center (the nervous system in an animal) is typically connected to the part of the nervous system that directs the organism toward or away from the odor.

406. (C) This is an example of **classical conditioning**, a form of associative learning, where the animal relates a **neutral stimulus,** the sound of the bell, to food. The neutral stimulus eventually becomes associated with food (and becomes the **conditioned stimulus**). The dog will salivate in the presence of food—that was a fact exploited in the experiment. Salivation is a reflex, not something typically done on command. (However, thinking about food, particularly savory or sour food, may make you salivate, especially when you are hungry.) It is not possible to know *directly* what a dog is thinking. Choice D switches the relationship between the **unconditioned stimulus** (the food, the presence of which will naturally result in salivation) to the conditioned stimulus (the bell).

407. (B) The proportion of surviving hatchlings in species 1 is almost the same, with or without the filter.

408. (D) Almost 100% of the hatchlings survive in species 1 under both conditions. However, there is no data indicating the absolute number of eggs or the absolute number of hatchlings that survive.

409. (D) Both proteins with a "gray" N-terminus are targeted to the mitochondria. The two proteins with the "black" N-terminus are targeted to the ER. There is no indication of the function of any of the proteins.

410. If the sequence of a gene is known or the gene has been isolated from a genome and put into a plasmid (cloned), the polypeptide (or protein) can be expressed, usually in bacteria cells (as long as there are no introns, which are usually spliced out when cloning). By using restriction enzymes, a coding region of the gene can be spliced with another coding region that was cut with the same restriction enzyme.

There are other ways, too. For example, you could use PCR to "tape together" pieces of sequence and then have them replicated. *Electrophoresis* is a method used to separate a mixture of DNA molecules according to size.

To be expressed, the sequences need to be ligated into plasmids.

411. (C) In order to make new ER, a segment of existing ER is needed. Normally, this occurs during cytokinesis, when the contents of the cytoplasm are divided up between the daughter cells. *The fact that the second cell, whose ER was not entirely removed, was able to make new ER while the first cell could not indicates that the information in the genome is not sufficient to construct the ER "from scratch."*

412. (B) The two cells were taken at different stages of development but from the same organism, a mouse (a diploid animal). Tumor cells do not undergo meiosis (choice A), and although mouse embryo cells are diploid, it does not explain the findings. *The results show* *that the variable and constant regions are connected on one fragment in the tumor cells, cells that have differentiated.* In embryo cells, the two sequences are present on different fragments of DNA, indicating that during development, "cutting and pasting" of the genome occurred.

413. (B) At 20 mm Hg adult hemoglobin has a saturation of approximately 20% (which means that 20% of the available oxygen binding sites are occupied with oxygen) as compared to over 30% saturation of fetal hemoglobin and approximately 70% saturation of myoglobin.

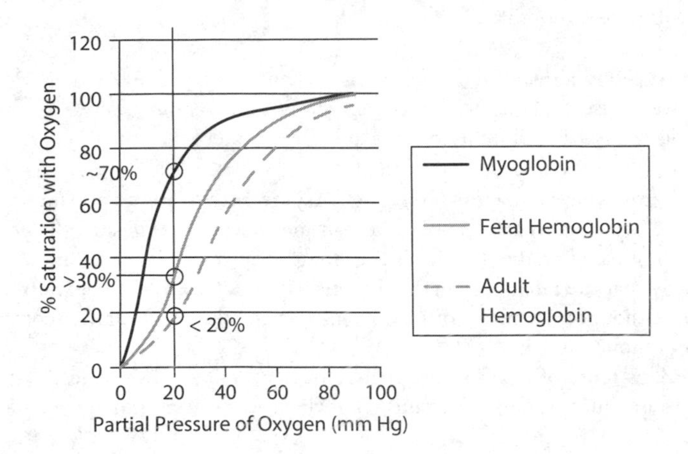

414. (D) The fetal hemoglobin binding curve was arbitrarily chosen. The black line shows the "right" shift of the curve at a lower pH. At a lower pH (greater acidity), there is a lower percent saturation, indicating that less oxygen is binding (or more oxygen is "falling off"). Either way, the result is the same: a lower affinity of the oxygen-binding protein.

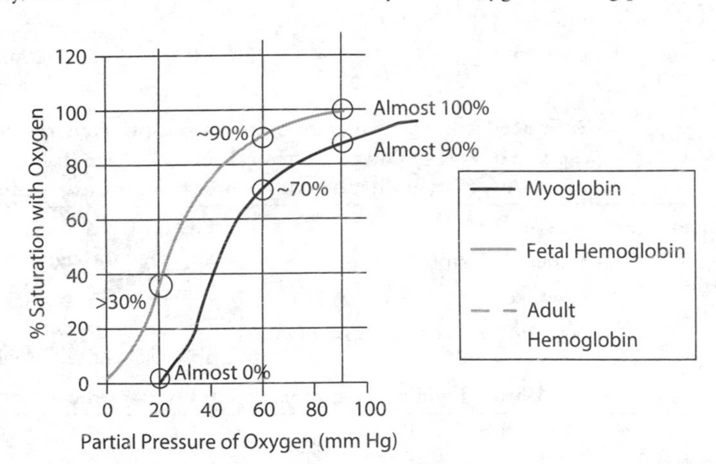

There are important physiological consequences to the affinity change at lower pH. Muscle cells secrete lactic acid into the bloodstream when working anaerobically, which locally decreases blood pH. One of the many effects this has is to increase the "drop off" of oxygen at the tissue by hemoglobin.

415. **(C)** The shape of the myoglobin curve shows that even at low partial pressures of oxygen, myoglobin saturation is significant. **The high affinity of oxygen for myoglobin means that myoglobin has oxygen bound to it unless the partial pressure of oxygen is very low.** The result of this is that myoglobin will keep its oxygen until there is very little oxygen available from hemoglobin.

416. **(D) In order for fetal oxygen to get oxygen from maternal hemoglobin, the affinity of oxygen for fetal hemoglobin must be greater than the affinity for hemoglobin.** Once a baby is born, the fetal hemoglobin gets replaced with adult hemoglobin as red blood cells turn over within 3–4 months.

417. **(A) Oxygen** is fairly non-polar, meaning it **is not very soluble in water**. The solubility of a gas in water also depends on the temperature. As temperature increases, the solubility of oxygen decreases. The solubility of oxygen in 37°C water is less than 7 mg/L.

418. **(D) Tyrosinase is the rate-limiting enzyme in the pathway that synthesizes melanin, the pigment in fox fur** (and human melanocytes). Typically, the coat color of an animal is not based on diet (choice B). Although changes in daylight can affect animals, it is not likely that the rate of pigment synthesis will be affected by the pattern of daylight (choice A). Although that may not be obvious to you, choice D makes sense and (uses) biology you (know).

Choice C is more of a reason the coat color may change. Theoretically, it can be an "ultimate" cause, but it is not a mechanism by which the fur can undergo a color change.

419.

	Animals	Plants
Requirements	• Obtain organic carbon and many molecules from the environment • Oxygen	• CO_2, typically gaseous, and oxygen • Minerals from the soil • Sunlight
	Both need water and other substances from the environment, a source of energy and gases from the atmosphere (for aquatic organisms, oxygen dissolves into water from the atmosphere).	
Mechanism of nutrient acquisition	• Seek and inject food • Sometimes hunt • Filter feed • Action/behavior typically required	• CO_2 from atmosphere absorbed through leaves • Liquid water • Dissolved minerals in soil absorbed through roots
	Both rely on active and passive transport processes.	

Metabolic processes	• Ingested food must be digested, absorbed, and assimilated. • Some organic molecules are oxidized for fuel while some are used as building blocks.	• Synthesis of complex compounds from simple substances in the chloroplast • Energy of sunlight powers photosynthetic cells • Oxidation of organic fuel occurs in the mitochondria other cells.
	Both require processing of absorbed substances and the oxidation of organic fuel as a source of energy.	
Growth	• Determinate • Cell types and body plan well developed at "birth" • Growth usually involves cell division but may involve increase in cell size (e.g., fat cell and muscle cell growth, for example).	• Indeterminate • Lifelong growth • Can produce all cell types in the zone of differentiation at meristems. • Cells divide and most increase dramatically in size.
	Both rely on cell division and cell growth to grow and develop.	
Reproduction	• Sexual reproduction • Many species can reproduce asexually. • Sexual partners usually have direct interactions.	• Most species can reproduce sexually and asexually. • Sexual partners are usually secured through the intervention by a biotic (bees or bats, for example) or abiotic (wind and water) factors.
	Both produce gametes by meiosis and produce zygotes by fertilization.	

420. There are several strategies you can use to estimate the leaf surface area. A simple and fairly accurate way is as follows:

(1) Count the number of boxes that are at least $^2/_3$ of the area covered by the leaf.

$$\sim 25 \therefore \sim 25 \text{ cm}^2$$

(2) Count the number boxes where approximately ½ of the box is covered; then divide by 2 (2 boxes will contain approximately 1 cm^2 of surface).

$$^8/_2 = 4 \text{ cm}^2$$

(3) Count the number of boxes that have ¼ or less covered; then divide by 4.

$$^{20}/_4 = 5 \text{ cm}^2$$

$$\textbf{TOTAL} = 25 \text{ cm}^2 + 4 \text{ cm}^2 + 5 \text{ cm}^2 = \mathbf{34 \text{ cm}^2}$$

421. The section contains 9 full stomata and 2 stomata that look like there is 50% or more in the section. Depending on your method you may count each as a full stomata or combine the two into one based on surface of stomata. Either method is acceptable. Here, assume each stomata individually ∴ 11 stomata.

The **area of the leaf** shown is 2 mm × 1.5 mm = 3 mm^2.

$$\frac{11 \text{ stomata}}{3 \text{ mm}^2} = \frac{\mathbf{3.7 \text{ stomata}}}{\mathbf{mm^2}}$$

The leaf has **34 cm^2** of area (from answer 420).
Assume only bottom surface has stomata.

$$1 \text{ cm}^2 = 10 \text{ mm} \times 10 \text{ mm} = 100 \text{ mm}^2$$

$$34 \text{ cm}^2 \times \frac{100 \text{ mm}^2}{\text{cm}^2} = \mathbf{3{,}400 \text{ mm}^2}$$

$$\frac{3.7 \text{ stomata}}{\text{mm}^2} \times 3{,}400 \text{ mm}^2 = \mathbf{12{,}580 \text{ stomata}}$$

422. The stomata are openings in the leaf that allow for gas exchange between the atmosphere and the inside of the leaf. The **guard cells** regulate the opening of the stomata. Stomata allow water to leave by **transpiration. Transpiration is a major mechanism of water transport in the plant and is responsible for the majority of water transport in the xylem.** Stomata allow the oxygen gas generated during the light reactions to escape into the atmosphere, inhibiting photorespiration. Stomata allow the uptake of carbon dioxide from the atmosphere, providing carbon atoms for the Calvin cycle. The evaporation of water from open stomata can help cool the plant but at the expense of water. Closing the stomata can reduce water lost but may cause an increase in temperature, an accumulation of oxygen, and a reduced partial pressure of CO_2.

423. (C) The pea plant has the second highest stem-weight-to-root-weight ratio. For a given root weight, cotton has a greater weight of stem compared to the pea plant and carrot has a lower weight of stem.

424. Determinate growth:

- *Advantage:* greater complexity of development can be achieved because a more complex sequence of events can unfold for a longer period of time. Limited size means limited energy requirements.
- *Disadvantage:* many structures cannot be regenerated if lost during life.

Indeterminate growth:

- *Advantage:* The rate of growth can change in response to the environment. However, increased growth requires increased energy, and nutrient and water requirements for both the initial growth and maintenance of the larger body size. In plants, a larger body size can translate into the increased capacity to make energy if the right proportion of the growth is photosynthetic surface. The ability to grow continuously allows certain plants to "take over" an area. In addition, plants can generate and regenerate many parts of their bodies because they retain stem cells at meristems.

- *Disadvantage:* Growth can incur requirements for energy, nutrients, and water that may be lacking after growth has already occurred.

425. (B) The promoter of many archae and eukaryotic genes contains a TATA box, a binding site for proteins involved in gene expression (transcription factors). The TATA box is named for the high number of thymine and adenine base pairs in the region. Although you don't have to know the specific fact, regions where transcription begins usually have more A-T base pairs because there are only 2 hydrogen bonds per base pair, so less energy is needed to separate the DNA strands when transcription factors and RNA polymerase bind to the promoter. DNA sequences with a high G-C content require more energy to separate because G-C pairs are held together by 3 hydrogen bonds.

426. (C) The amino group of the incoming amino acid gets added to the carboxyl group carbon of the amino acid at the end of the chain to form a peptide bond. Polypeptides, like the mRNA they were translated from and the DNA that encoded them, are synthesized in a particular direction. **mRNA is read from 5′→3′ by the ribosome, while the polypeptide is synthesized from the N-terminus to the C-terminus.** The first amino acid will always have its amino group on one amino end of the chain, and the last amino acid will have its carboxyl group free at the opposite end.

427. (D) The mussel shell thickness was measured in the experiment (so choice A is incorrect). The mussels that were bred from the original northern population of mussels had no previous exposure to *Carcinus*, yet their shells thickened the most when exposed to *Carcinus.*

The mussels bred from the southern population of mussels also showed increased shell thickness with exposure to *Carcinus* and *Hemigrapsus* even though those specific mussels had never been exposed to either type of crab.

428. (A) *The mussels in the experiment were naïve to crabs;* in other words, they were never exposed to crabs. **The fact that the shell thickness increased only when exposed to the type of crab the native population of the mussels had been exposed to indicates that the population from which they were derived had adapted to the presence of the specific crab.** This requires the ability to detect the presence of the specific type of crab that was a threat and effect a response (thicken their shell).

However, the increase in shell thickness was technically an acclimation, made possible by the adaptation of the ability to sense and respond to the presence of specific crabs that the mussels were never exposed to.

Choice B is incorrect because *the ability to respond to the presence of a specific crab was an adaptation,* not the inheritance of a learned or acquired trait. Choice C is incorrect because there is no indication of the mechanism of shell thickening. It is likely that the mussels can chemically detect the presence of the crabs, but there is nothing to indicate that mussels must be near other mussels to have the response. Choice D is incorrect because there is no indication that mussels feel fear, although they can clearly perceive certain threats.

429. (D) The two species compete when cultured together, but they are *not* parasites. Parasitism is a specific case of **symbiosis**, which is when two different species live in very close association (usually one lives in or on the other). These two species compete when cultured together, and their competition reduces their growth rate and final population density.

430. (B) *P. caudatum* has the highest growth rate and maintains the highest population density when grown alone. When *P. caudatum* is grown with *P. bursaria*, the population density is cut in half and the initial growth rate decreases. *P. bursaria* alone has a lower growth rate and maintains a lower population density when grown alone, but when grown with *P. caudatum* the population density is cut by at least two-thirds. The initial rate of growth is reduced drastically.

431. (B) Choices A, C, and D are incorrect because *they assume competition always has negative consequences.* **Competition can have positive effects** on one (or more) of the competing populations.

432. (A) Population density can change for a variety of reasons and the effect is determined by a number of factors. Choice B and D do not account for the multitude of possibilities. Choice C is not true.

433. (B) Populations of short-lived organisms can fluctuate wildly, even over many orders of magnitudes, within short periods of time. They can rise rapidly and crash because there are often many that are born at once and almost as many that die soon after. **High turnover rates are associated with instability in a population.**

434. (C) The most direct consequence of rain forest destruction is the permanent loss of massive biodiversity. There are numerous other consequences, including an **increase in atmospheric carbon dioxide,** because there are less plants available to fix carbon into organic forms (and if the land is used for grazing cattle, which are more animals that produce methane, another greenhouse gas).

Carbon dioxide is a greenhouse gas, but ozone depletion occurs when pollutants like chlorofluorocarbons (CFCs) are allowed to escape into the atmosphere. The rain forest is not the major producer of oxygen gas for the planet.

435. (A) Silicon makes up 33% of the mass of the elements in the soil.

436. (D) The availability of the elements varies with its chemical form in the soil and is affected by temperature, soil pH, and the presence of other ions.

437. (C) The soil content represents the mass of the mineral per square meter of soil. The annual plant uptake is how much the plants take up the mineral each year. If the soil content is high and uptake is low, the soil will take a long time to get depleted. If the soil content is low and uptake is high, the soil will get depleted quickly if the mineral is not replenished. For example, with an uptake of 30 kg hectare^{-1} year^{-1} it takes only 40 years for nitrogen to be depleted (nitrogen is likely being cycled, so it is at least, in part, being replenished).

438. (A) Compare the soil content of these elements to the annual plant uptake. Elements like **calcium, potassium, and nitrogen** are taken up in much greater amounts than their concentration in the soil. These elements, along with phosphorus, have the lowest soil content per annual plant uptake as shown in the far-right column.

439. (A) Plants absorb dissolved nutrients as solubilized ions and molecules. **Highly mobile ions are very soluble, so their uptake from the soil is limited to the absorptive capacity of the roots.**

440. (B) Increasing absorptive surface area is the only choice in which a plant can adjust to a low mineral soil. Increasing stomatal density on the leaves wouldn't increase absorption of minerals from the soil. The enzymes of biochemical pathways are adaptations that require major changes in the cell that would likely require the evolution of new biochemical systems in the cell (highly improbable in one generation).

Plants *can* switch from passive to active transport of certain ions depending on their abundance in the soil. For example, when phosphorus concentrations in the soil are less than those in cells of the roots, plants can take up phosphorus by active transport (at the expense of cellular energy, but phosphorus is important in DNA, RNA, ATP, and phospholipids).

441. Smaller plant bodies require less energy to grow and maintain. The same is not exactly true of animal bodies.

Root growth and maintenance require a continuous input of energy, but roots can only obtain energy through cellular respiration. Roots are not photochemical cells (they can't convert light into chemical energy), so they make ATP solely through cellular respiration. **Photochemical cells in the leaves use ATP from photophosphorylation, so the photosynthetic shoot system provides the plant with organic compounds *and* can power itself with energy from the sun.**

The roots absorb essential water and minerals, and store energy in proportion to their surface area and mass, while the **shoots capture sunlight for energy in proportion to the amount of surface by which they can capture sunlight**. Large shoot systems may require a larger root system for anchorage. Root growth is often prioritized over shoot growth when soil nutrients are low.

442. (B) Phosphorus, usually in the form of phosphate, is required for all life. This is an example of *a specific, abiotic factor that influences the biomass of the plant population.*

Arriving at the correct answer choice may require the process of elimination, since you may not know that **phosphorus forms insoluble complexes (and therefore becomes unavailable to plants) in acidic or alkaline soils.**

443. (C) Plants, like animal cells, can perform exo- and endocytosis. **All eukaryotic cell membranes are highly dynamic and constantly turn over via the vesicle cycling within the endomembrane system.**

444. (A) Although species adapted to live in nutrient-poor soil *may* also have high rates of transpiration, it is *not a characteristic* of them. They do allocate a higher percentage of their biomass to their root system relative to species that are not adapted to nutrient poor soils. They tend to form associations with fungus (**mycorrhizae**) to increase their absorptive capacity, and they grow slowly as their access to nutrients is limited and larger plant bodies require more nutrients to build and maintain.

445. (B) Species adapted to nutrient-poor soil will typically absorb and store the excess nutrients. They are much more conservative in increasing their growth rate even when nutrients are available. **All plant cells store minerals in their central vacuole.**

446. (B) Plants usually take up K^+ from the soil by active transport because it is present at much lower concentrations in the soil than in cells. The transporter is not too specific for the uptake of K^+. **Because Na^+ is present at a higher concentration in the soil, it is more likely to be taken up than K^+. Na^+ accumulation is toxic to all cells.**

Ion	Plant intracellular (mM)	Animal intracellular (mM)	Plant extracellular (mM)	Animal extracellular (mM)
Na^+		5–15		145
K^+	22	140	1.0	5
Ca^{2+}	5	10^{-4}	0.2	1–2
Cl^-	27	5–15	1.2	110

447. (B) Only seven ions are listed in the table, but there are dozens to hundreds dissolved in the solutions compared in the table.

448. (D) If there is less solute dissolved, there will be a smaller number of ions, a lower mass of solutes, and a lower ion concentration.

449.

	2	3			

23 to 24.

$$\text{Total mass} = 42.1 \text{ grams}$$
$$\text{Dry mass (without water)} = 9.7 \text{ grams}$$
$$\frac{9.7}{42.1} \times 100 = 23\%$$

450.

	2	.	0	2	

The masses given are for 10 plants, so each plant has an average mass of one-tenth the mass in the table.

$$\frac{20.2 \text{ g}}{10} = 2.02 \text{ grams}$$

451.

Bees and hummingbirds: As pollinators, **bees and hummingbirds contribute to plant reproduction.** The composition of the plant community is likely to change if the number of bees or hummingbirds declines because there is less pollination. Plant species that don't use bees or hummingbirds as pollinators may obtain a reproductive advantage in

this situation. The plants that the bees and hummingbirds pollinate serve as shelter and food for other members of the community, so the entire community may be affected.
Beavers: Beavers are **habitat engineers**. Their **dam building provides still water for many species.**
Sea stars: Mussels have no natural predators, so **sea stars help keep the mussel population in check.** If the only natural predator is removed, mussel populations can explode.
Elephants: **Elephants consume small trees,** preventing them from growing into large trees, which can shade out the grasses. Large trees also require more water from the soil.

452. (D) The increased biomass relative to control in the nitrogen and phosphorus groups show that **phosphorus does not inhibit plant growth, at least when in the presence of nitrogen.**

453. (B) Excess nitrogen in the soil slightly reduced biomass relative to the control plants, even when the nitrogen is added with phosphorus. Phosphorus is limiting to *Ceanothus* growth as shown by an approximate 25% increase in biomass relative to control in the phosphorus only supplemented group. The decreased biomass indicates that the excess nitrogen somehow inhibited the nitrogen fixing bacteria and/or supplied a form of nitrogen that was not as easily accessible to the plant than that supplied by the bacteria.

Nitrogen shouldn't poison plants, unless it's provided in extremely high concentrations. Organisms typically don't produce excess proteins, so the protein composition of the cell isn't the most likely possibility for the reduced biomass upon addition of nitrogen. (The amount of nitrogen needed for nucleic acid synthesis can vary greatly, however. Many plants are polyploid and can have as many as eight sets of chromosomes!)

454. (D) The biomass of all the control groups is 100, as in 100%, because **the biomass of the control relative to the control is unity (1).** The *y*-axis clearly states that the biomass is "percent of control."

455. (C) Substances X and Y clearly *inhibit* the activity of the enzyme. The reaction rate decreases as their concentrations increase (notice the *x*-axis of the bottom two graphs are not substrate concentration). However, *the reaction rate would be expected to slow at a constant concentration of substrate* anyway—but adding more substrate does not increase the rate in the presence of substance X, rather it transiently increases the rate in the presence of substance Y.

456. (B) The inhibition of a **competitive inhibitor**—an inhibitor that resembles the substrate and competes for the active site—would be inhibited by the addition of more substrate. In other words, adding excess substrate should overwhelm the inhibitor by outnumbering it, at least transiently while the substrate concentration is high enough. Because the inhibition is competitive, the enzyme can bind to either, and assuming they have equal affinities for the active site, the one that is present in the highest concentrations is the most likely to get "chosen" by the enzyme, simply because there's more present which increases the likelihood the enzyme will bump into it. **Irreversible binding would prevent the substrate from ever binding to the active site again.** It functionally kills or permanently inactivates the enzyme.

457. (A) If a substance reversibly binds to the active site, the inhibition can be reversed by the addition of more substrate. Note that the *y*-axis is reaction *rate,* not the amount of product formed, so as time passes, the substrate concentration will decrease and the reaction rate will decrease, as well. The top graph shows that the rate stabilizes because the enzymes are saturated (working as fast as they can) once the substrate concentration gets too high. The addition of the substrate in the presence of substance Y was able to increase the rate until it was converted to product or the concentration was too low to compete with substance Y.

458. (B) Variations in population density depend on the magnitude of the fluctuations in the environment and the stability of the population. *Populations with short-lived individuals are inherently unstable, because they don't live long enough to sustain changes in birth and death rates.* For example, a long-lived animal in a population may survive through many fluctuations in birth and death rates in its population over the course of many years. This tends to have a stabilizing effect on the population.

Larger organisms are typically more resistant to environmental change, as well. The amount of available sunlight is expected to vary with the season, and yet high population densities are observed when days are short (January has higher population densities than June, choice C). Choice D is true but incorrect because although the density of water does change with temperature (maximum at 4°C), the change is very small relative to the changes in cell number.

459. (B) Cell centrifugation separates cell components by size and density. Typically, the largest parts of the cell are subject to the greatest forces and settle to the bottom of the centrifuge tube most quickly.

460. (A) Pellet A was obtained by the *lowest-speed centrifugation* and contains the largest parts of the cell, including the *nucleus*, where DNA and RNA polymerase would be found.

461. (D) The enzymes of glycolysis are soluble in the cytosol (the liquid part of the cytoplasm), so they are too small and soluble to be subject to the centrifugal force enough to form a pellet. They would be present in the supernatant in all the fractions.

462. (B) The **mitochondria are much larger than the ribosomes**. The mitochondria contain ribosomes! The ribosomes would be found in the last pellet after a very high-speed centrifugation. The ribosomes attached to ER (in rough ER) would be found in pellet C, the fraction that contains **microsomes**, the vesicles derived from the ER. These membranes are often disrupted by the homogenization and centrifugation processes and tend to form smaller vesicles.

463. (C) The process is called **combinatorial diversification** and is an important mechanism for diversifying the antigen-binding sites of antibodies.

464. (A) Gene duplication and mutation of the duplicates are the primary source of new alleles in eukaryotes.

465. The greater the diversity of the antibody segments in an organism, the greater the diversity of antibodies it can generate and the greater the number of pathogens it can recognize. A pathogen's surface usually has more than one recognition site (antigenic determinant, epitope), so that with a broad diversity of antibodies there may be several kinds than can bind to one particular pathogen, increasing the effectiveness of the immune response.

Pathogens evolve and change to evade host immune systems, so the greater the diversity of the host immune system, the less likely the evolution of the pathogen will allow it to escape detection.

466. (A) A baseline is used to establish a standard for the reaction by providing a reference to which other conditions can be compared. Relative reaction rates require baselines, but absolute reactions rates do not—they require only numbers.

467. **(D)** The reaction rate for the baseline was the same in both experiments.

468. **(B)** **The enzyme concentration is the limiting factor because when additional substrate is added, the rate does not increase.** However, the reaction does continue because the absorbance keeps increasing. *Increasing the enzyme concentration increases the rate, but the same final absorbance is reached because the same amount of substrate has been converted to product.*

469.

	0	.	6		

0.5–0.6

	pH 5	pH 6
Absorbance at T = 0	0	0
Absorbance at T = 4 min	~0.12	~0.2
$\frac{\textbf{Absorbance}}{\textbf{Time}}$	$\frac{0.03}{\text{min}}$	$\frac{0.05}{\text{min}}$

$$\textbf{Relative rate to pH 5 versus pH} = \frac{\textbf{reaction rate at pH 5}}{\textbf{reaction rate at pH 6}} = \frac{\mathbf{0.03}}{\mathbf{0.05}} = \mathbf{0.6}$$

470. **(A)** The extra substrate isn't converted into product much faster. Usually increasing concentration increases reaction rate, but in this case it doesn't happen. Instead, **the enzymes just keep converting substrate to product. When the absorbance doesn't change, it means that all substrate had been converted into product.** The absorbance continues to increase (although it looks like the reaction rate is beginning to slow down at 10 minutes), indicating that product continues to get formed.

471. **(B)** There isn't data to indicate the difference in reaction rate with substrate concentrations at the concentrations tested. The enzyme seems to be present far in excess of what is needed to consume the substrate. *Adding more substrate does not increase reaction rate, although adding more enzyme does.*

The reaction rate is clearly sensitive to pH, although only acidic conditions were tested.

472. In science, **a model is a theoretical framework.** A model **can take on a variety of forms** including (but not limited to) an analogy (lock and key), a formula, or a simulation. The purpose of models is **to make a particular system easier to visualize and understand.** It **reveals testable questions and predictions.** Most models will ultimately fail in their representation of the system, but their continued refinement is fundamental to the process of science.

Observations that supported the **Lock and Key Model of enzyme action** include *enzyme specificity*—the fact that enzymes have one or a few substrates. The increased reaction rate observed with increasing substrate concentration supports that the enzyme and substrate must collide and interact to increase reaction rate. The ability to inhibit enzymes with molecules of specific structures that resemble the substrate but are not able to be acted on by enzymes further supports the lock and key model. Many protein structures have been solved, and the shape of their active sites is typically complementary to the shape of the substrate.

Induced fit is like a "secret handshake" between the enzyme and the substrate (which is itself a model . . . see how useful they are!). The ability of enzyme and substrate to "form into one another" **enhances the fidelity of molecular recognition**, especially in the presence of molecules of similar structures ("background noise"). This is called "**conformational proofreading**," a general mechanism of molecular recognition that is now known to operate in many situations that require protein binding (and maybe all of them). Conformational proofreading allows for the detection of a structural mismatch between the molecular "reorganizer," the enzyme in this case, and its target, the substrate.

473. The best way to determine enzyme (or protein) homology is to compare the amino acid and/or DNA sequences.

474. (D) Bacteria have several types of metabolism not observed in eukaryotes. They include chemoautotrophy and photoheterotrophy. Autotrophs can fix carbon, but not all carbon fixation occurs via the Calvin cycle in photosynthesis. These chemolithoautotrophic bacteria are found in many places on the surface of the earth but are the producers of the hydrothermal vent communities at the bottom of the ocean where there is no light (so no photosynthesis). They oxidize compounds like sulfur, sulfides, hydrogen, and iron as people oxidize glucose. They use oxygen as their final electron acceptor (they're aerobes, **all aerobic organisms require oxygen as a final electron acceptor**).

475. (C) Temperature is proportional to the average kinetic energy of the particles in a substance. High temperatures correspond to greater thermal motion, which destabilizes molecular interactions, particularly weak ones, like hydrogen bonds. Proteins that function at high temperatures need a greater number of strong, electrostatic forces to maintain their structures. Choices A, B, and D all increase the structural stability of the polypeptide, like using a stronger glue. **Polar parts of molecules have partial charges that allow them to form attractive electrostatic interactions with chemical groups of opposite charges or partial charge.** *Non-polar groups do not have permanent dipoles that create stable electrostatic interactions.*

Chapter 5: Lab-Based Questions

476.

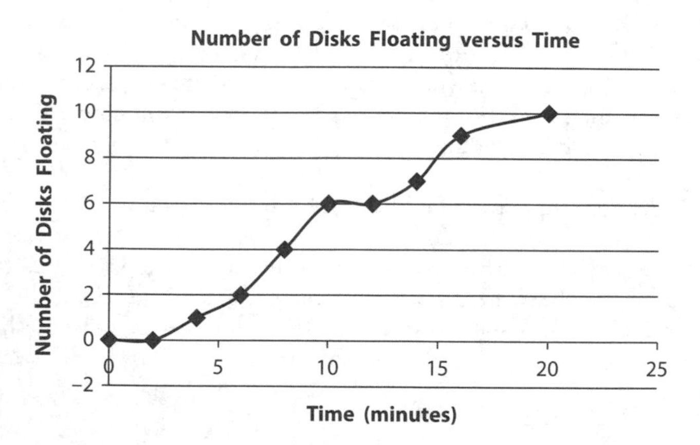

477. The ET_{50} is 9 minutes.

- It takes 20 minutes for 10 disks to float $\therefore$ $^{10}/_2$ = **5 disks**
- It takes ~**9 minutes** for 5 disks to float.

478.

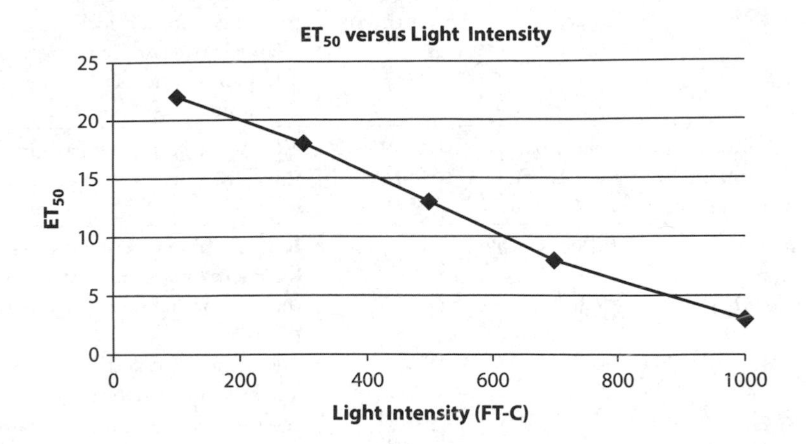

479. The slope of the line can be calculated from any two data points on the line. I will use the first and last:

$$\frac{\Delta y}{\Delta x} = \frac{(3-22)}{(1{,}000-100)} = -\frac{19}{900} = \mathbf{-0.021}$$

The slope represents the rate of photosynthesis (as represented by the ET_{50}) relative to the light intensity. The value of the slope is *negative* because the value of the ET_{50} *decreases with an increased rate* of photosynthesis. The *faster* the disk floats, the higher the rate of photosynthesis but the *lower* the quantity of time on the *y*-axis.

480. The oxygen produced by the light reactions accumulates in the mesophyll, causing it to float. *The carbon dioxide is in the form of soluble bicarbonate ions, so it does not contribute to the leaf disks buoyancy.* If the gases were *not* removed prior to immersion in bicarbonate solution, they would float due to the gases already present in the disk. *The technique requires the disks be evacuated of gases before the experiment.*

481.

Environmental Variable	Increase or decrease photosynthetic rate?	Explanation
Availability of water	• Too little water would *decrease* photosynthetic rate.	• **Transpiration** brings water up from the roots and into the leaves. • **Water is the source of electrons and protons** needed for the light reactions.
Partial pressure of carbon dioxide in the leaf	• Low partial pressure of CO_2 would *decrease* rate.	• CO_2 is the **source of carbon** for the Calvin cycle.
Quality of light (specific wavelengths)	• Red and blue wavelengths *increase* photosynthetic rate.	• **Red and blue** wavelengths provide the **energy to excite electrons** for the electron transport chain in the thylakoid. • The electron transport chain provides the energy to build the proton gradient that powers **ATP synthesis.** • ATP is needed by some reactions in the Calvin cycle.
The amount of plant pigment	• Lack of pigment *decreases* photosynthetic rate.	• Lack of adequate pigment would **limit the quantity and quality of light that could be absorbed.** • Light energy is needed to excite electrons for the electron transport chain in the thylakoid membrane.
Partial pressure of oxygen in the leaf	• High partial pressure of O_2 in the leaf *decreases* photosynthetic rate.	• High partial pressure of O_2 promotes **photorespiration**. • The process in which **RuBisCO**, the carbon-fixing enzyme of the Calvin cycle, uses O_2 instead of CO_2 as a substrate, **diminishing the output of the Calvin cycle**.
Temperature	• Too high or too low temperature *decreases* photosynthetic rate.	• **High temperature can denature enzymes** or **cause stomata to close** (which reduces CO_2 availability, increases O_2 partial pressures in the leaves, and reduces transpiration). • **Low temperatures reduce enzyme activity**.

482. The state in the cell cycle, mitosis or interphase, may not always be clear, so different people may estimate differently. One technique is shown as follows (for the untreated group only). Cells in interphase are crossed out. The left column (on the right of the diagram) shows the estimated number of cells in interphase. The right column shows the estimated number of cells in mitosis.

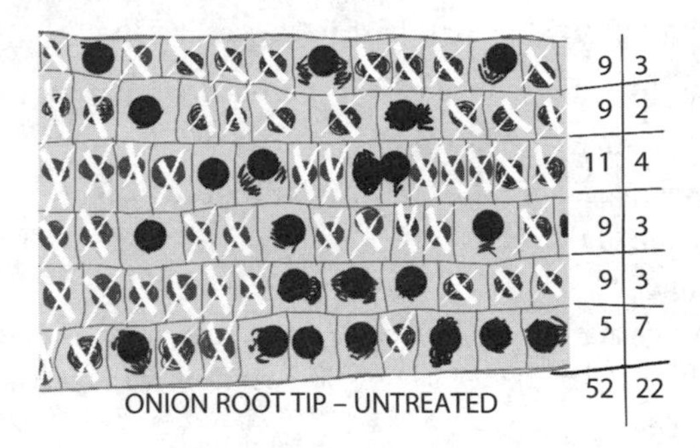

	Interphase	Mitosis	Total	Percentage in mitosis
Untreated	52	22	74	**29.7**
Treated	28	46	74	**62.1**

483. The **null hypothesis** is that *there will be same percentage of cells undergoing mitosis in both the treated and untreated groups.*

Observed		Expected	(Observed – Expected)^2	(Observed – Expected)^2 ÷ Expected
Interphase	28	52	576	11.1
Mitosis	46	22	576	11.1

There are two states possible, interphase and mitosis; therefore, $2 - 1 =$ **1 degree of freedom.**

$$\chi^2 = 11.1 + 11.1 = \mathbf{22.2}$$

A χ^2 value of 3.84 would allow you to reject the null hypothesis because it means there is less than a 5% probability that the deviation of observed results from the expected results are due to chance. **A value of 6.64 means there is a 1% probability** that the "unexpected" results are due to chance.

484. There is little information with which to formulate a hypothesis, so you have a lot of room to *imagine* as long as your hypothesis is **reasonable** and **based on sound biology**.

When you are asked to design an experiment, choose your hypothesis carefully. You'll design a better experiment with a solid hypothesis.

Hypothesis: Rapidly dividing cells have weaker cell walls.

Other reasonable hypotheses could relate poor root developments with:

- a sudden increase in energy requirements
- abnormal development of tissues when cells divide too rapidly
- lack of regulation of the cell cycle
- ectopic cell division
- uncoordinated enlargement of only some tissues, particularly at the expense of others
- weakened cell walls due to the inability of the rate of cell wall synthesis to keep pace with cell division

This experiment would compare the structural and mechanical properties of the cell walls of mitogen-treated and untreated onion root cells.

Examples of experiments that would test this hypothesis:

- Provide a hypotonic environment. Compare increases in cell expansion (microscopy), water uptake (mass). If cell walls are significantly weakened, some cells may lyse.
- Cellulase is an enzyme that digests cellulose. Cells could be immersed in various concentrations of cellulose and the time to lysis measured.

485.

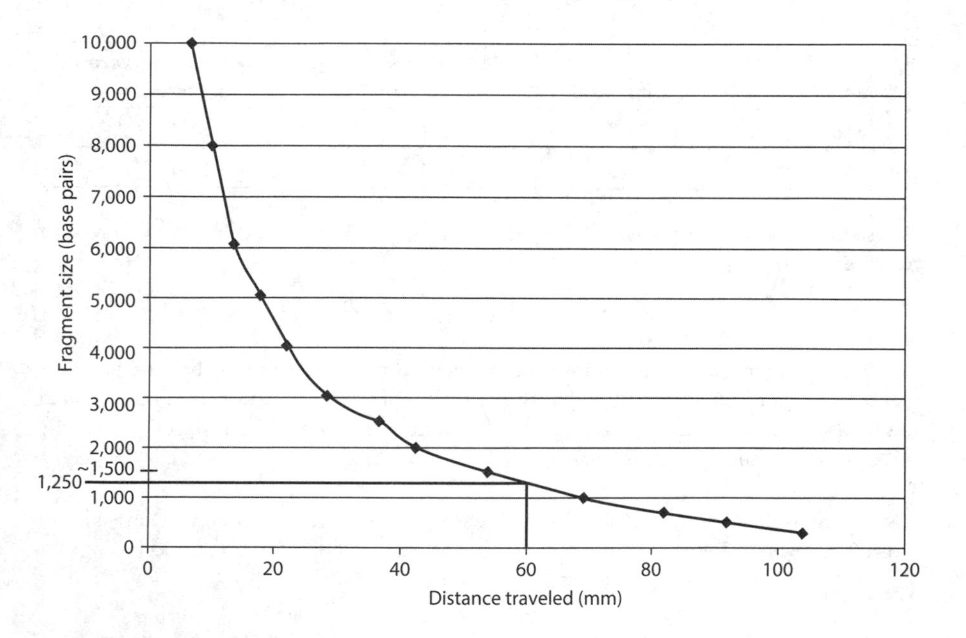

486. A band that traveled 60 mm on the gel would contain DNA fragment lengths of approximately 1,250 base pairs (see the preceding graph of fragment size versus distance traveled).

487. The size of the fragment size is **inversely related** to the distance traveled.

488. The purpose of the pebbles in the respirometer is to indicate any changes in external, or atmospheric, pressure that may have occurred during the experiment. If so, the pressure changes are used to correct the data generated by the respirometers with the dry and germinating peas.

489.

	0 min (mm)	5 min (mm)	10 min (mm)	15 min (mm)	20 min (mm)	25 min (mm)	30 min (mm)
Pebbles at 22°C	0	1	1	2	2	3	3
Germinating peas at 22°C	0	3	6	9	14	19	23
Germinating peas at 22°C Adjusted	0	2	5	7	12	16	20

	0 min (mm)	5 min (mm)	10 min (mm)	15 min (mm)	20 min (mm)	25 min (mm)	30 min (mm)
Pebbles at 30°C	0	1	1	2	3	3	2
Germinating peas at 30°C	0	6	11	17	22	28	36
Germinating peas at 30°C Adjusted	0	5	10	15	19	25	34

	0 min (mm)	5 min (mm)	10 min (mm)	15 min (mm)	20 min (mm)	25 min (mm)	30 min (mm)
Pebbles at 10°C	0	1	2	3	3	3	4
Germinating peas at 10°C	0	2	4	7	11	14	17
Germinating peas at 10°C Adjusted	0	1	2	4	7	11	13

490.

10°C

$$\frac{13\text{ mm}}{30\text{ min}} = \frac{\mathbf{0.43\ mm}}{\mathbf{min}}$$

22°C

$$\frac{20\text{ mm}}{30\text{ min}} = \frac{\mathbf{0.67\text{ mm}}}{\mathbf{min}}$$

30°C

$$\frac{34\text{ mm}}{30\text{ min}} = \frac{\mathbf{1.13\text{ mm}}}{\mathbf{min}}$$

491.

	Time = 30 minutes	Time = 0 minutes	Total distance / time
10°C dry peas	5	0	
10°C pebbles	4	0	
10°C adjusted dry peas	**1**	**0**	$\frac{\mathbf{0.03\text{ mm}}}{\mathbf{min}}$
22°C dry peas	5	0	
22°C pebbles	3	0	
22°C adjusted dry peas	**2**	**0**	$\frac{\mathbf{0.067\text{ mm}}}{\mathbf{min}}$
30°C dry peas	8	0	
30°C pebbles	2	0	
30°C adjusted dry peas	**6**	**0**	$\frac{\mathbf{0.2\text{ mm}}}{\mathbf{min}}$

	10°C	22°C	30°C
Dry peas	$\frac{\mathbf{0.03\text{ mm}}}{\mathbf{min}}$	$\frac{\mathbf{0.067\text{ mm}}}{\mathbf{min}}$	$\frac{\mathbf{0.2\text{ mm}}}{\mathbf{min}}$
Germinating peas	$\frac{\mathbf{0.43\text{ mm}}}{\mathbf{min}}$	$\frac{\mathbf{0.067\text{ mm}}}{\mathbf{min}}$	$\frac{\mathbf{1.13\text{ mm}}}{\mathbf{min}}$

492. The amount of energy an aerobic organism is using is proportional to the amount of oxygen it consumes. The *dry peas* contain fewer cells in a dormant embryo. The few cells it the embryo of the dry seed are not undergoing any cell division. They are using the minimum energy needed to sustain life processes. The *germinating peas* contain more cells (because the embryo cells have divided), and the cells are continuing to divide (and rapidly!), so they are using a great deal of energy.

493.

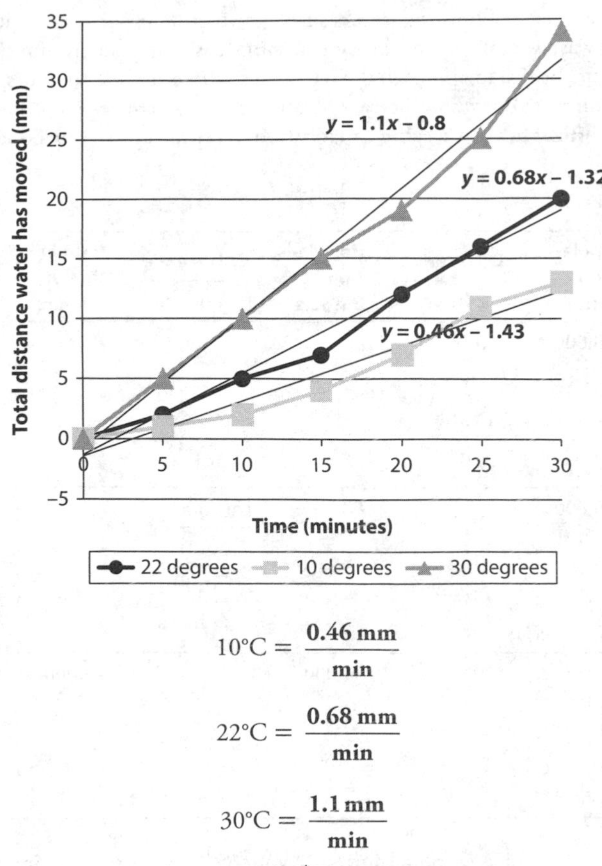

$$10°\text{C} = \frac{\mathbf{0.46\ mm}}{\mathbf{min}}$$

$$22°\text{C} = \frac{\mathbf{0.68\ mm}}{\mathbf{min}}$$

$$30°\text{C} = \frac{\mathbf{1.1\ mm}}{\mathbf{min}}$$

494. The slope of the line is calculated by $\frac{\Delta y}{\Delta x}$ or, in this specific case, $\frac{\text{distance traveled (mm)}}{\text{time (min)}}$.

In this lab, the respiration rate is measured by the distance the water traveled per minute in this experiment.

The rates differ based on method of calculation because the algebraic method uses *only* the final minus initial data points. The *best fit* line is an average of all the data points.

495. The growth rates of the bacteria depend on biotic and abiotic factors. From the first data point (around 1.5 hours) until about 5 hours, only a slight increase in population is observed as there are very few bacteria initially present. From 5 to 10 hours, there is a logarithmic pattern of growth when the bacteria population has reached a critical density where the doubling of the population is a significant increase in absolute number of bacteria (if there are 4 bacteria and the population doubles, there's 8, but if there are a million bacteria and the population doubles, that's 2 million bacteria).

Although the population growth is rapid in the 5–10 hour time period, it is not exponential, due to lack of limiting factors such as reduction in the availability of nutrients and buildup of wastes, examples of the influence of density-dependent limiting factors.

The death rate begins to approach the reproductive rate around 10 hours. At this point, the accumulation of toxic wastes increases death rate and decreases reproductive rate. **The population stabilizes at the carrying capacity when reproductive rate equals death rate.**

496.

Population at 12 hours	Population at 10 hours	$\frac{\text{Difference}}{\text{2 hours}}$	Population at 8 hours	$\frac{\text{Difference}}{\text{2 hours}}$	Population at 6 hours	$\frac{\text{Difference}}{\text{2 hours}}$
250,000	220,000	$\frac{30,000}{\text{2 hours}}$ $\therefore \frac{15,000}{\text{hr}}$				
	220,000		80,000	$\frac{140,000}{\text{2 hours}}$ $\therefore \frac{70,000}{\text{hr}}$		
			80,000		40,000	$\frac{40,000}{\text{2 hours}}$ $\therefore \frac{20,000}{\text{hr}}$

497. Bacteria are useful as a model for eukaryotic cells because they are **cells**, however simple, that share the **same genetic code** and basic mechanisms of transcription and translation as eukaryotic cells. They are single-celled with **one chromosome**, so there is no recombination of homologous chromosomes or heterozygous states that mask recessive alleles. They are **quick, easy, and inexpensive to grow** in a laboratory and **can easily be manipulated in a massive variety of ways**. The first models of the regulation of gene expression, the *trp* and *lac* operons, were worked out in bacteria. Because they have no intracellular membrane-bound organelles, their cells are **easy to fractionate** (by homogenization and centrifugation).

498. One of the most complex undertakings of life is the development of a multicellular organism with multiple cell types from a single zygote. Development requires coordinated differential gene expression. Bacteria are single-celled so there is **no development and differentiation**. Bacteria have their **genes arranged in operons**, so there are multiple genes under the control of one promoter. In addition, their promoters have **operators**, which function as simple on/off switches, whereas **eukaryotes have more complex gene expression** requiring multiple types of proximal and distal regulatory sequences, alternate RNA splicing, and histone proteins, which assist in coordination of gene expression by making certain regions of the DNA accessible for transcription (and making some regions inaccessible).

499.

Number of 4:4 (parental types)	Number of 2:2:2:2 or 2:4:2 (recombinants)	Total	% Asci showing crossover / 2	Gene to centromere distance
240	260	500	$\frac{52}{2} = 26$	26

$$\frac{260}{500} \times 100 = 52\%$$

$$\frac{52}{2} = \textbf{26 map units}$$

In **diploid organisms**, *map distances can be calculated from the crossover rates (recombination frequencies) between two genes.* In **haploid organisms**, *the crossover rate can be calculated by the percentage of crossovers and the centromere.*

You divide the crossover percentage by 2 because **only two of the four chromatids actually cross over**, so only half the chromatids are involved. The other half are parental. Notice in the picture that accompanies the question, after mitosis in the cross-over group, only half the chromatids have actually crossed over. Those are the **recombinants**. The solid color **chromatids** are the parental types.

500. Recombination frequencies are greater for loci that are far apart on the **chromosome** because **crossing over is more likely to separate distant loci than those that are close together.**